£38.50
Nett

D1426754

AERODYNAMIC DRAG MECHANISMS OF BLUFF BODIES AND ROAD VEHICLES

PUBLISHED SYMPOSIA

Held at the
General Motors Research Laboratories
Warren, Michigan

Friction and Wear, 1959
Robert Davies, *Editor*

Internal Stresses and Fatigue in Metals, 1959
Gerald M. Rassweiler and William L. Grube, *Editors*

Theory of Traffic Flow, 1961
Robert Herman, *Editor*

Rolling Contact Phenomena, 1962
Joseph B. Bidwell, *Editor*

Adhesion and Cohesion, 1962
Philip Weiss, *Editor*

Cavitation in Real Liquids, 1964
Robert Davies, *Editor*

Liquids: Structure, Properties, Solid Interactions, 1965
Thomas J. Hughel, *Editor*

Approximation of Functions, 1965
Henry L. Garabedian, *Editor*

Fluid Mechanics of Internal Flow, 1967
Gino Sovran, *Editor*

Ferroelectricity, 1967
Edward F. Weller, *Editor*

Interface Conversion for Polymer Coatings, 1968
Philip Weiss and G. Dale Cheever, *Editors*

Associative Information Techniques, 1971
Edwin L. Jacks, *Editor*

Chemical Reactions in the Urban Atmosphere, 1971
Charles S. Tuesday, *Editor*

The Physics of Opto-Electronic Materials, 1971
Walter A. Albers, Jr., *Editor*

Emissions From Continuous Combustion Systems, 1972
Walter Cornelius and William G. Agnew, *Editors*

Human Impact Response, Measurement and Simulation, 1973
William F. King and Harold J. Mertz, *Editors*

The Physics of Tire Traction, Theory and Experiment, 1974
Donald F. Hays and Allan L. Browne, *Editors*

The Catalytic Chemistry of Nitrogen Oxides, 1975
Richard L. Klimisch and John G. Larson, *Editors*

Future Automotive Fuels – Prospects, Performance, Perspective, 1977
Joseph M. Colucci and Nicholas E. Gallopoulos, *Editors*

Aerodynamic Drag Mechanisms of Bluff Bodies and Road Vehicles, 1978
Gino Sovran, Thomas Morel and William T. Mason, Jr., *Editors*

AERODYNAMIC DRAG MECHANISMS OF BLUFF BODIES AND ROAD VEHICLES

Edited by
GINO SOVRAN, THOMAS MOREL and WILLIAM T. MASON, Jr.
General Motors Research Laboratories

PLENUM PRESS • NEW YORK – LONDON • 1978

Library of Congress Cataloging in Publication Data

Symposium on Aerodynamic Drag Mechanisms of Bluff Bodies and Road Vehicles, General Motors Research Laboratories, 1976.
Aerodynamic drag mechanisms of bluff bodies and road vehicles.

"Proceedings of the Symposium on Aerodynamic Drag Mechanisms of Bluff Bodies and Road Vehicles held at the General Motors Research Laboratories, Warren, Michigan, September 27-28, 1976."
Includes bibliographical references and index.
1. Motor vehicles–Aerodynamics–Congresses. 2. Drag (Aerodynamics)–Congresses. I. Sovran, Gino. II. Morel, Thomas. III. Mason, William T. IV. General Motors Corporation. Research Laboratories. V. Title.
TL245.S95 1976 629.2'31 77-16545
ISBN 0-306-31119-4

Proceedings of the Symposium on Aerodynamic Drag Mechanisms of Bluff Bodies and Road Vehicles held at the General Motors Research Laboratories, Warren, Michigan, September 27-28, 1976

A Division of Plenum Publishing Corporation
227 West 17th Street, New York, N.Y. 10011

Printed in the United States of America

PREFACE

These Proceedings contain the papers and oral discussions presented at the Symposium on AERODYNAMIC DRAG MECHANISMS of Bluff Bodies and Road Vehicles held at the General Motors Research Laboratories in Warren, Michigan, on September 27 and 28, 1976. This international, invitational Symposium was the twentieth in an annual series, each one having been in a different technical discipline. The Symposia provide a forum for areas of science and technology that are of timely interest to the Research Laboratories as well as the technical community at large, and in which personnel of the Laboratories are actively involved. The Symposia furnish an opportunity for the exchange of ideas and current knowledge between participating research specialists from educational, industrial and governmental institutions and serve to stimulate future research activity.

The present world-wide energy situation makes it highly desirable to reduce the force required to move road vehicles through the atmosphere. A significant amount of the total energy consumed for transportation is expended in overcoming the aerodynamic resistance to motion of these vehicles. Reductions in this aerodynamic drag can therefore have a large impact on ground transportation energy requirements.

Although aerodynamic development work on road vehicles has been performed for many years, it has not been widely reported or accompanied by much basic research. Certainly, the research has not been on a scale of the comprehensive efforts in the aeronautical field. This is partly attributable to the complexity of the flows, but has been more the consequence of a less pressing need to know.

The primary objective of the Symposium was to explore the basic mechanisms by which drag is generated on bluff bodies. The configurations and flow fields of interest shared the following major characteristics: three-dimensional bodies (symmetric and asymmetric) with flow fields containing numerous regions of quasi-two-dimensional and three-dimensional flow separation both in front as well as in the rear, with discrete vortices in the wake, and a lift-to-drag ratio of about unity. The schematic representation, shown below, of some of the key flow features thought to exist around automobiles, excluding those of the rotating wheels and of the underbody, was drawn to visually convey these characteristics to prospective contributors of the technical program.

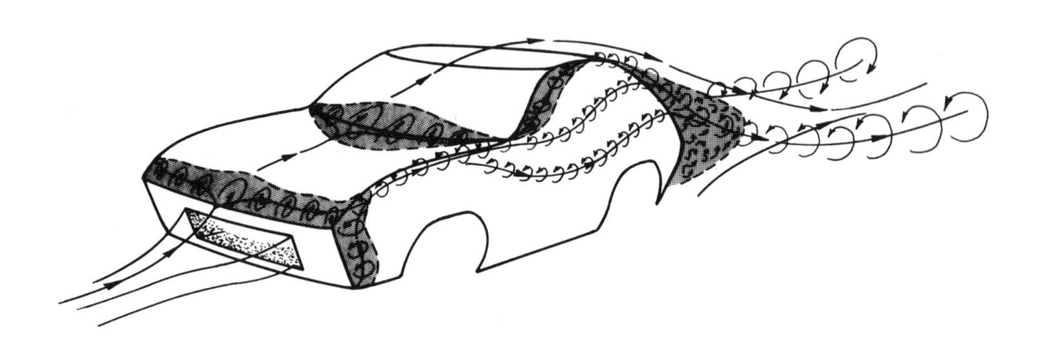

For some of the authors not previously involved with road-vehicle aerodynamics, and therefore working outside their field of direct experience and knowledge, this constituted a principal exposure to the type of flow field of interest and to its most significant features. To whatever degree the schematic is inaccurate or unrepresentative, the responsibility is that of the Symposium Chairmen. Another purpose of the sketch was the portrayal of general geometrical characteristics relevant to road vehicles, of which the most significant are: complexity of body shape due to functional constraints, planview aspect ratio less than 0.5, length-to-diameter ratio of about 2. Ground clearance about 10-15% of body "diameter", motion relative to the ground, and an engine-compartment cavity with through flow are other distinctive features. In addition, in the case of tractor-trailer trucks, there is the problem of two-body configurations.

The program of the Symposium was structured to produce a Proceedings that constituted the present state of understanding of bluff body fluid mechanics relevant to road vehicles, defined the most critical unknown areas, and delineated promising avenues for future research. The situation that has existed, and largely still exists, is one where the major expertise in fluid mechanics resides in people having little knowledge of, or interest in, the flow fields generated by road vehicles. As a result, those actively involved in vehicle aerodynamics have not generally had the advantage of inputs from these experts while exploring the lower *limits* of aerodynamic drag.

Increased communication between these two groups was a goal of the Symposium, and we hope that at least some success is reflected in the contents of these Proceedings.

Active and intensive discussion of the technical presentations was vigorously promoted. Also, all attendees were invited to bring prepared discussions. These were cleared by the particular session Chairperson and the organizers for appropriateness and length. Although the discussions were recorded, verbatim transcripts were not necessarily used. As required, discussions were edited to aid brevity and clarity. Some reordering of the material was done in several discussion periods so as to effectively group all comments on a given subject. We gratefully acknowledge the assistance of a number of the attendees whose on-the-spot evaluations were very helpful in the discussion editing process.

A significant highlight of the symposium not recorded in the Proceedings was the banquet presentation by MR. BOBBY ALLISON. His account of personal experiences with the aerodynamics involved in stock-car racing was fascinating, and effectively cut through the varied background of the Symposium participants. He "told it like it is" on the track, and his descriptions of "drafting" and "sling-shotting" challenge translation to basic aerodynamic terms.

We would like to express appreciation to the Management of both the Fluid Dynamics Research Department and the Research Laboratories for their expression of confidence in permitting complete freedom of expression with regard to the scope and detailed structure of the technical program. We also wish to recognize the very important contributions of our two technical advisers – PROFS. MARK V. MORKOVIN and ANATOL ROSHKO. Their technical knowledge, experience, wisdom and judgment provided large and invaluable inputs to the detailed program planning and execution.

The smooth functioning of the meeting and its pleasant atmosphere were principally the work of MR. KURT T. ANTONIUS of the Technical Information Department, who handled the many details of the physical arrangements in an excellent manner. Many thanks go to several of our colleagues in the Fluid Dynamics Research Department who worked as session aides and provided valuable assistance in many different ways. Finally, the overall Symposium experience could never have been translated into this final copy without the very able and extensive efforts of MR. DAVID N. HAVELOCK and his Graphic Arts Group of the Technical Information Department, all of whom have not yet learned how to say no.

GINO SOVRAN
THOMAS MOREL
WILLIAM T. MASON, JR.

CONTENTS

INTRODUCTION

G. SOVRAN

General Motors Research Laboratories, Warren, Michigan

The purpose of these introductory comments is to set the stage and the context for the technical program. Their objective is to help the deliberations associated with it proceed in an effective manner, and to permit the limited available time to be used to best advantage. To do this, we need to focus attention on the most significant aspects of our particular subject, minimizing any diversions to peripheral questions, regardless of how interesting they are, or how important they may be in other contexts.

First of all, the scope of the aerodynamic characteristics being considered is limited to that of drag. Discussion of other aerodynamic effects, unless directly related to drag, is not in order. This emphasis on drag is timely because of the present day concern about energy conservation. From a technological point of view, this focusing of attention also permits depth of treatment, and increases the possibility of significant technical progress being made.

While the drag mechanisms of simple bodies will be discussed, a major thrust of this Symposium is to specifically consider the drag of road vehicles (i.e. cars, trucks, buses). We are seeking a systematic, critical appraisal of the means by which pressure drag is generated on these bodies. We need to inquire into the basic drag mechanisms for aeronautical-type bluff bodies. We should then consider the degree to which these are relevant to road vehicles. What are the drag mechanisms that are peculiar to road vehicles? What are the directions for future research leading to the better understanding that will permit design control of the drag of these vehicles?

By purposeful intent, an aim of the Symposium has been to bring together basic researchers of fluid mechanics and applied researchers of road vehicle aerodynamics. One group has the in-depth knowledge of fluid mechanics, the other knows the geometrical constraints that must be imposed on road vehicles, and the general nature of their aerodynamic characteristics; it also has familiarity with the real-world vehicle applications, and the needs relevant to them.

The practitioners amongst us will attempt to define the current state of understanding of road vehicle aerodynamics in terms of key fluid mechanical modules; the researchers should search their experience for applicable technology, and also give consideration to the needs for new research. Hopefully, dialogue and mutual understanding will be established between our two groups, with the cross-fertilization of expertise hopefully becoming a significant stimulus to the advancement of both technologies.

The outline and organization of the technical program itself represents the result of a comprehensive analysis of road vehicle aerodynamics and related bluff body research. It is based on *our* interpretation of the state of the two arts – the nature of the flows, the unknowns, the needs – and we at the General Motors Research Laboratories take full responsibility for it. If you feel it is incomplete or off the mark, the criticism should be directed toward us.

To implement the program we have solicited and carefully chosen "command performances" to raise and cover the key issues. We wanted this done adequately and effectively and with a minimum of duplication.

The opening session will start with two descriptions of the general and known characteristics of road vehicle flow fields. These two presentations are not intended to be exhaustive; rather, they will seek to focus on data which is representative and illustrative of the drag mechanisms. The particular presenters are not the only qualified ones in their fields; they do, however, represent some of the more advanced technological work being done. Their task is to present the technologies, and to define the problems sufficiently well that we, the assembled participants, can get on with fruitful discussions.

These opening papers will be followed by: an analysis of drag generating mechanisms on simple bluff bodies, and an assessment of the degree to which they are applicable and transferable to road vehicles; new research on three-dimensional bluff bodies having road-vehicle-like characteristics; and finally, an evaluation of the prospects for qualitative analysis and quantitative computation of the aerodynamic forces.

Primarily for the benefit of those not active in vehicle aerodynamics, it will be useful at this point to define some important characteristics peculiar to road vehicle drag.

On the average, there is always an ambient wind. Consequently, a vehicle's air speed is different than its ground speed. Furthermore, since the direction of travel of road vehicles is controlled by tire-to-road forces rather than by aerodynamics, all ambient winds except those blowing along the roadway cause the relative wind to approach a vehicle at some angle of yaw (Fig. 1). This angle depends on the magnitude and direction of the wind, and on the ground speed of the vehicle. The drag force relevant to propulsion requirements is that which is aligned with the path of *travel*, rather than with the *relative wind* as in aeronautical practice.

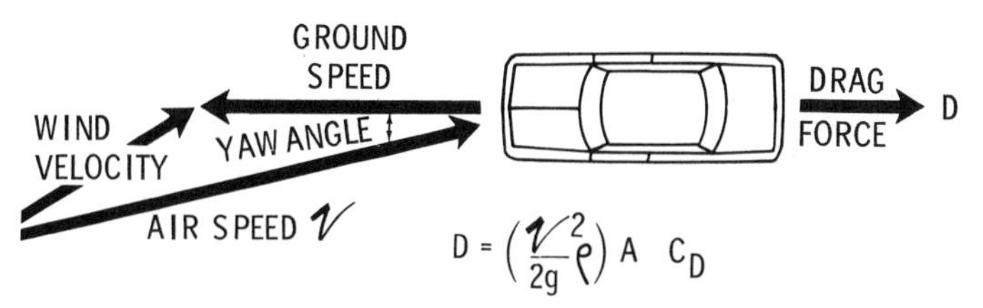

Fig. 1. Air speed and drag force for a road vehicle.

The typical effect of yaw on the drag coefficient defined in this way is shown in Fig. 2, C_D being normalized with respect to its value at zero yaw. For a range of car and truck data, a maximum drag over-run of 55 percent is not uncommon. For tractor-trailer trucks the maximum occurs near 20° of yaw; for cars, near 40°. The *shape* of this drag characteristic must be considered when evaluating the on-road performance of vehicles.

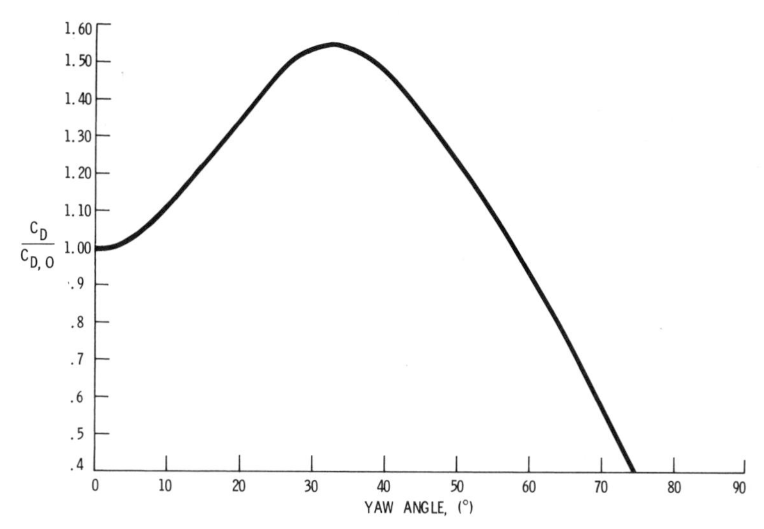

Fig. 2. Typical drag coefficient characteristic for a road vehicle.

Due to ambient wind, operating vehicles experience a yaw angle history as they travel, and their drag coefficient therefore moves back and forth along such a curve. As a result, the average effect of wind is to increase the power required for propulsion over actual driving routes. (A corollary is that winds make it even more desirable to reduce drag.) The shape of the drag characteristic is important to on-road fuel economy considerations. For example, a configurational change which reduces the zero-yaw drag of a particular vehicle but also produces a more steeply rising characteristic, crossing the original curve and peaking at a smaller ψ, will not necessarily reduce the power, and hence the fuel, required in on-road operation. The fluid mechanics which determine the shape of this yaw characteristic *are* within the scope of the symposium; techniques for using it in establishing a wind-averaged drag criterion are not.

Some cautionary comments on the absolute level of the drag coefficient values that will be discussed are in order. As is shown in Fig. 1, reference values of area and velocity head are used to establish the drag coefficient corresponding to a given drag force. Due to the complex geometry on the underside of most road vehicles, there is a degree of subjectivity and arbitrariness in defining A; different procedures are used by different practitioners. The corresponding arbitrariness in C_D must be recognized when comparing values from different sources. Primary considerations should therefore be given to the drag *changes* resulting from configurational modifications that are reported by any given investigator, rather than to any small differences in the absolute levels reported by different experimenters.

The velocity head or q used for non-dimensionalization is also something of a loose junction. It is not within the scope of this Symposium to consider wind tunnel corrections for q. More broadly, test technique *per se*, on-road as well as in-tunnel, is out of bounds. The one exception is the ground plane question in wind tunnel testing, and this will be treated by Dr. Bearman in the first session. Discussion on the subject should be deferred to that paper. The presence of a moving ground plane, and one that is also in "close" proximity to the body under investigation, is an inherent difference between the flow configuration of a road vehicle and that of either an aeronautical or an architectural bluff body. The possible fluid mechanical ramifications of this *have* to be discussed.

In summary, then, our objectives at this Symposium are the following:

(1) A critical review and appraisal of the mechanisms by which drag is generated on bluff bodies in general, and on road vehicles in particular.

(2) An exploration of means for breaking through to significantly lower levels of drag for these vehicles.

(3) A definition of the most critical areas of unknown, and a delineation of promising avenues for profitable research.

(4) The identification, if possible, of practical lower limits for road vehicle drag.

Because of the general shortage of specific data on three-dimensional bluff bodies, hypothesis and speculation are not only in order, but encouraged. They should, however, be clearly labelled as such.

SESSION I

Session Chairperson
P.B.S. LISSAMAN

AeroVironment, Incorporated
Pasadena, California

THE AERODYNAMIC DRAG OF CARS CURRENT UNDERSTANDING, UNRESOLVED PROBLEMS, AND FUTURE PROSPECTS

W. -H. HUCHO

Volkswagenwerk AG, Wolfsburg, Germany

ABSTRACT

An introduction to the drag-related aerodynamics of cars is presented from a practitioner's point of view. Both current understanding and unresolved problems are elucidated with the aid of examples from design work carried out on cars now on the road. Prior to discussing aerodynamic drag in detail, a brief outline is given of the entire field of vehicle aerodynamics, of the economic significance of drag, and of the constraints that have to be observed during design.

Subsequently, aerodynamic drag is classified by type. The mechanisms of external and internal drag are discussed. Attention is drawn to gaps in related knowledge, especially to induced drag. These more or less fundamental considerations are followed by the description of a method which has proved to be very effective during actual design work. This so-called optimization technique has been developed at Volkswagenwerk AG over the recent years. Some results achieved with the technique are presented.

Finally, today's state of the art is placed in the context of a time history of vehicle aerodynamics. What has been achieved to date on production automobiles is compared with what is known to be attainable. This contrast should be viewed as a challenge, one that must be met by increased levels of research and development in vehicle aerodynamics.

1. INTRODUCTION

1.1. Scope of Vehicle Aerodynamics – The energy crisis has generated strong public interest in the fuel economy of cars. In those countries, where fuel prices were

References pp. 39-40.

high, even before the oil embargo of 1973/74, good mileage always has been an important quality of road vehicles. However the current world energy situation has caused increased effort in *all* countries to improve the efficiency of ground transportation. While, in the recent past, great progress has been made in improving the safety of cars and in reducing their emissions to a level ensuring an acceptable air quality, there is still a good ways to go before the existing potential for improving the fuel economy of cars will have been converted into practical solutions. Together with this technological task, a scientific challenge must be recognized. It has two objectives. The one is to make already-known phenomena understandable to vehicle designers, and also to aerodynamicists, and thus applicable in engineering practice; the other is to generate entirely new knowledge.

In the literature there exists a well-documented variety of possibilities for improving the fuel economy of cars; see, for example, Schmidt (1938), Kamm (1969), and Hurter & Lee (1975). The only factor that will be discussed here is the influence of aerodynamic drag. But before dwelling upon this subject in detail, attention must be drawn to the fact that drag is not the only aerodynamic phenomenon a car experiences. The meaning of this is illustrated with the aid of Fig. 1, which shows a vehicle in its natural environment. The flow field of a car is the result of its driving speed and the speed and direction of the ambient wind. The air around the car may be wet with rain and contaminated with soil whirled up from the ground. Closely related to the external flow field are two internal flow systems – engine-cooling and passenger compartment. They are at temperatures different, but not independent, from ambient. The various objectives of vehicle aerodynamics can be classified into four categories, as shown in Fig. 2.

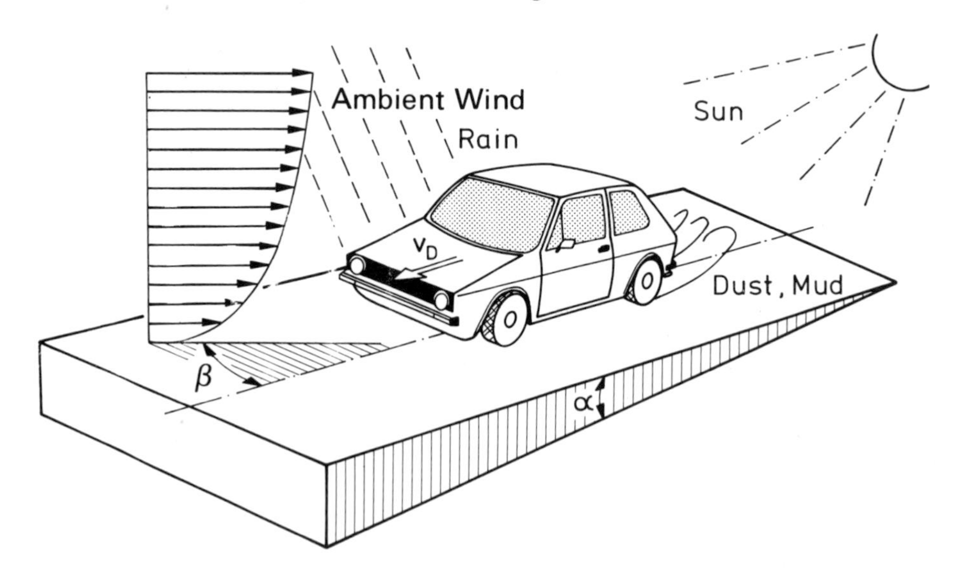

Fig. 1. Vehicle in its natural environment.

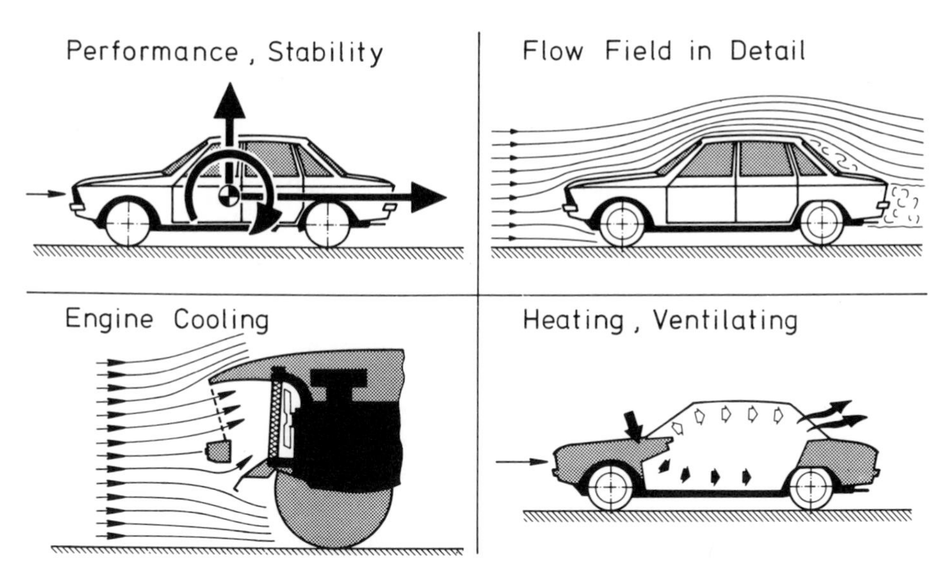

Fig. 2. The four main categories of vehicle aerodynamics.

The first area of concern deals with the forces and moments imposed on a car by its flow field. In many cases, e.g. for almost all passenger cars, only one of the six components, namely drag, is relevant to the aerodynamic quality of the car. This is not true for fast sports cars and racing cars. For the latter, for instance, a compromise has to be found between low drag and high negative lift. The optimum balance of these forces is not only dependent on the specific car but also on the nature of the race track; for example, see Flegl (1974) and Braess *et al* (1975). A different kind of compromise has to be made for light-duty vans, where a trade off between drag, yawing moment and side force is necessary.

The second objective of vehicle aerodynamics is the flow field of a car in its details. The flow path of rain drops; the mechanisms of soil deposition, wind noise, and panel flutter; the cooling air flow to the brakes; and the forces on windshield wipers are all determined by the external flow. The quality of a car to a large extent depends on the success in tailoring the flow field with respect to these problems. It will be shown that at least one of these items is closely related to a car's drag.

Engine cooling is the third area where aerodynamics influences the design of a car. A properly-guided ram air flow can save radiator material; see Olson (1976). On the other hand, cooling air flow can contribute significantly to drag if it is not carefully matched to the external flow and the requirements of the cooling system; see Janssen & Hucho (1973).

Finally, vehicle aerodynamics is decisive with regard to the climate inside a car since the mass flow of ventilating air is related to the pressure distribution over the

References pp. 39-40.

car; see, for example, Wallis (1971). When the flow system inside a passenger compartment is dealt with, the local temperature has to be taken into account as an additional variable. Fortunately, however, the internal flow field at least can be treated without regard to external drag.

All four categories of vehicles aerodynamics outlined in Fig. 2 require roughly equal attention during the design of a car. In other words, they are of almost the same importance to its quality. This should be borne in mind when, for the rest of this paper, only drag is considered.

1.2. Scope of this paper – It is the intent of this paper to draw a picture of the aerodynamic drag of cars from a practitioner's point of view. In order to do so, it will first be shown how important air drag really is when improvement in fuel economy is desired. Attention will be focused on the two main properties of the body, namely air drag and weight. Realistic driving conditions will be used as a baseline. From the constraints which must be observed during the design of a car, the degree of freedom left for the aerodynamicist will be deduced. The components of air drag will then be discussed in order to indicate present understanding of the mechanisms involved. More questions will be raised than answered. It will then be shown how present knowledge about air drag is applied during the course of actual car design and development.

In the final section an attempt is made to evaluate the true potential of vehicle aerodynamics. Some groping steps are taken which may help to overcome the drag limits of today, and lead to aerodynamic drag far lower than that of contemporary cars.

The purpose of this paper is to prepare the ground for the following contributions, and to establish a mutual basis of understanding between the participants of this symposium. Because of the wide scope of the subject, the depth in which details can be treated is necessarily limited.

2. AERODYNAMIC DRAG AND FUEL ECONOMY

2.1. Purpose of Analysis and Car Characteristics Considered – In the recent past a number of papers have been published on the subject of the fuel economy of cars. Generally speaking, they have left the impression that only little can be gained by aerodynamic improvements. What can really be achieved by reducing aerodynamic drag will be demonstrated in this section. The following results are based on data deduced from the car population in Europe.

In the context of fuel economy, a car can be defined by its size, its weight, the properties of the installed engine, and the drivetrain characteristics. There is a correlation among the first three of these parameters, as can be seen in Fig. 3. The left diagram indicates an approximately linear growth of frontal area, A, which is a good measure of a car's size, with weight, m. The scatter of the data is relatively small. The right diagram shows a parabolic increase in installed engine power, P, with car weight. The somewhat larger scatter at the small-car end is due to the fact that, especially in this range, engines of a broad power spectrum are very often an option for a given car. Of course the fuel economy of a given car depends on the kind of engine installed. For the following calculations, similarity among the specific-fuel-consumption maps of all engines is assumed; the SFC map used can be found in Hucho, Janssen, & Emmelmann (1976). The efficiency of the drivetrain is kept constant ($\eta_G = 0.90$). A 4-speed manual transmission with the same ratio spreads is used for all cars. The rear-axle ratio is selected such that top car speed is reached at 100 rpm beyond the point of maximum engine power. The tires are radials, and an identical rolling-resistance coefficient curve, $f_R(V_D)$, is assumed for all cars.

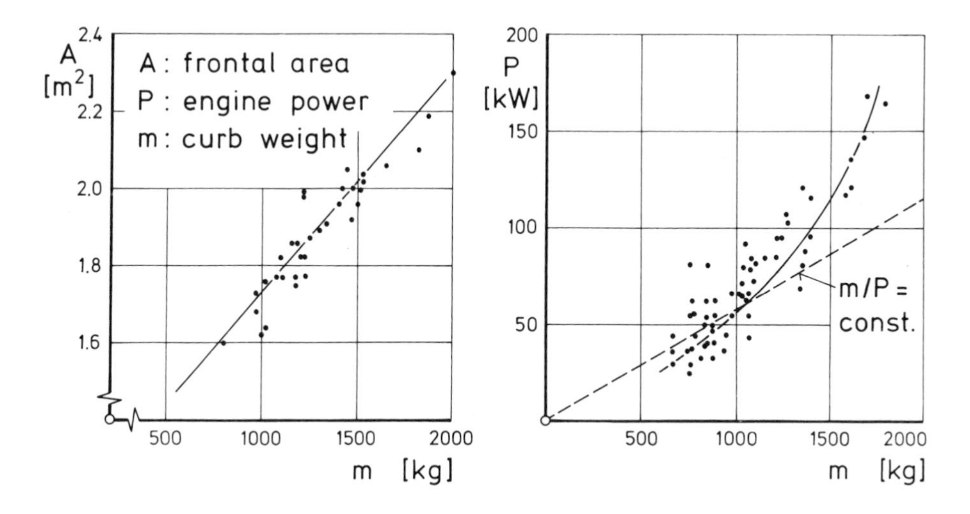

Fig. 3. Correlation between the data characterizing European passenger cars: frontal area, curb weight and installed engine power.

2.2. Driving Resistances — Under level road conditions and steady speeds there are two driving resistances that must be overcome. The first is rolling resistance, R:

$$R = f_R(V_D)\, m_T \qquad (1)$$

The coefficient f_R is a function of driving speed, V_D. The function is given in Fig. 4, and is taken as invariant. The second resistance is aerodynamic drag, D:

$$D = C_D \frac{\rho}{2} V_D^2 A \tag{2}$$

In this well-known relation C_D is the nondimensional drag coefficient, ρ the density of the ambient air, V_D the driving speed, and A the frontal area, which, as has been shown in Fig. 3, is also a good measure of car weight. Ambient wind speed is assumed to be zero.

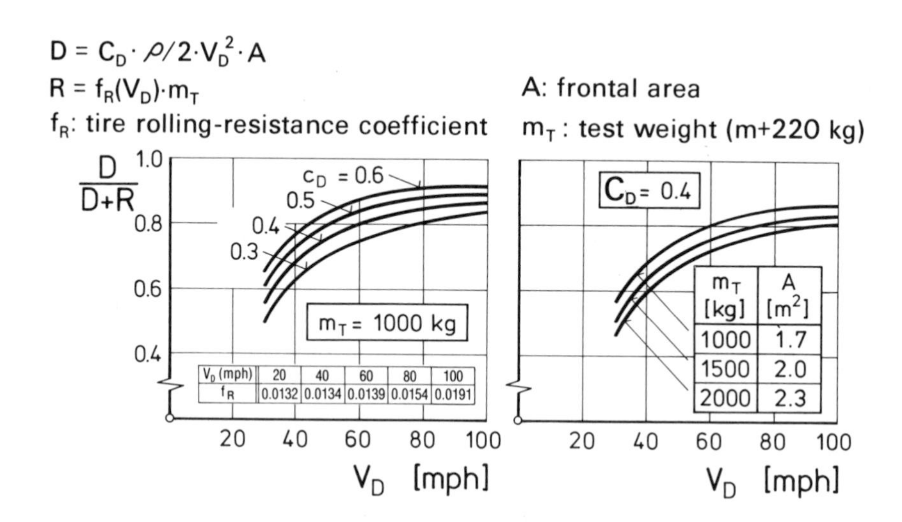

Fig. 4. Ratio of aerodynamic drag, D, to total external resistance, D+R, the sum of aerodynamic drag, D, and rolling resistance, R.

The ratio of aerodynamic drag, D, to total external resistance to motion, D + R, during constant-speed driving is plotted as a function of driving speed in Fig. 4. In the left diagram the weight of a car is kept constant (m_T means test weight, which in all cases is curb weight, m, plus 220 kg). Above speeds as low as 25 mph the air drag starts to exceed the rolling resistance. Depending on the drag coefficient, air drag amounts to 80 to 90 percent of the total drag at higher speeds. The weight itself, see right plot of Fig. 4, changes this picture only slightly, at least as long as the relationship A = A (m) from Fig. 3 is valid.

2.3. Driving Cycles for Fuel Economy Determination – In the United States the fuel economy of cars is currently evaluated over standardized driving cycles. The EPA-Composite Cycle, Fig. 5, is made up from the urban driving cycle imposed during Federally-mandated exhaust emissions tests (Federal Test Procedure) and a highway cycle. Up to now, nothing comparable has been officially introduced in Europe, where fuel economy figures are still quoted for steady-state driving conditions; see, for instance, DIN 70030 (1968). From the consumer's point of view, the influence of aerodynamic drag on fuel economy, which is substantial anyhow, is much overestimated by such steady-state conditions. This has been demonstrated with the aid of a specific example by Hucho (1976). For European purposes a procedure similar to the one shown in Fig. 5, but extended by a portion of high-speed driving on the Autobahn, has been proposed (Hucho, 1976). Due to its higher average velocity this proposed cycle yields a stronger influence of air drag on fuel economy than the U.S. cycle.

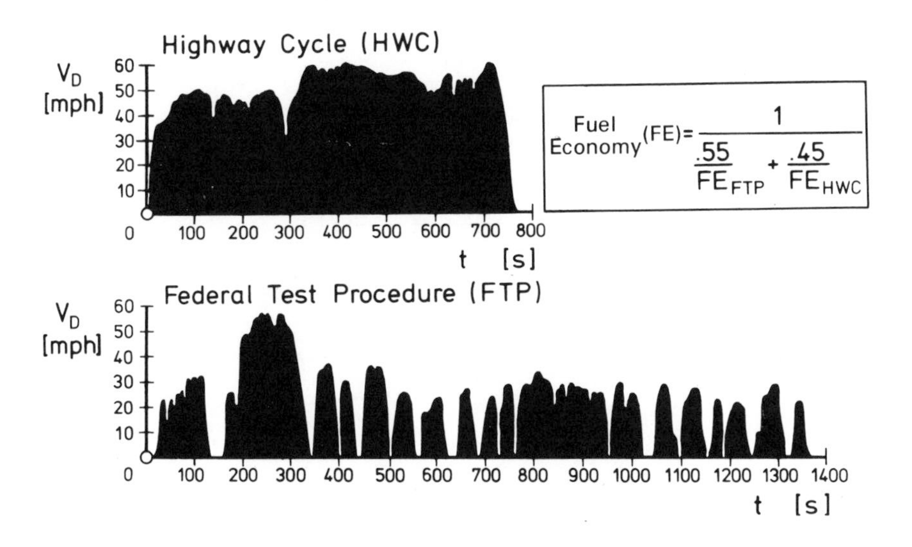

Fig. 5. Evaluation of the "Composite Fuel Economy", FE.

In the following, fuel economy is evaluated on the EPA-Composite Cycle. During the calculations throttle-related transients were taken into account. Where comparison was possible, calculated fuel economy was well within the repeatability of experimental data for all examples.

Calculations were made for three different test weights. Together with the weight variation, the frontal area was altered according to the correlation in Fig. 3. The

References pp. 39-40.

rolling resistance coefficient was varied with speed as shown in Fig. 4. Engine power was rated such that the EPA acceleration recommendation ($t_{0\text{-}60\ mph}$ = 13.5 sec) was met.

2.4. Air Drag and Fuel Economy – The results of the analysis are shown in Fig. 6. Fuel economy is plotted versus drag coefficient for the three different car families; they are characterized by their test weight, frontal area and installed engine power. On an absolute basis, the improvement in fuel economy by reducing air drag is most pronounced for light cars. On the other hand, the relative or percentage improvement, referred to the fuel economy at the highest drag, is roughly the same for all three cars considered. If one starts from C_D = 0.5, which is at the high-drag end of the European car population (see Hucho, Janssen, & Schwarz, 1975), fuel economy can be improved by almost 15% if air drag can be brought down to C_D = 0.3. A drag value in this range is below that of today's cars, but, as will be shown in the final section, it is judged to be attainable.

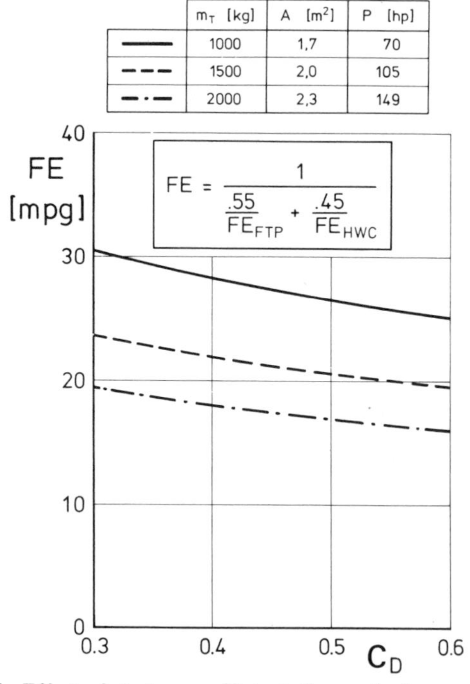

Fig. 6. Effect of air drag coefficient, C_D, on fuel economy, FE.

3. BODY SHAPE DESIGN

3.1. Design Considerations and Constraints – Instead of turning directly to the aerodynamic task of creating bodies of low drag in order to make use of the potential shown in Fig. 6, we should first consider the unalterable constraints which must be observed during body design. Fig. 7 gives a summary of the major considerations. Without going into detail – which can be found in a proper textbook – the following can be stated: the overall dimensions, e.g. length l, height h, and width b of the body, which lead to the aerodynamically relevant fineness ratios h/l and b/l, or a combination thereof, as well as the wheelbase, a, and track, s, are determined by the class of the car, the number of passengers to be carried, the power and location of the engine and, finally, by the size of the luggage compartment. In other words, the enveloping box shown by the dotted line in Fig. 7 is determined when the aforementioned design tragets have been set. For European passenger cars the height is almost invariant for all classes. The width, to some extent, and the length grow with weight. But in a given class, the scatter of these data among competitors' cars is surprisingly small.

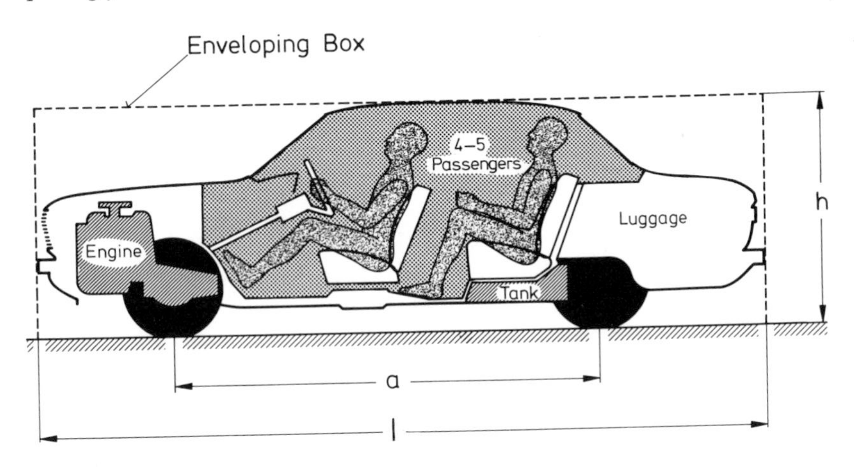

Fig. 7. Design considerations and constraints for a passenger car.

This close coincidence of the data must be recognized as the result of continuous evolution. Technical, biomechanical, and economic arguments were the forces behind the scene which have led to the convergence that has developed with time. All attempts of the past to swim against this stream have failed.

3.2. Comparison of Body Shapes – The shapes of today's cars are far from making use of the entire space within the enveloping box shown in Fig. 7. Surprisingly or not, however, the contours of cars from different manufacturers are far more alike than their visual appearance would suggest. The contours of cars within the three main European classes are compared in the next three figures. In Fig. 8, which includes cars of the size of the VW Rabbit, two different kinds of contour can be distinguished – hatchback and notchback. But within these two categories the contours of different cars are very much alike. The same is true for Fig. 9, which encompasses cars like the VW Dasher and the Audi Fox. Quite recently the fastback has come back into favor

References pp. 39-40.

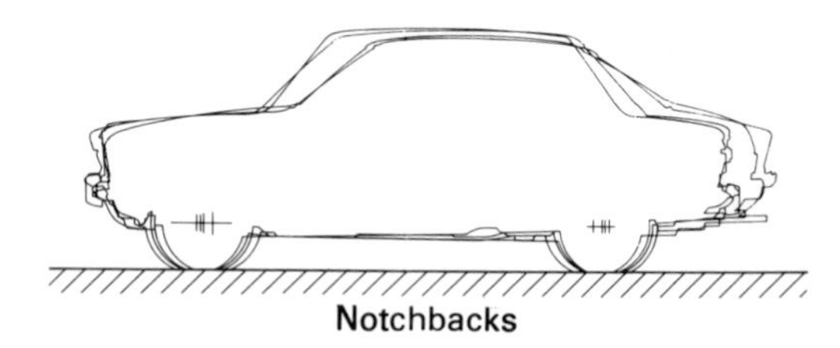

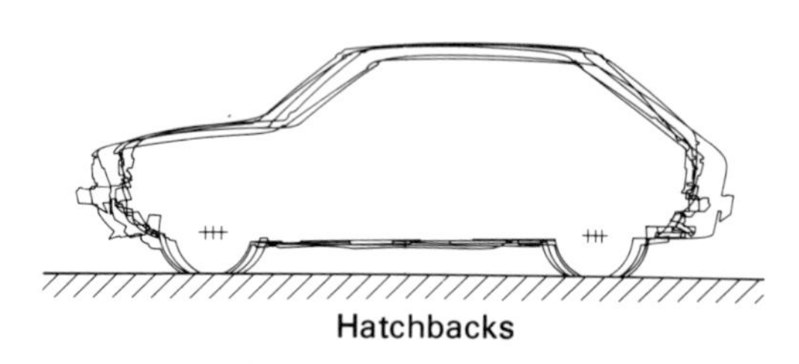

Fig. 8. Centerline cross-sections of small European cars.

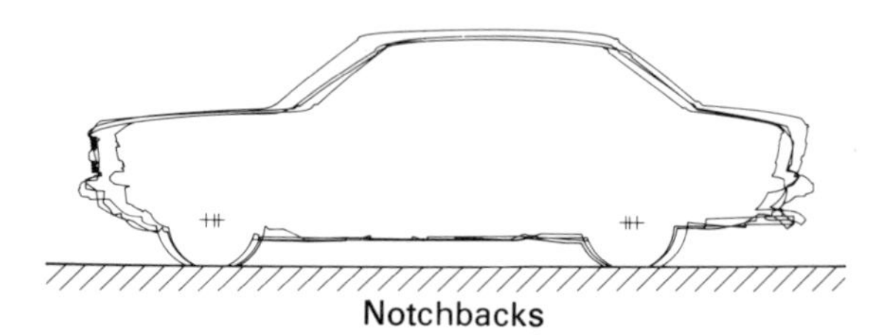

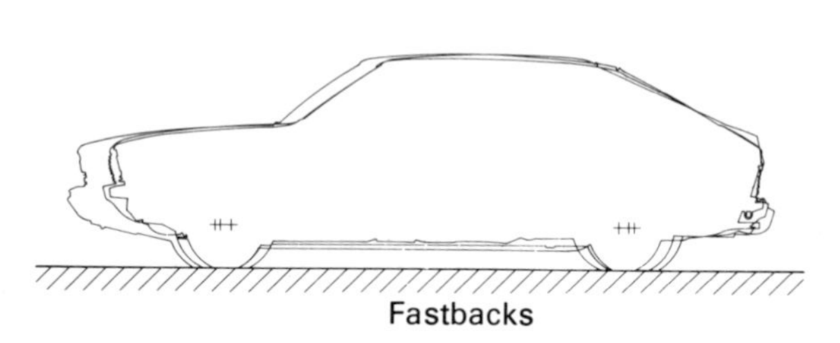

Fig. 9. Centerline cross-sections of medium European cars.

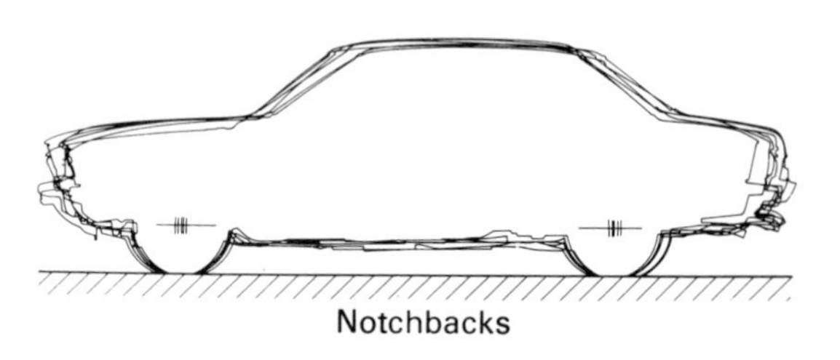

Fig. 10. Centerline cross-sections of larger European cars.

in this class. Among the various shapes, a close similarity can again be seen. Finally, Fig. 10 shows larger cars; this is the class of the Audi 100 and Mercedes 240 (W 123). Even here only comparatively small differences are observed, a fact which Goetz (1971) has already identified.

Of course the contour of the centerline does not define the entire body shape. It is only suited for characterizing its main proportions. But these are very much alike within the different classes and basic shapes. As was the case in the discussion of overall dimensions in the preceding section, here also it is technical and economic arguments that are the reason for the similarities.

Despite the similarities of shape, the variance of drag among cars is remarkable. As can be seen from Fig. 11, which is an updated version of the one published by Hucho, Janssen, & Schwarz (1975) and contains only European cars (excluding sports cars), the drag coefficient of cars varies between 0.37 and 0.52.

It will be shown in section 5 how a drag coefficient at the low end of the distribution in Fig. 11, say $C_D = 0.4$, can be realized with shapes of today. However, on the basis of our present knowledge of vehicle aerodynamics, drag coefficients significantly better than this, say $C_D = 0.3$ or below, will require new body shapes. Section 6 will consider this question.

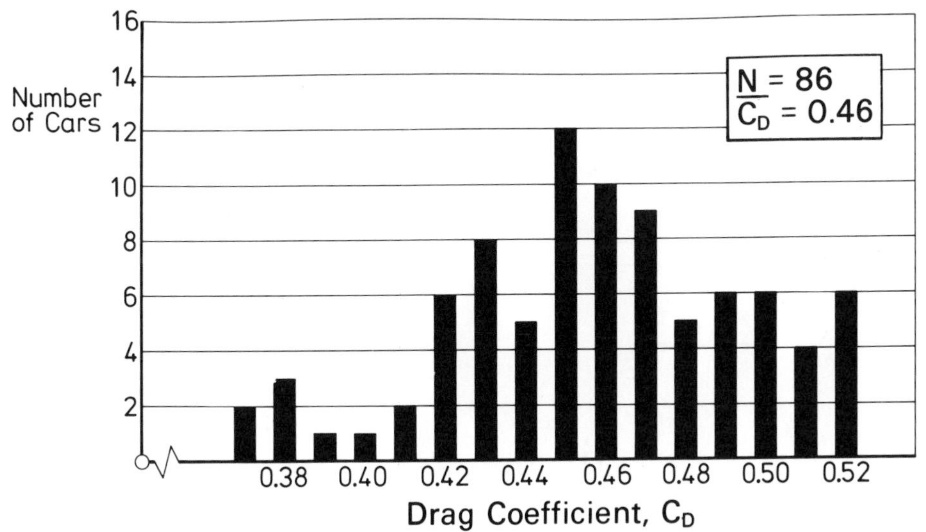

Fig. 11. Compilation of drag coefficients of N = 86 European passenger cars, model year 1968 through 1976.

4. MECHANISMS OF AIR DRAG

4.1. Flow Field Around a Car – The classification of aerodynamic drag by type, which will be carried out in this section, must be related to the flow field surrounding a car. A lot of detailed information about this flow field is available from visualization techniques, which are widely used in studying vehicular aerodynamics. The flow pattern on the surface of a body can be made visible by wool tufts or a variety of surface coatings. The remainder of the flow field, attached as well as separated flow regions, can be visualized by smoke (see, for example, Oda & Hoshino, 1974) or neutral-buoyancy gas bubbles.

All these techniques yield principally qualitative information. Flow-field measurements which would give quantitative details of the velocity and pressure fields around cars are scarcely to be found. Fig. 12 contains one of the rare cases. It has been produced by Howell (1974). The pitot-pressure traverse behind a notchback car shows a distorted vortex sheet; surveys further downstream show that the distortion is rapidly dissipated. Nevertheless a downwash field *is* induced, which is similar to one generated from a pair of longitudinal contrarotating vortices. The lack of such quantitative results and the fact that even the qualitative information is full of gaps makes it very difficult to draw a flow picture of general validity.

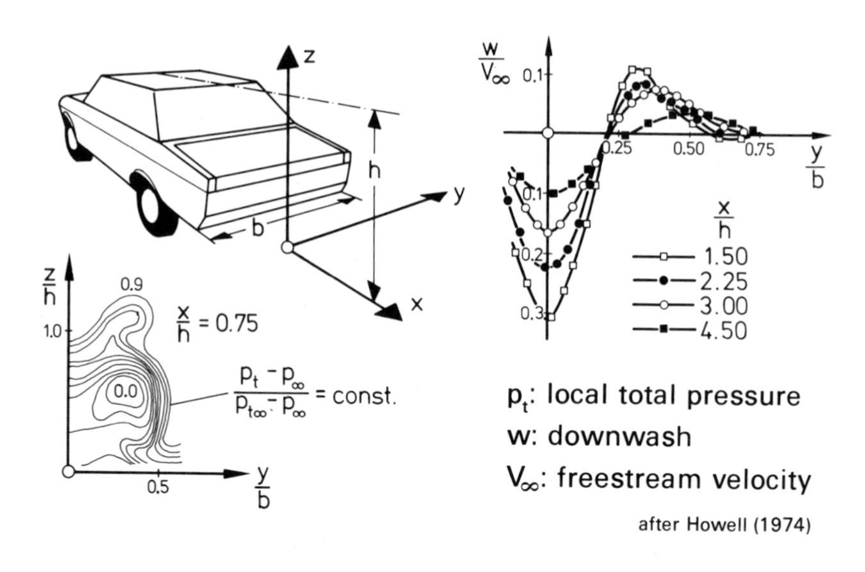

Fig. 12. Flow field behind a notchback car.

The complexity of the flow field around a car will be discussed for fastback cars. Figs. 13 and 14 show the main features. Only the zero-yaw condition is presented. Both figures point out where the flow is attached and where it might be separated. This is not to say that separation *must* take place at the indicated positions in any specific example. The avoidance, or the appropriate positioning, of separation is one of the means for creating shapes of low drag.

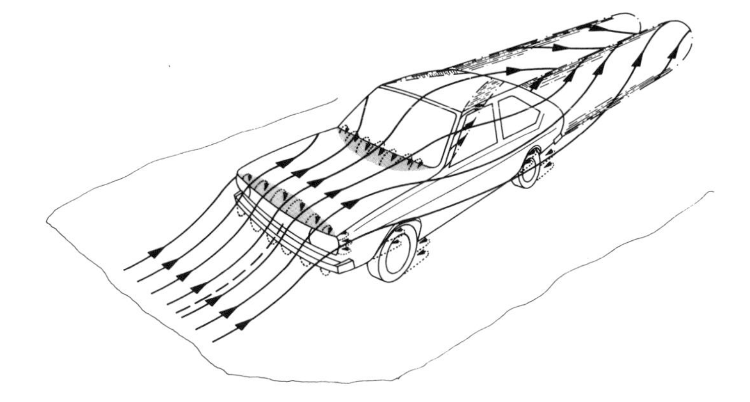

Fig. 13. Flow field around a fastback car, front three-quarter view (schematic).

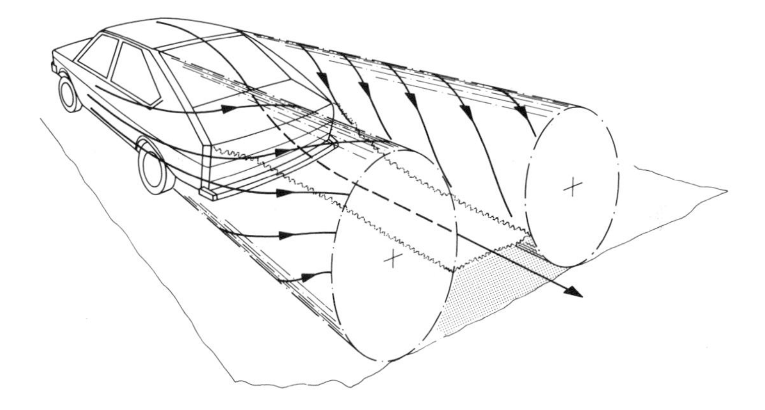

Fig. 14. Flow field around a fastback car, rear three-quarter view (schematic).

Two different kinds of separation can be distinguished. In those cases where the streamlines are drawn as dotted lines the detachment of the flow takes place as in a two-dimensional flow. Bubbles are generated if the flow reattaches further downstream. The flow at the leading edge of the front hood serves as an illustration. Due to friction losses the pitot pressure within such bubbles is below that of the undisturbed flow. The separation bubbles contain vorticity; the axis of the vortices tends to be perpendicular to the undisturbed flow and parallel to the line of separation. Vortices cannot end at the sides of a separation bubble; their vorticity must be shed into the outer flow in the form of longitudinal vortices, presumably weak ones. However, none of the flow visualization studies made by, or known to, the author have yet verified their existence. At those locations where no reattachment occurs (at the rear, for example) the orientation of the vortices seems to be irregular. Large-scale eddies are generated, and have been observed.

The other kind of separation is three-dimensional in nature. Vortex filaments are shed from yawed edges, forming vortex patterns similar to the one on a delta wing at angle of attack. Small vortices of this kind originate from the A-pillars (i.e. the windshield-sideglass junction); another pair of vortices originates from the rear-end, as shown by the broken lines and shaded areas. The axis of the A-pillar vortex is oblique to this pillar, having a larger angle of inclination. There is indication that the A-pillar vortices are bent rearwards over the roof and converge toward the centerline. The manner in which they are entrained into the rear flow field is not known.

The formation of the pair of trailing vortices is similar to that of the tip vortices of a wing. The pressure on the roof of the car is below that on the underbody, resulting in a lift. This pressure difference induces an upward flow on both sides of the car which forms the pair of vortices that trail behind it. The axis of these vortices is approximately parallel to the freestream. Their vorticity and the location of their axes depend on the shape of the rear-end of the car body. The pair of trailing vortices induces a downwash in the space between them, as seen in Fig. 12. The magnitude of the downwash velocity is determined by the strength of the vorticity within these trailing vortices. The location of the quasi two-dimensional separation is dependent on the strength of the downwash. For the car geometry shown in Fig. 14, the presence of the trailing vortices causes separation to take place at the rear edge of the trunk. If, by some means, the generation of the pair of longitudinal vortices is prevented, then the two-dimensional separation will move forward and occur at the rear edge of the roof. This interaction between the two kinds of separation can be observed during actual aerodynamic development work on cars, as will be illustrated in section 5. Photographic evidence of these separation patterns is shown in Figs. 15 and 16, where the separation bubbles have been made visible by introducing smoke into the near wake. In Fig. 15, separation takes place at the rear edge of the hatch door. The wake is short and, due to the downwash induced by the trailing vortices, the separation bubble is squeezed downwards. In Fig. 16, separation has been artificially fixed at the rear edge of the roof, and the wake extends far downstream.

Flow field features not detailed in Figs. 13 and 14, but typical of the overall flow past cars, are those of the underbody flow, the inlet and outlet of air for the purposes of engine cooling and ventilation of the passenger compartment, and the flow around the rotating wheels and inside the wheelhouses. These fields, with the exception of the latter, will be discussed in conjunction with the following breakdown of drag into its main categories.

4.2. Drag Breakdown – First of all, we can distinguish between external and internal drag. The latter is due to losses in through-flow passages such as, for instance, the engine cooling-air duct. It will be discussed in section 4.5. The external drag of

Fig. 15. Wake flow for a fastback-type flow regime. Smoke has been introduced into the wake at the rear.

Fig. 16. Wake flow for a squareback-type flow regime. Separation has been artificially fixed at the roof.

vehicles usually is treated in the literature in the same manner as in aircraft aerodynamics. It will be shown that this is feasible only with limitations.

The drag coefficient, C_D, of a wing is made up as follows:

$$C_D = C_{Do} + C_{Di} \tag{3}$$

The first term on the right side, C_{Do}, is called the profile drag. It is measured (or calculated) under two-dimensional conditions, and accounts for all effects due to

References pp. 39-40.

the viscosity of the air. The second term, C_{Di}, is called the induced drag. Its calculation is based on an inviscid flow model. Induced drag is generated by a system of free vortices shed from the wing. It can be explained and related to this vortex field, and thus to the lift of the wing, by either momentum or energy considerations, using an inviscid flow model. The above equation only holds if the effect of viscosity in two-dimensional flow and the effect of vorticity in inviscid three-dimensional flow are strictly separable. This implies that the flow regimes do not interfere with each other. This means that the character of the *two-dimensional* flow past any given section of the wing is not changed by the three-dimensional vortex effects, and vice versa. It is well known that this is valid only if the aspect ratio of the wing – i.e. the ratio of the square of the wing span to the planform area – is not too small. Obviously, this aspect-ratio condition is not met for cars. Nevertheless the external drag of cars *is* generated both by viscous effects and a vortex field. However, the two influences can hardly be separated as in (3) because of strong interaction between the viscous flow field and the vortices.

4.3. Pressure Drag and Skin Friction Drag – The nature of pressure drag and skin friction drag, which are both due to viscosity, is well known. Fig. 17 recalls the underlying mechanisms. The surrounding flow field exerts a pressure distribution and a field of shear stress on the surface of the body. If the local adverse pressure gradient exceeds a given steepness, the flow separates from the surface. In the sketch

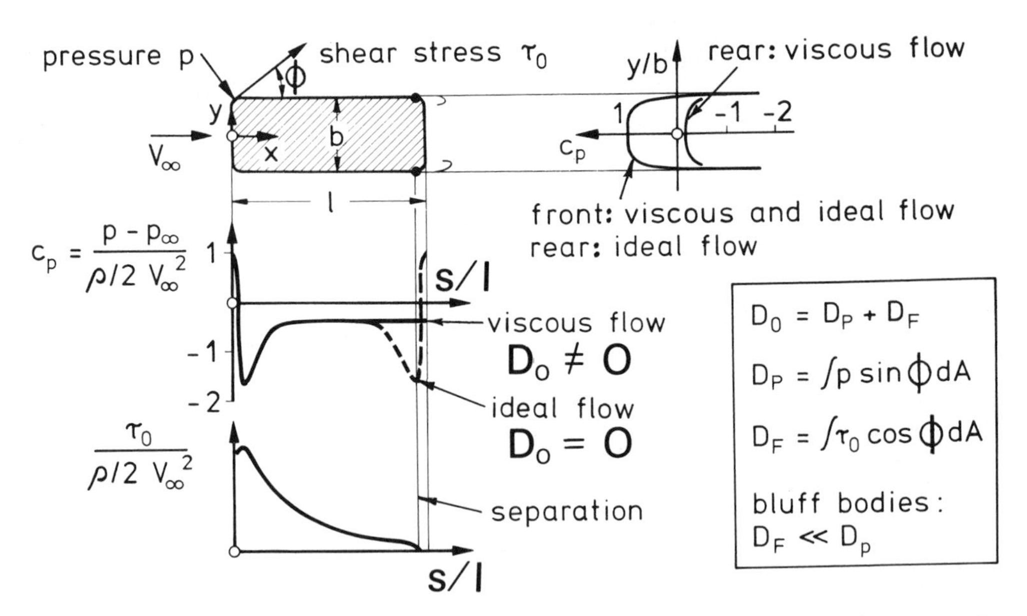

Fig. 17. Definition of pressure drag, D_P, and skin friction drag, D_F; distribution of pressure and surface shear stress on a car (schematic).

shown in Fig. 17 this is assumed to take place at the rear end. Thus the pressure distribution in the rear, shown schematically in both the longitudinal and crosswise directions, deviates significantly from the one in non-viscous flow. With the increase in boundary layer thickness that occurs, the local shear stress decreases, until it is zero at the point of separation. By integrating the pressure and the shear-stress distribution over the entire surface, both pressure drag and skin friction drag can be determined using the integrals in Fig. 12. Their sum yields the entire external drag of a car, *including* the induced drag. In principle, drag can be evaluated in this manner for *smooth* bodies. In fact, however, it is not that simple for cars because of the many roughnesses and protuberances, especially on the underbody. In this region, suspension, drivetrain, and other components contribute mainly to the pressure drag. However, for the sake of analytical tractability, it might be advisable to treat the entire underbody as an extremely rough wall and count its drag as skin friction drag.

As with other bluff bodies, the pressure drag of cars generally exceeds the friction drag significantly. To the author's knowledge the relationship between the two is not known precisely. The information that can be found in the literature (see, for example, Barth, 1956) is based on estimates rather than on measurements. Friction drag is said to be about 10 to 15 percent of the total drag. It is doubtful whether this figure, if valid at all, is applicable to all cars. During the discussion of practical results in section 5, it will be shown that it is worthwhile to deal with *both* drag components when a car is tailored for low drag. However, up to now the major overall progress has been achieved by reducing the pressure drag.

4.4. Induced Drag – As shown in Fig. 14, the flow behind a car contains longitudinal vorticies generated by the pressure difference between the car's roof and underbody. This vortex field represents an amount of kinetic energy which is equivalent to the work that must be performed against a portion of the car's total drag. This portion of drag is called induced drag. The said vortex field seems to be related to the car's overall lift.

In many cases a close correlation between drag and lift *has* been observed for modifications performed on a given car. Fig. 18 serves as an example. Both the drag and the lift coefficient of a small-size car are plotted versus the inclination angle, ϕ, of the hatch door. Without going into detail here – it will be given in section 5 – the following conclusion can be drawn: for certain modifications, any means which leads to a reduction of drag is accompanied by a reduction of lift, and vice versa.

All attempts of different authors to create a formula for induced drag are based on such a relationship between lift and drag (see, for example, Morelli, Fioravanti & Cogotti, 1976). The present author himself went in the same direction *until* he obtained the results which are reproduced in Fig. 19. The car is a Type 3 Volkswagen, prepared as a full-scale calibration model for comparison purposes between different wind tunnels. Various add-on parts can be fitted to the body to produce a broad range of drag and lift coefficients with one and the same car. Configuration A, the

References pp. 39-40.

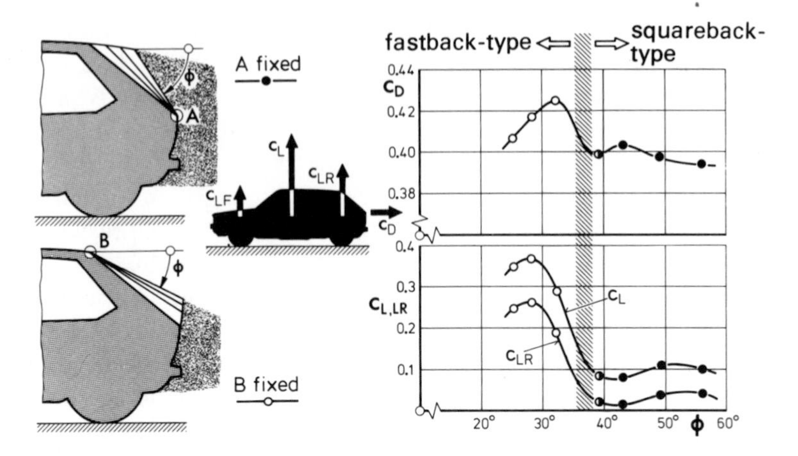

Fig. 18. Effect of rear-end inclination angle, ϕ, on drag and lift of a car.

	Drag C_D	Lift C_L
 (A) Basic Configuration (no add-on parts)	0.34	0.38
 (B) Rear Spoiler	0.33	0.18
 (C) Side Spoiler	0.38	0.48
 (D) Front Spoiler	0.38	0.29

Fig. 19. Pairs of drag and lift coefficients produced by different add-on parts to a basic car.

baseline vehicle, has essentially the same drag coefficient as configuration B with a rear spoiler. But for the latter, the lift is less than half the value of the former. Configurations C and D, the one with front-fender spoilers, the other with a front spoiler, have exactly the same drag coefficient, and again a big difference in lift is observed. Finally, if configuration A is compared with D, it can be seen that car A has a lower drag but a higher lift than D.

In summary it is apparent that in some cases drag is increased when lift grows, but that in others the contrary is true. It is obvious that *both* trends can not be described by a single formula linking "induced drag" and overall lift. Thus, a drag breakdown following the concept of Eq. (3), which has proved to be very helpful for the aerodynamics of wings, is not useful for bluff bodies like cars. However, trailing vortices do induce low pressure on adjacent surfaces and thus may contribute to pressure drag significantly. The portion of drag caused by these vortices might be called "vortex drag" if a special name is at all needed.

4.5 Internal Drag – The several through-flow systems cause an additional drag. This drag is due to internal losses and to interaction between the internal air flows and the external flow field. These drag components, the second of the two might even be negative, really can't be separated. For passenger cars, only the engine-cooling air flow has to be dealt with. The mass flow through the other systems, like those for passenger compartment ventilation or brake cooling, is an order of magnitude smaller than the engine-cooling air flow and thus can be neglected.

The relationship between the internal losses of a ducted system and its external drag has been investigated in depth in connection with the optimization of aircraft-engine oil coolers, and is well understood. A compilation of these results was given by Hoerner (1965). For the limiting case where the entire cooling air flow leaves the system with zero absolute momentum in the longitudinal direction, the well-known Betz formula characterizing the so-called sink drag can be applied to calculate the drag due to radiator flow:

$$D_R = \rho \dot{Q}_R V_D \tag{4}$$

where D_R is the drag due to the radiator air flow, ρ is the density of the ambient air, $\dot{Q}_R$ is the volume flow rate of cooling air, and V_D is the driving speed. If this drag component is made nondimensional in the same manner as the external air drag, the following expression is obtained:

$$C_{DR} = \frac{D_R}{\frac{\rho}{2} V_D^2 A} = 2 \frac{V_R}{V_D} \frac{A_R}{A} \tag{5}$$

where V_R is the average velocity through the radiator and A_R is the frontal area of the radiator. Using actual data for V_R/V_D and A_R/A (see Emmenthal & Hucho, 1974) this equation yields:

$$0.02 \leq C_{DR} \leq 0.06 \tag{6}$$

The drag increase due to the engine-cooling system has been measured on a great

References pp. 39-40.

number of cars, and is compiled and classified in Fig. 20. These results are well within the limits of (6). The average drag increase due to radiator air flow is $\Delta C_{DR} = 0.03$. With a properly designed cooling air duct, $\Delta C_{DR} = 0.01$ can be achieved, which is about 2 percent of the total drag of a car. On the other hand, the drag increase associated with the cooling system can easily be as high as 10 percent of the overall drag if careful attention is not paid to this problem.

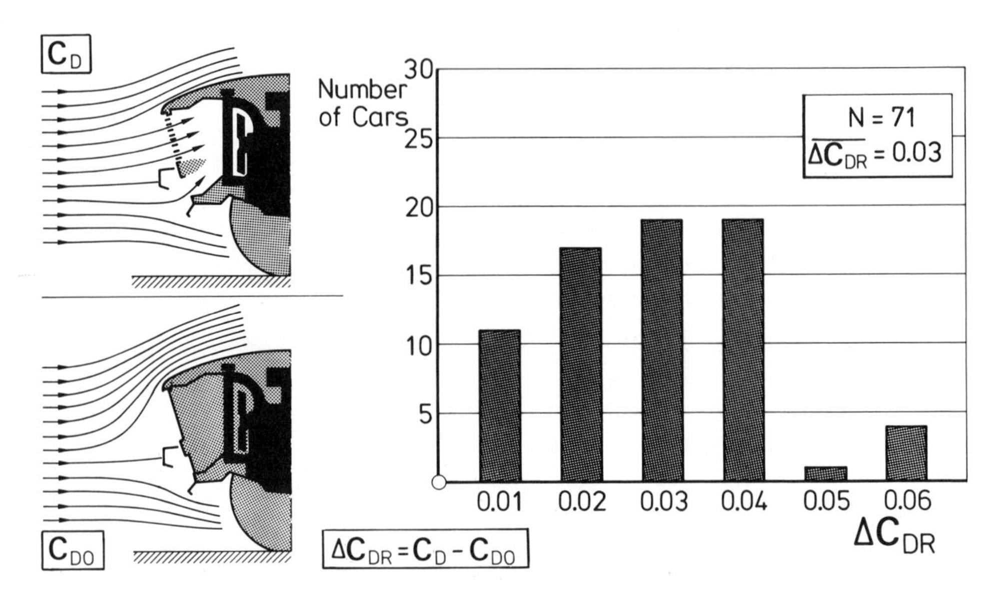

Fig. 20. Compilation of drag-increase coefficients, ΔC_{DR}, due to radiator air flow, European cars, model year 1968 through 1976.

5. BODY OPTIMIZATION METHOD

5.1. Philosophy of the Optimization Method — The optimization method employed by Volkswagenwerk has evolved along with the development of new VW car programs. It has been described in great detail quite recently (Hucho, Janssen, & Emmelmann, 1976). Only its characterizing essentials will be described here.

The optimization method is based on the postulate that the styling concept of a vehicle must be accepted as it stands. Bodies which are designed around the idea of low drag (so-called streamlined cars) are the exception today. The overall dimensions and the main proportions of most passenger-car bodies are within the limits shown in Figs. 8 through 10. Aerodynamic improvements can only be sought by changing body *details*, such that the vehicle's appearance is not altered. Within this limitation all those body details which are important with respect to drag are modified step by step in order to establish a relationship between the car's drag and the geometric parameters describing the specific detail under consideration.

It has turned out that many of these relationships have a saturation characteristic. Fig. 21, taken from Hucho (1972), serves to illustrate this. The vertical leading edges of the VW-van were rounded in several steps. Starting from a sharp corner, the drag is reduced considerably when the radius, r, is increased. But beyond a specific radius, no further improvement is gained. That radius with which the lowest drag is first achieved is called the optimum one. The mechanism governing this behavior is well known. The photographs in Fig. 21 show the different flow regimes. The flow separates at the leading edge when it is sharp. Reattachment can be observed somewhere downstream, its location depending on the curvature of the face and on the width-to-length ratio of the van. With the small radius, r/b = 0.024, separation can still be detected, but the point of reattachment has moved upstream. At the optimum radius separation no longer exists, nor at any larger radii. This interpretation of the flow pictures is confirmed by the pressure-distribution measurements shown in Fig. 22. The separation occurring with the sharp edge and with small radii is characterized by the comparatively moderate suction peaks near the leading edge. As has been shown by Hucho, Janssen, & Emmelmann (1976), the flow phenomenon discussed above also depends on Reynolds number. This has to be considered when small-scale models are investigated.

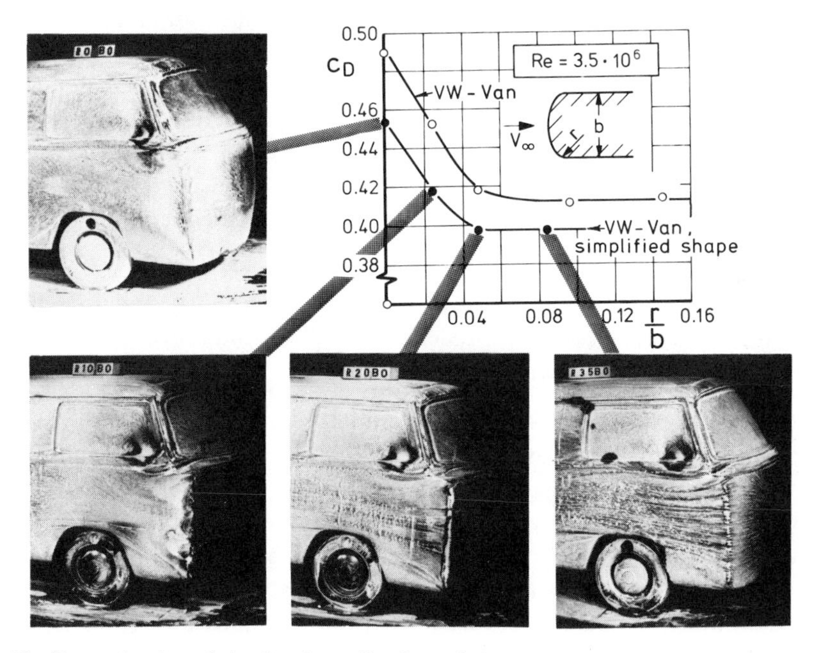

Fig. 21. Determination of the "optimum" radius of the leading edges of VW-van. Flow pictures showing separation from sharp leading edge and attached flow with larger radii. Reynolds number is based on model length.

Many details of a car's body can be optimized in accordance with the procedure illustrated by the example of the van. In fact, curves of similar shape to the ones

References pp. 39-40.

shown in Fig. 21 are obtained as long as only one geometric parameter is modified at a time. Such curves form a rational basis for discussions with stylists and body engineers.

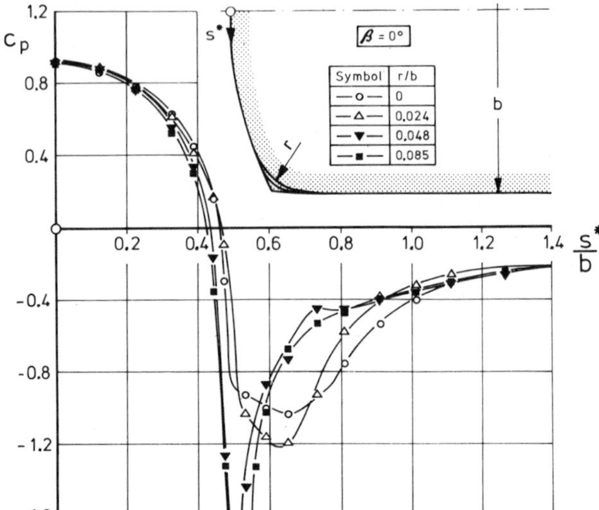

Fig. 22. Pressure distributions for various leading-edge radii, VW-van.

Fig. 23 (see Janssen & Hucho, 1973) shows the final result from one of these discussions. The shape changes on the left side were necessary to achieve the drag reductions compiled in the diagram on the right. Each individual detail was systematically modified in steps. In this example a drag reduction of 21 percent could be realized without changing the overall styling of the car. When shape modifications altering the styling were made, a drag reduction of 33 percent was obtained. This example indicates the capabilities of the optimization method, and the potential of vehicle aerodynamics beyond this method.

The way in which individual details are optimized will be illustrated using the front end, the rear end, and spoilers as specific examples. Additional examples can be found in Hucho, Janssen, & Emmelmann (1976).

5.2. Optimization of the Front End – Prior to proceeding in the way described above it is helpful to investigate what the ultimate drag reduction would be if the front end of the car did not have any separation. This separation-free flow field can be produced by means of an add-on nose section designed purely on aerodynamic principles with no consideration given to the car's appearance. This technique is illustrated in Fig. 24, which is a result of the optimization of the new Audi 100. The add-on nose, which is marked by the dotted line, leads to a drag reduction of 7 percent. This same drag reduction could be realized by rounding-off or chamfering the leading edge of the hood. For styling reasons the front end was finally changed to shape 2, which is almost as good as either the add-on nose or the optimized shape of front-end 1. The optimization is done without cooling air. Later the cooling-flow air inlet is added. Generally, this does not change the *geometry* resulting from the

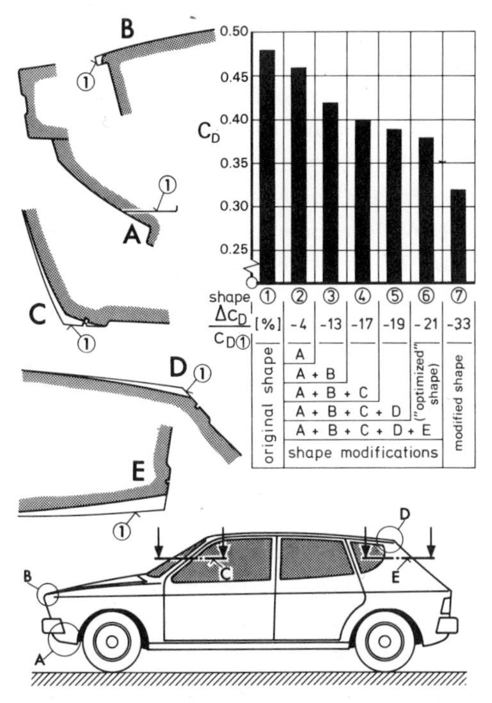

Fig. 23. Example of the optimization process of a passenger car.

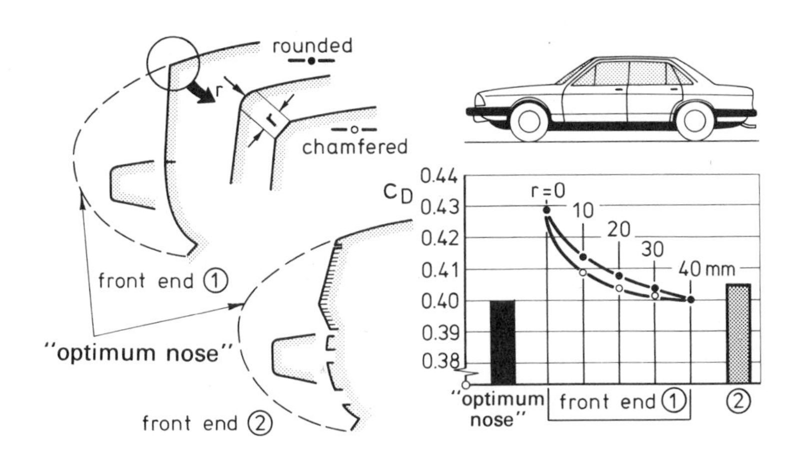

Fig. 24. Optimization of the front end of the Audi 100, model year 1977.

optimization, although the level of drag will be higher. This example together with others published by Hucho, Janssen, & Emmelmann (1976), demonstrates that with almost any front-end configuration of current styling, the drag of the optimum nose can be achieved within close limits through optimization of local geometry.

References pp. 39-40.

However, it should be pointed out that the application of this procedure has one potential drawback. The flow which has passed an optimized corner, although still attached, may be close to separation. As a result it might separate at the *next* steep adverse pressure gradient further downstream. Consequently, the drag improvement achieved at the front-end might be partly or entirely offset by a drag increase at the rear due to premature separation there. It is not known to the author whether such interaction is significant for contemporary cars. However, the more progress that is made in realizing low drag on cars, the more important this kind of interaction will become.

5.3. Alternative Rear-End Shapes — The rear end of a car is characterized geometrically by a variety of shape parameters. In the following only one of them, namely the slant of the rear end, will be discussed. Fig. 25, from Hucho, Janssen & Emmelmann (1976), shows the influence of the hatch-door inclination angle, ϕ, on the drag of a hatchback car and the character of its wake flow. Similar results were presented in Fig. 18. Two flow regimes can be distinguished. If the angle of the hatch panel is steep, say $\phi = 40$ to $50°$, the flow separates at the rear edge of the roof. The advantage of the comparatively low drag of this flow regime is accompanied by the disadvantage of rain droplet and soil deposition on the rear window. When the angle is reduced, a certain value is reached at which the line of separation moves down to the lower apex of the door. Along with this downshift of the separation line the drag is increased and so is the lift. Two strong trailing vortices like those sketched in Fig. 14 can be observed. The rear window remains clean, and water droplets are blown away during cruise.

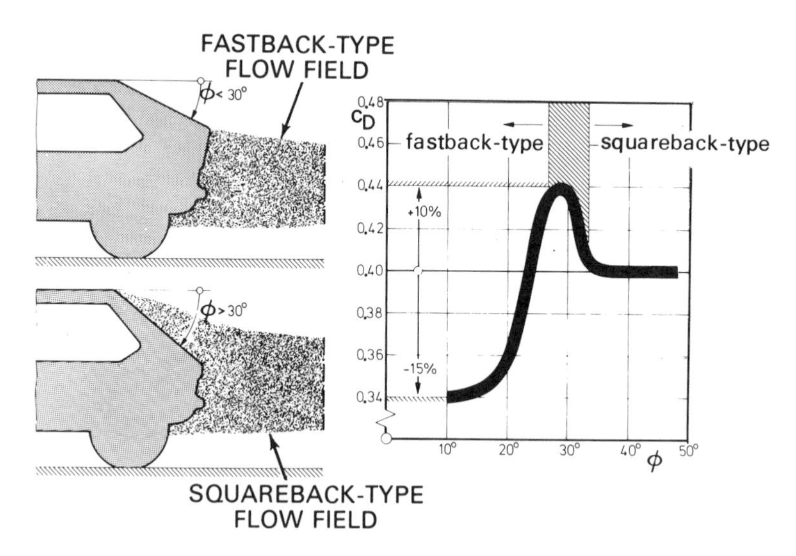

Fig. 25. Influence of rear-end inclination angle on drag coefficient, C_D, separation line, and wake formation.

In both examples shown in Figs. 18 and 25 it was observed that, in the range of slant angles marked by the shaded bands, the flow pattern oscillated randomly between the squareback-type flow regime and the so-called fastback flow regime. The nature of this oscillation has not been investigated in depth, but it must be assumed that it is influenced by the curvature of the roof at its rear end. Furthermore, the average slant angle at which transition takes place seems to be dependent on the upstream flow history.

If the slant angle is reduced further, the drag falls off again. Up to now the lowest drag values have been achieved with an angle as flat as 15°, justifying the term "fastback" used for such configurations. However, such a small angle causes problems with rearward vision. With $\phi = 0$ the squareback configuration is reached again. The drag should be the same as for large slant angles, in this specific example $C_D = 0.40$. This has been confirmed by tests on other car configurations.

Both flow regimes have been visualized with smoke. Fig. 26 shows the squareback pattern. The separation at the rear end of the Rabbit's roof is clearly seen. Fig. 27 is a representative example of a fastback pattern. The flow at the rear of the Scirocco is maintained attached until the apex of the rear door.

Fig. 26. Streamlines in the center cross-section of a VW Rabbit.

References pp. 39-40.

Fig. 27. Streamlines in the center cross-section of a VW Scirocco.

5.4. Spoiler Development – The use of spoilers to improve the aerodynamic quality of cars has increased rapidly in the recent past. Front spoilers – sometimes called air dams – and rear spoilers are either incorporated in the body shell or mounted as add-on parts. The function of both front and rear spoilers seems to be well understood.

A front spoiler reduces the velocity through the underbody region. Velocity profiles underneath a car which have been measured by Kramer *et al.* (1974) and which are reproduced in Fig. 28 clearly indicate this effect. By reducing the velocity adjacent to the extremely rough underside surface, the friction drag of the underbody region is reduced. If the front spoiler is properly matched to the car, this drag reduction can overcompensate for the additional drag originating from the spoiler itself.

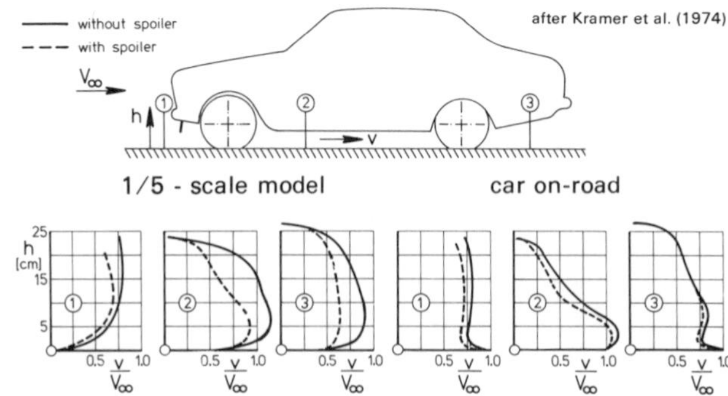

Fig. 28. Velocity distribution underneath a car with and without spoiler; 1/5-scale model tests made in wind tunnel, full-scale tests made on road.

Additionally, a front spoiler causes a reduction in pressure on the front underbody panel, see Fig. 29, and under the engine compartment's hood. The latter effect is welcome when the engine-cooling air flow has to be increased. In general, these reductions in pressure help to reduce the front-axle lift. In Fig. 30 the results from an actual spoiler optimization are compiled. The height of the spoiler is increased in steps. Very often, but not in the present example, if lowest drag is desired, a shorter spoiler results than when maximum lift reduction is the target. More results on front-spoiler design can be found in Hucho, Janssen, & Emmelmann (1976).

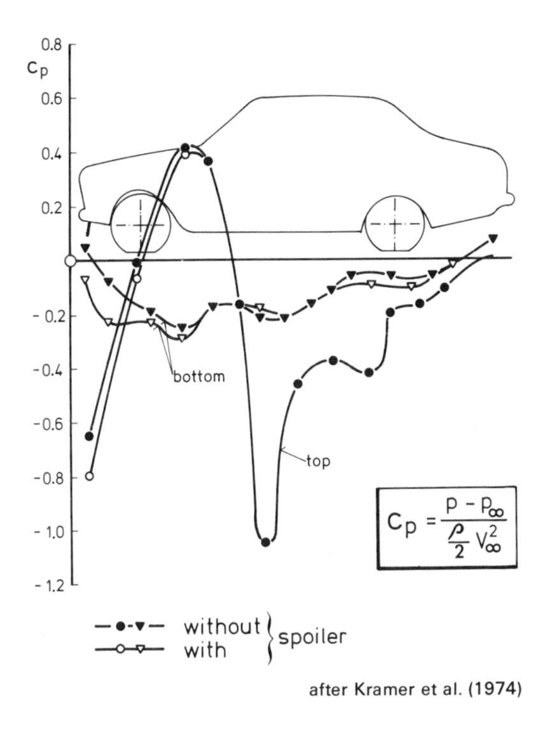

Fig. 29. Effect of a front spoiler on pressure distribution, center cross-section, 1/5-scale model.

The effect of a rear spoiler has been enlightened by Ohtani, Takei, & Sakamoto (1972). Similar to a flap on an airfoil, a rear spoiler increases the pressure on the body surface immediately upstream of it. Depending on the slant of this upstream region, both drag and rear-axle lift can be reduced. In Fig. 31 a result achieved by Ohtani *et al.* (1972) is reproduced. The pressure at the forward part of the car remains unchanged, while at the rear the pressure increase due to the spoiler is clearly visible. Our own measurements have confirmed these findings. They are well explained by a theoretical model based on the flow past a wing section with a trailing-edge flap (see Ohtani *et al.* 1972).

References pp. 39-40.

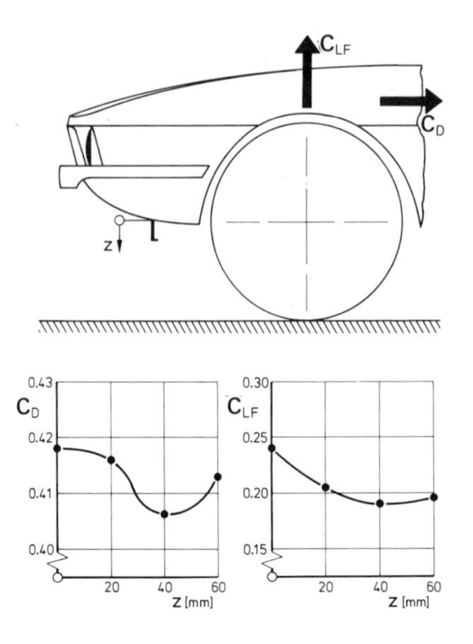

Fig. 30. Optimization of front spoiler.

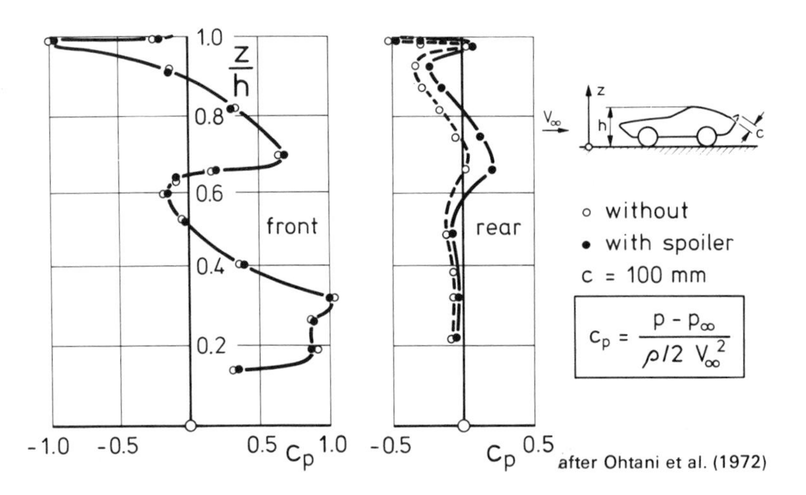

Fig. 31. Effect of a rear spoiler on pressure distribution, center cross-section.

The procedure of optimizing a rear spoiler can be seen in Fig. 32. In this specific example a significant drag reduction was achieved by this device. The saturation character of the resulting relation $C_D(z)$ is typical for the optimization of such a spoiler.

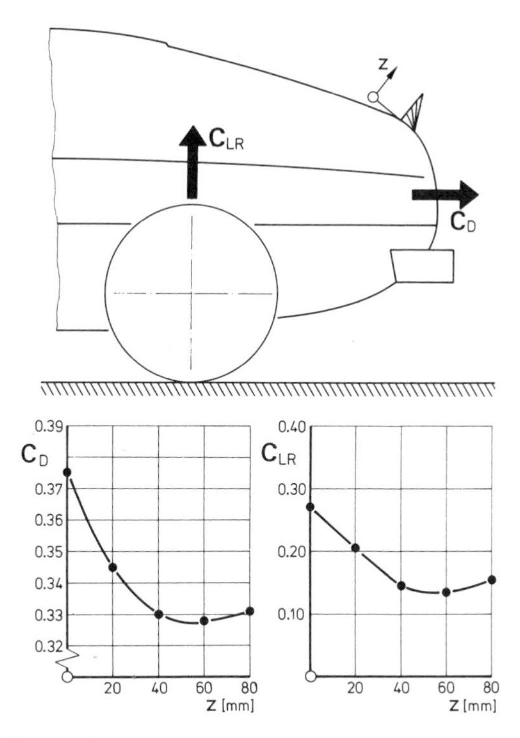

Fig. 32. Optimization of a rear spoiler on a sports car.

6. PHYSICAL POTENTIAL AND FUTURE PROSPECTS

6.1. Drag-Related Potential of Car Aerodynamics – The application of the optimization technique results in a remarkable drag reduction with any given body shape. However in the course of using this method its limitations have become evident. The many results achieved in the past have led to the recognition that drag coefficients below 0.4 are very difficult to realize with the method. If, nevertheless, a drag lower than $C_D = 0.4$ *is* desired by this procedure, it comes at great expense and with high risk of failure. Up to now, drag values notably better than $C_D = 0.4$ have only been achieved with shapes of limited public acceptance, e.g. with so-called streamlined bodies.

The question the practitioner of vehicle aerodynamics is subsequently faced with is the following: have we already arrived at the limits of car aerodynamics? If one contemplates the history of vehicular aerodynamics, one tends to answer this question with a yes. The drag-related history of car aerodynamics is summarized in Fig. 33, taken from Hucho, Janssen, & Emmelmann (1976). The drop in drag coefficient from 0.8, for cars of the Twenties, to an average value of 0.46, for today's European automobiles, occurred in two phases. In the first, the years between the World Wars, cars became longer, lower and smoother in detail. Despite free-standing

References pp. 39-40.

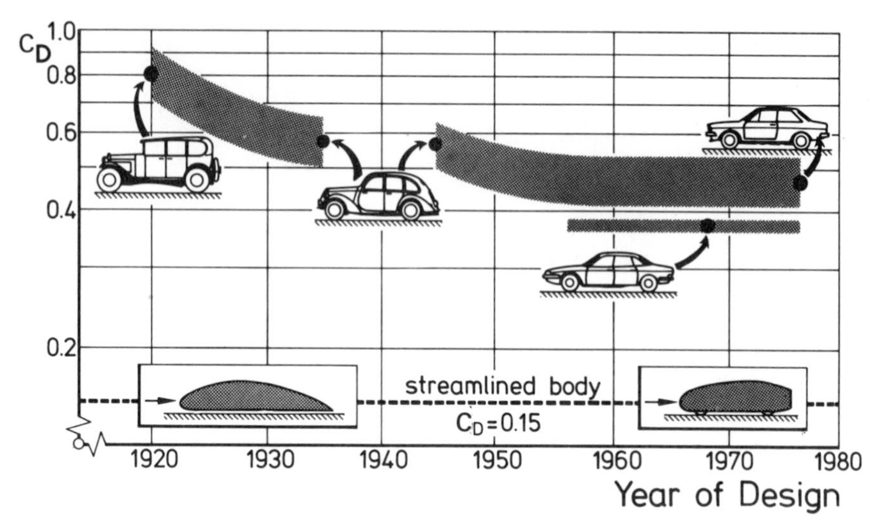

Fig. 33. Time history of the aerodynamic drag of cars in comparison with streamlined bodies.

headlamps and fenders an average drag coefficient of 0.55 was obtained. Together with a reduction of the frontal area, this led to a significant reduction in air drag. The second phase was accomplished with the fully integrated body. However, since 1960 no further tendency toward a systematic drag reduction can be detected.

On the other hand, from fundamental investigations, it has long been known that a drag coefficient far lower than those realized on today's cars is feasible for bodies close to the ground. As early as 1922, Klemperer reported a model having a drag coefficient of 0.15. This model is seen in the lower left part of Fig. 33. The same drag coefficient was achieved at Volkswagenwerk quite recently with the body shown in the lower right corner of the diagram. In contrast to Klemperer's model, the dimensions of the VW-model were maintained within the enveloping box of a modern passenger car (see Fig. 7). Both models were quarter scale, of smooth contour on all surfaces, and without cooling air flow. The VW model had wheels, but no wheelhouses.

For the practitioner, these results lead to the following question: how can use be made of the potential hidden between $C_D = 0.46$, as the average of present cars, and the $C_D = 0.15$ that has been shown to be feasible? Prior to discussing this question another fact should be mentioned. A body of revolution, having a fineness ratio between 3 and 5 has a drag coefficient in the range of 0.04 to 0.05. If such a body is brought close to the ground, its drag is increased by a factor of two or three, even when its shape is modified and "matched" to the local flow field in ground proximity. Without doubt, an explanation of this effect would contribute significantly to the understanding of vehicle aerodynamics. Scientists are called upon to help resolve the related problems. At the moment, the question of how the proven low-drag potential can be utilized has first priority for the practitioner.

By its nature, the optimization method described in section 5 can provide only partial answers to this question. A complete answer can only be expected from systematic research work in car aerodynamics. As has been shown, most of the knowledge in vehicle aerodynamics is qualitative in nature. Some attempts to transfer results from other fields of aerodynamics to the flow around cars have been quite successful. Nevertheless, what can be deduced from them is guided empiricism rather than systematic design procedures. Unlike other disciplines of fluid dynamics, as for instance turbomachinery or aeronautics, little quantitative information is available on which a rational design procedure for road vehicles can be based.

What is needed first is a complete picture of the flow field surrounding a car. The various flow modules have to be identified, and their contributions to drag elaborated. The only way to do this is by experiment, and appropriate techniques are available. However, in order to generate complete understanding and to develop design procedures which will convert vehicle aerodynamics from an art to a subject of rational engineering, the experiments have to be supported by theory.

In the course of this Symposium it will be indicated what can be expected from the application of fundamental theoretical flow models, namely: potential flow, boundary layer theory, wake flows and vortex flow-fields. Despite the fact that a complete theoretical model can not be expected in the near future, this theoretical penetration of the subject is not meaningless. Theory must serve to both guide and interpret the experiments. This should bring a better understanding of the flow around cars. This deeper understanding would be of great help in solving the problems related to all four categories of vehicle aerodynamics outlined in Fig. 2.

6.2. Geometry Analysis – The knowledge generated by the experimental and theoretical investigations that have been suggested would be applied in creating new body shapes. Along with this, two questions arise: how much will these new shapes differ from today's shapes? Is low drag only possible with streamlined bodies of the type seen in Fig. 33, which we already know have almost no chance of being accepted? Of course, no reliable answers can be given prior to having these still-unknown shapes. Nevertheless, one should try to establish some reference points.

In Fig. 34 an attempt is made to identify the formal differences between bodies of different air drag. The complexity of the body geometries of existing cars and of low-drag configurations has been characterized by a so-called contour parameter, κ. This parameter κ is defined, as can be seen in Fig. 34, as the line integral of the rate of change of curvature, k, of the surface contour. For the sake of simplicity, this integral is taken for the centerline cross-section only. If it would be applied for the entire body surface, an even better distinction would be possible. The same is valid for more sophisticated methods to characterize surfaces.

For a streamlined body the rate of change of curvature along its contour is only moderate. Such a body has no abrupt changes in surface curvature, and thus the

References pp. 39-40.

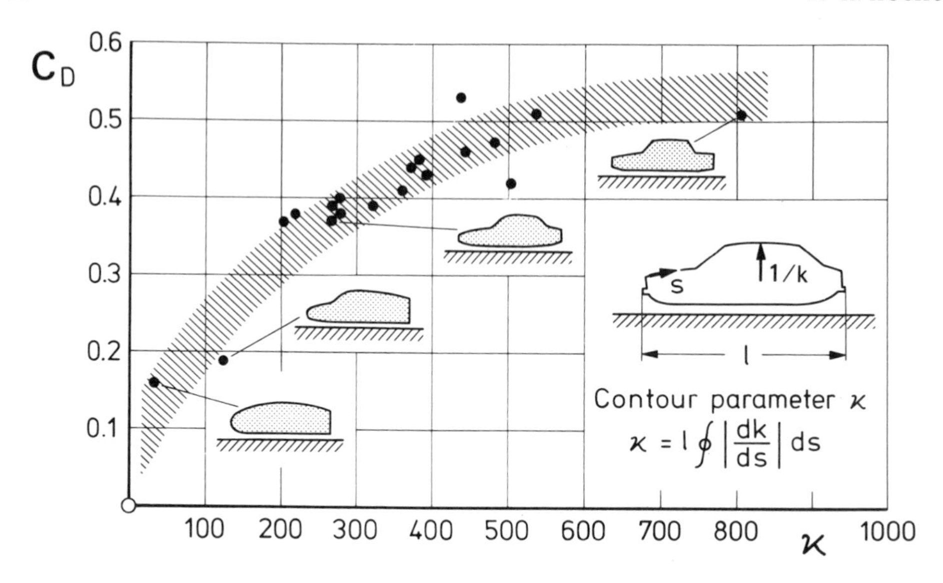

Fig. 34. Correlation of body complexity and drag coefficient; contour parameter, κ, determined for center cross-section only.

contour parameter, κ, is small. In contrast, a standard car is characterized by a number of steep gradients in curvature, thus leading to a large value for the parameter κ. For the sake of completeness it must be added that a small value of κ is a necessary condition only; it is not sufficient for low air drag.

In Fig. 34 the measured drag coefficient is plotted versus the contour parameter, κ. During the numerical evaluation of κ, bumpers and underbody protuberances were neglected. A reasonable correlation between drag and the contour parameter was obtained. Cars in the range of $200 < \kappa < 300$ were styled under strong aerodynamic guidelines. Related examples are the NSU-Ro 80, the Porsche 924 and cars from Citroen. With bodies only slightly less complex than these, a drag coefficient of 0.3 should be possible. This drag value, which was used as the lower limit in the calculation of fuel economy, see Fig. 5, seems to be a reasonable target for the not too distant future.

However, as long as this target can only be hit by cut and try, it will not be accepted as being realistic. Design tools, equivalent to those in other fields of aerodynamics, have to be developed from both experiment and theory before a drag coefficient of 0.3 or less can become a design criterion in the engineering specification of cars to be developed.

7. CONCLUDING REMARKS

In the light of the severe energy situation, the feasibility of improving the fuel economy of cars by means of aerodynamics has been considered. It turns out that great improvement can be gained when aerodynamic drag is reduced. This is valid not only for high-speed steady-state cruise, but also for representative driving cycles. In the latter case, a 15 percent reduction in fuel consumption can be achieved when the drag coefficient is reduced from 0.50 to 0.30. Within existing constraints and the knowledge available today, $C_D = 0.30$ is out of reach for cars designed under usual time and cost limitations. Most of the phenomena associated with the flow past a car seem to be well understood in qualitative terms. Application of this knowledge as an empirical guideline has led to an optimization technique which has been able to book considerable success. The low-drag potential beyond the limits of this method can only be made use of through quantitative relationships. These have to be developed using both experiment and theory.

ACKNOWLEDGEMENT

The author is particularly indebted to two of his colleagues for their assistance during the preparation of this paper. Mr. H.-J. Emmelmann performed the fuel economy calculations. Mr. H.-J. Hochfeld detailed the definition of the contour parameter and carried out its numerical evaluation. The author expresses his gratitude to Volkswagenwerk AG for permission to publish this paper.

REFERENCES

Barth, R. (1956), Einfluss der Form und der Umstroemung von Kraftfahrzeugen auf Widerstand, Bodenhaftung und Fahrtrichtungshaltung, VDI-Zeitschrift, 98, pp. 1265-1312.

Braess, H.-H., Burst, H., Hamm, L. & Hannes, R. (1975), Verbesserung der Fahreigenschaften durch Verringerung des aerodynamischen Auftriebes, ATZ, 77, pp. 119-124.

Emmenthal, K.-D., & Hucho, W.-H. (1974), A Rational Approach to Automotive Radiator Systems Design, SAE 740088.

Flegl, H. (1974), Die Fahrleistungsgrenzen heutiger Rennwagen – elaeutert am Beispiel des Porsche 917/10 CAN-AM, Koll. Industrie Aerodynamik, Aachen, Part 3: Aerodynamik von Strassenfahrzeugen, pp. 141-152.

Goetz, H. (1971), The Influence of Wind Tunnel Tests on Body Design, Ventilation and Surface Deposits of Sedans and Sportcars, SAE 710212.

Hoerner, S. (1965), Fluid-Dynamic Drag, Published by the Author, Brick Town, N.J.

Howell, J. (1974), Wake Properties of a Saloon Car, Koll. Industrie Aerodynamik, Aachen, Part 3: Aerodynamik von Strassenfahrzeugen, pp. 85-95.

Hucho, W.-H. (1972), Einfluss der Vorderwagenform auf Widerstand, Giermoment und Seitenkraft von Kastenwagen, Zeitschrift fuer Flugwissenschaft, 20, pp. 341-351.

Hucho, W.-H., Janssen, L. J., & Schwarz, G. (1975), The Wind Tunnel's Ground Plane Boundary Layer – Its Interference with the Flow Underneath Cars, SAE 750066.

Hucho, W.-H., Janssen, L. J. & Emmelmann, H.-J. (1976), The Optimization of Body Details – A Method for Reducing the Aerodynamic Drag of Road Vehicles, SAE 760185.

Hucho, W.-H. (1976), Steigerung der Wirtschaftlichkeit von Kraftfahrzeugen durch geeignete Formgebung, to be published in Automobil Revue, Bern.

Hurter, D. A. & Lee, W. D. (1975), A Study of Technological Improvements in Automobile Fuel Consumption, SAE 750005.

Janssen, L. J. & Hucho, W.-H. (1973), Effect of Various Parameters on the Aerodynamic Drag of Passenger Cars, Advances in Road Vehicle Aerodynamics, Cranfield, UK, pp. 223-253.

Kamm, W. (1969), Der Weg zum wirtschaftlichen autobahn- und strassentuechtigen Wagen, Strasse, 6, pp. 104-109.

Klemperer, W. (1922), Luftwiderstandsuntersuchungen an Automobilmodellen, Zeitschrift f. Flugtechnik u. Motorluftschiffahrt, 13, pp. 201-206.

Kramer, C., Gerhardt, H. J., Jaeger, E. & Stein, H. (1974), Windkanalstudien zur Aerodynamik der Fahrzeugunterseite, Koll. Industrie Aerodynamik, Aachen, Part 3: Aerodynamik von Strassenfahrzeugen, pp. 71-83.

Morelli, A., Fioravanti, L., & Cogotti, A. (1976), The Body Shape of Minimum Drag, SAE 760186.

Oda, N. & Hoshino, T. (1974), Three-Dimensional Airflow Visualization by Smoke Tunnel, SAE 741029.

Ohtani, K., Takei, M. & Sakamoto, H. (1972), Nissan Full-Scale Wind Tunnel – Its Application to Passenger Car Design, SAE 720100.

Olson, M. E. (1976), Aerodynamic Effects of Front End Design on Automobile Engine Cooling Systems, SAE 760188.

Schmidt, C. (1939), Fahrwiderstaende beim Kraftfahrzeug und die Mittel ihrer Verringerung, ATZ, 41, pp. 465-477 and pp. 498-510.

Wallis, S. B. (1971), Ventilation System Aerodynamics – A New Design Method, SAE 710036.

DIN 70030 (1968), Ermittlung des Kraftstoffverbrauch von Kraftfahrzeugen, Beuth Vertrieb, Koeln.

DISCUSSION

R. T. Jones *(NASA-Ames Research Center)*

I'm going to make a comment which I think may add some insight to a conclusion about the relation of induced drag to lift that I believe the speaker has already drawn. The thrust of my comment is that I think there is no essential relation between lift and induced drag for a ground vehicle, as there is for an airplane. For example, you can have a ground vehicle that develops lots of lift but no drag. Consider the *potential* flow around a sphere where the bisecting plane of symmetry represents the ground. The average pressure coefficient over the upper surface is a negative 11/16, so there is an upward force, or lift, without trailing vortices, and without induced drag.

As another example consider a flat plate mounted perpendicularly to the ground in a *real* flow. Clearly, the plate can develop no lift, but there will be a terrible wake with lots of negative pressure and a pair of trailing vortices with downflow between them. With this flat plate there is drag and trailing vortices, but without any lift.

R. T. Jones

In summary, these two examples show that there can be lift without vortices and drag, and there can be drag and vortices without lift. I think that reinforces the idea that Dr. Hucho expressed in his presentation – that there is no essential relation between lift and drag.

W. H. Bettes *(California Institute of Technology)*

I think it's important to point out that the rear flow field you described in Fig. 14 contains a very idealized representation of the trailing vortex system that is not necessarily general.

W.-H. Hucho

I agree; I tried to make that clear.

W. H. Bettes

In our work at Cal Tech we've put a tuft grid at various stations behind ground vehicles and it's been very difficult to detect any predominant rotational flow in such planes. Rather, there seems to be a recirculating system with high flow angularities behind the centerline of the vehicle, with no discernible longitudinal vortex motion.

R. T. Jones

Doesn't the combination of the negative wake pressure together with the restricted underbody flow almost always produce such longitudinal vortices?

W. H. Bettes

Not that we've been able to determine. It may take better experimental tools than a tuft grid to find the vortices, but they certainly don't seem to be comparable to other motions in the wake.

S. J. Kline *(Stanford University)*

I'm going to make a comment on using smoke for flow visualization. The natural thing that most of us do is to inject smoke into the airstream ahead of the test object. This often leaves some of the most interesting separated regions unenlightened. It's a simple matter to correct that situation by injecting smoke directly into regions of suspected separation. In this manner, recirculating regions can be clearly outlined. If you then use combinations of both methods of smoke flow visualization, but one at a time, I think you can develop a better understanding of the rear flow field and, for example, determine the effects of rear spoilers on fastback cars.

W.-H. Hucho

That is a good comment, and we do employ both techniques at VW. That may not have been evident in some of the flow visualization photos I showed because of their composite nature. To minimize the number of slides for our presentation we used a multiple exposure photographic technique; we don't ever use both methods of smoke injection at the same time. In fact, we have found the technique of injecting smoke directly into the near wake to be a very useful technique for determining the location of the rear line of separation. This helps generate needed information about soil deposition, etc.

W.-H. Hucho

B. Pershing *(The Aerospace Corporation)*

I noticed that you said nothing about the effect of a rotating wheel in a wheelhouse. Have you done any research on that subject?

W.-H. Hucho

The answer is no, but I can comment. From all we know, the rotation of wheels enclosed by wheelhouses adds only slightly to drag. However, we are concerned about the rotational effects of the wheels with respect to soil deposition, but these experiments are done on-road. Open-wheeled racing cars are another matter. Claude Williams from Lockheed-Georgia has shown* how to cope with this problem based on the work of Fackrell & Harvey.** You have to ensure that the point of flow separation from the fixed wheel in the wind tunnel is the same as that from the rotating wheel on the road.

W. H. Bettes

I have some supplementary information regarding the 0.15 drag coefficients of the Klemperer and VW low-drag ground vehicles that you presented in Fig. 33. As far as I know, the minimum drag coefficient that has been reported for a real ground vehicle is 0.12 based on frontal area.*† The vehicle is the Summers brothers' Goldenrod, the wheel-driven land speed record holder.

Editors' Comment

In the mid-Sixties General Motors Corporation designed and fabricated wind tunnel models of a "simplified" fastback car in four different scales (1/8, 1/6, 1/4, and 1/3). The body design was realistic in shape and dimension (see Fig. 35), and featured a blunt nose and rear end, a well-defined hood and windshield, slab sides which were perpendicular to the ground, and a planar underbody which produced a front contraction of the flow and a rear diffusion. There were realistic wheels and wheelhouses and the front underbody was equipped with a protuberance for roughness, but there was no engine compartment or cooling-air flow path. There was a small spoiler at the trailing edge of the fastback to fix the line of flow separation. Over the years these models were tested in three different wind tunnels during investigations of groundplane,† and blockage and wall effects. For the last 5 years the 1/4-scale version has been used in General Motors' aerodynamic test facility as a calibration model. At a Reynolds number of 2×10^6 based on effective diameter the drag coefficient†† has been consistently measured at a very low value of 0.23 (see Fig. 35).

**Wise, C. E. (1977), Aerodynamics at Indy, Machine Design, Penton/IPC, May 12, pp. 20-26.*

***Fackrell, J. E., & Harvey, J. K. (1975), The Aerodynamics of an Isolated Road Wheel, Proceedings of the AIAA Second Symposium on Aerodynamics and Racing Cars, edited by B. Pershing, Vol. 16 AIAA Lecture Series, Western Periodicals Co., California, pp. 119-125.*

**†Korff, W. H. (1966), The Aerodynamic Design of the Goldenrod – to Increase Stability, Traction, and Speed, SAE 660390.*

†Mason, Jr., W. T. & Sovran, G. (1973), Ground-Plane Effects on the Aerodynamic Characteristics of Automobile Models – an Examination of Wind-Tunnel Test Technique, Advances in Road Vehicle Aerodynamics, BHRA, England, pp. 291-309.

††Uncorrected for groundplane boundary layer or blockage/wall effects.

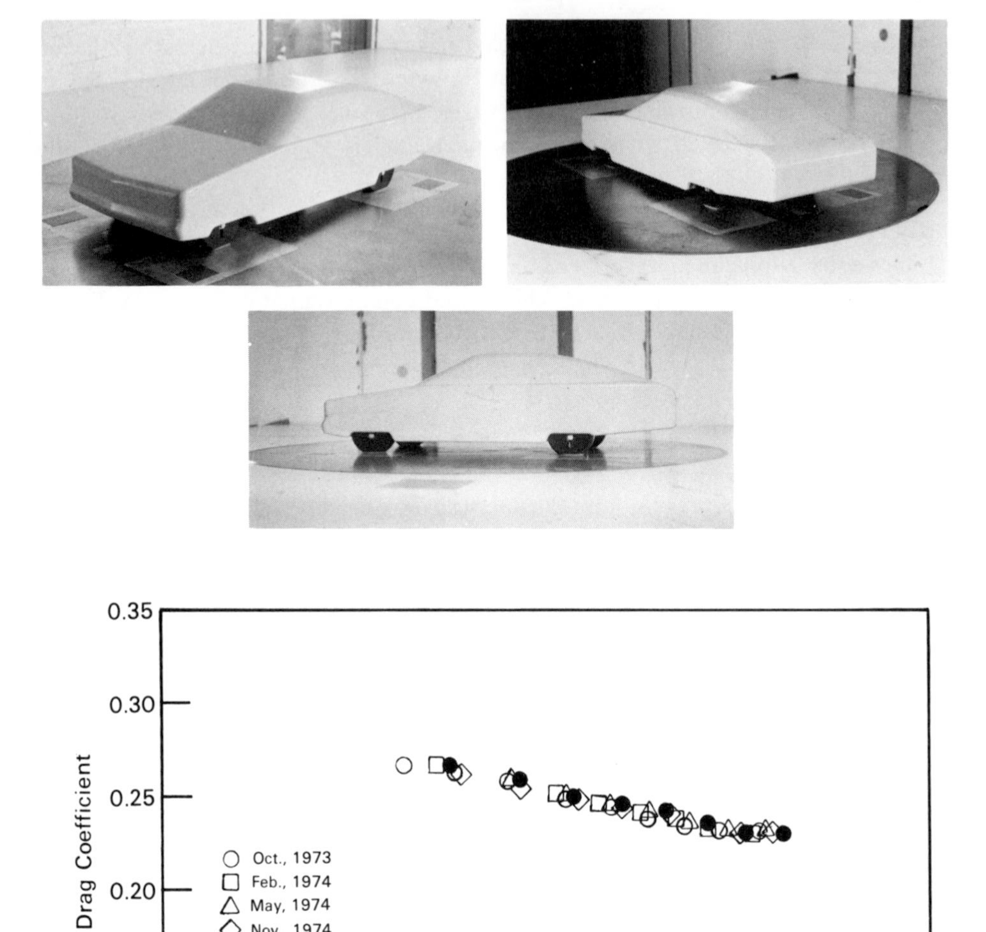

Fig. 35. Drag coefficient as function of Reynolds number (based on effective diameter) for 1/4-scale model of simplified fastback car.

THE DRAG RELATED FLOW FIELD CHARACTERISTICS OF TRUCKS AND BUSES

W. T. MASON, JR. and P. S. BEEBE

General Motors Research Laboratories, Warren, Michigan

ABSTRACT

Non-aerodynamic factors are largely responsible for the size and shape of contemporary trucks and buses. The results of wind tunnel experiments with 1/7-scale tractor-trailer and bus models are used to identify major drag producing regions of the flow fields, and to document some of the detailed characteristics. Some modifications of both the forebody and base flow fields are made in order to explore the practical potential for drag reduction. The largest drag reductions are shown to be achievable by changing the forebody flow field. By controlling flow separation from leading edges, either by modifying body contours or by employing add-on devices, apparent minimum drag limits have been identified. The possibility of even lower drag levels within existing constraints is analyzed. At the end, non-zero yaw drag characteristics are briefly discussed.

INTRODUCTION

The shape and size of contemporary trucks and buses are largely the result of functional and legal constraints; in addition, aesthetics and tradition have played a subtle, but important role. Aerodynamic drag levels have been somewhat of a by-product, rather than a specific design consideration. Load carrying efficiency is of utmost importance. The vehicle volume is maximized within the legal limits, and the payload, be it passengers or freight, is carried in a long body section of constant area. The forebody and base are necessarily blunt, and the overall shapes tend to be aerodynamically bluff rather than streamlined. Configurational differences between trucks and buses exist because of two distinct methods of payload management. Most

References pp. 89-90.

freight is understandably hauled by trucks comprised of two separate modules – a tractor and a trailer. A bus, on the other hand, does not utilize an independent payload module. The passenger section, engine, drivetrain, and driver are integrated into a single-chassis vehicle.

To establish the aerodynamically relevant geometric characteristics of trucks and buses it is helpful to consider their approximate, rather than exact, dimensions, as shown in Fig. 1. The overall length of a bus is less than that of a tractor-trailer, the width is the same, and the height is equal to that of the tractor. The length-to-diameter ratio (ℓ/d) is 4, where the diameter is an effective diameter based on projected frontal area. The aspect ratio of the body cross-section is one. The ℓ/d of a tractor-trailer ranges between 5 and 6, depending on the length of the trailer. The maximum length and height, which are legislated, are 40 percent larger than the corresponding bus dimensions. The ℓ/d of the trailer is between 4 and 5, and the aspect ratio of its cross-section is one. The trailer box dimensions are similar to those of a bus.

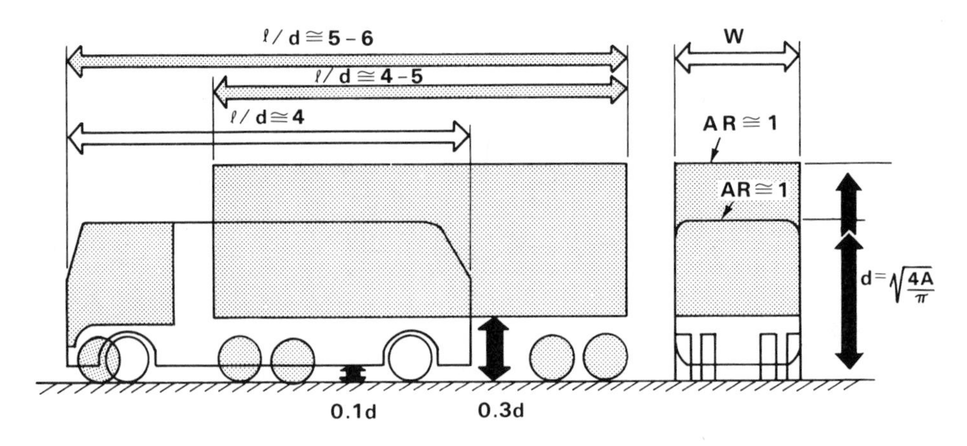

Fig. 1. Comparison between aerodynamically relevant geometries of tractor-trailers and buses.

The tractor and trailer are articulated for maneuverability, and an air gap is provided for small-radius turning. Many factors influence the length of the gap – tractor, trailer, and overall-vehicle lengths; vehicle weight distribution; geometry of the trailer front corners; and owner preference (accessibility, maintenance, etc.). Gaps tend to range anywhere from 20 to 100 percent of the effective diameter of the vehicle. Gaps between 30 and 60 percent are most common. Since trailers are a separate module and pulled by a tractor, their floor height is determined by the frame/suspension structure and the wheels, and is usually constant with length. A depressed center section is usually not practical because of palletized loading dock operations. The ground clearance is about 30 percent of the diameter. The clearance of the tractor and the bus is less – about 10 percent.

Two different types of tractor have evolved – the cab-behind-engine (conventional) and the cab-over-engine (C.O.E.); these are compared in Fig. 2. As the names infer, C.O.E. tractors are of less length and greater height than conventionals. The width of C.O.E. tractors is equal to that of the trailer and uniform with length. Conventional tractors are only 75 percent of the width of the trailer, and are non-uniform in cross-section. Tractor lengths range between 40 and 80 percent of the vehicle diameter. C.O.E. tractors are of two lengths – 40 percent (non-sleepers) and 60 percent (sleepers). Conventionals are longer than sleepers, and are twice the length of C.O.E. non-sleepers.

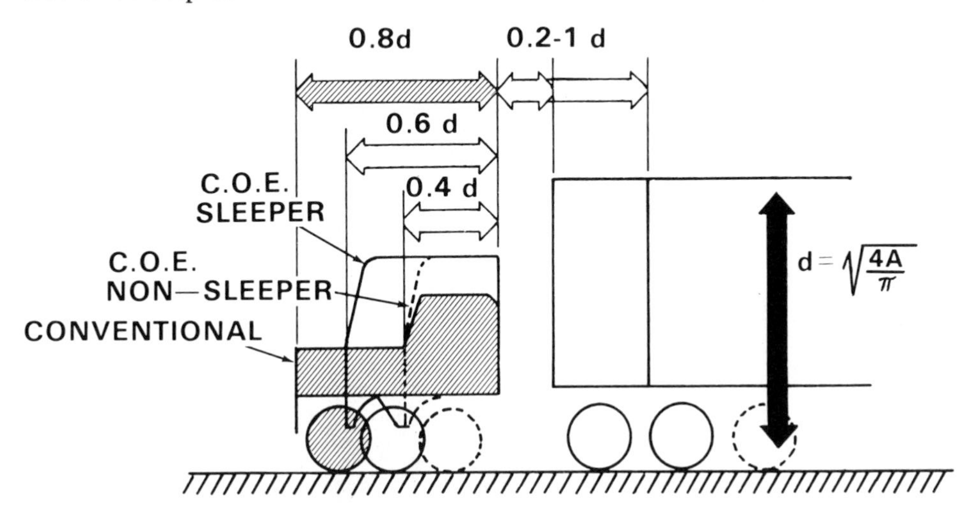

Fig. 2. Geometries of typical tractor-trailer combinations.

As the flow fields and drag mechanisms of tractor-trailers and buses are examined for the purpose of reducing aerodynamic drag, the functional and legal heritages which have produced the current geometries of these highly specialized vehicles must be constantly recognized as the primary design constraints.

SCOPE AND ORGANIZATION

The contributions of this paper deal primarily with zero-yaw flow fields. Non-zero yaw flows are certainly important with respect to real-world fuel economy, but in keeping with the objectives of the Symposium, coverage has been restricted to an in-depth analysis of zero-yaw flow modules. The zero-yaw modules are necessary building blocks for understanding the more complicated non-zero modules. Furthermore, the zero-yaw flow field establishes the level of drag from which the non-zero drag characteristics depart.

The forebody flow field characteristics are first discussed, followed by documentation of forebody drag reduction. The base flow characteristics are then presented, and

References pp. 89-90.

the results of some base drag reduction investigations are described. Some apparent minimum drag limits are identified, and the question of even lower drag levels is addressed. At the end, non-zero yaw drag characteristics are briefly discussed.

FACILITY AND TEST PROCEDURE

The research was conducted with scale-models in a wind tunnel. The overall test setup is shown in Fig. 3. Comparative results were the primary objective, not absolute drag magnitudes.

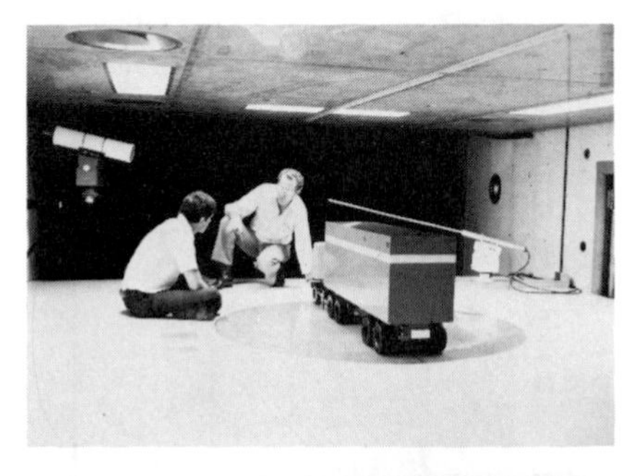

Fig. 3. Overall views of test section. For reference, diameter of yaw turntable is 1.8 m. Closed-circuit TV camera can be seen in upper-left of top photo; remotely controlled smoke generator in lower-right (camera and generator not installed during force and pressure testing).

Wind Tunnel – The aerodynamic test facility (see Mason, Beebe & Schenkel, 1973) was a closed test section, single-return wind tunnel with a fixed groundplane 6.10 m long, 4.57 m wide and 0.61 m above the test-section floor. The active test section was 1.35 m high. Models were situated on a 1.83 m diameter turntable located 2.13 m downstream from the groundplane leading edge. Test velocities were set using a calibration between the contraction pressure-drop and the empty test section dynamic pressure. No adjustments were made for any changes that may have been caused by the models. The maximum dynamic pressure was 2.7 kPa. The groundplane housed a six-component strain-gage platform-type balance under the turntable. All force coefficients were determined from 30-second simultaneous integrations of the balance output and the contraction pressure-drop. Surface pressure coefficients were obtained using 10-second simultaneous integrations of surface pressure and contraction pressure-drop. Surface pressures are reported relative to empty test section static pressure at the model location.

Modeling of On-Road Flow Field – All models were 1/7-scale, and semi-detailed. The full-scale frontal area of the two bus models was 7.2 m^2 (including open area between wheels). The length of each bus was 12.2m. The forebody of one of the bus models was designed to accept interchangeable sections. The frontal area of all tractor-trailers was 10 m^2 (maximum height times maximum width, 4.11 x 2.44 m). The length of the trailer was 9.15 m, and various tractor-to-trailer gaps were obtained by either adjusting the relative positions of the tractor and trailer or by adding segments to the trailer face. Tractor dimensions varied with tractor type, and are discussed in later sections.

Data were recorded at a nominal Reynolds number of 2×10^6 based on effective diameter; this was equivalent to 30 km/h full-scale road speed. The variation of drag coefficient over the available Reynolds number range of 0.9 to 2.1×10^6 was negligible. The turbulence intensity in the empty test section was 0.5 percent.

The bus blocked 2.4 percent of the active test section area; the tractor-trailer 3.3 percent. The test-section ceiling was 2.1 diameters above the bus roof and 1.4 diameters above the truck trailer. At zero yaw angle, the side walls were approximately 5.4 diameters to either side of the bus, and 4.4 diameters to either side of the tractor-trailer. With the models centered on the turntable, there were 2.9 diameters of groundplane upstream of the bus, and 2.4 diameters upstream of the tractor-trailer. The downstream extent of the groundplane was 7.1 diameters for the bus, and 5.9 for the truck. The groundplane boundary layer displacement thickness was 3.8 mm at the model location in an empty test section. The tractor-trailer wheels had a 1.3 mm gap between their flats and the groundplane. A smaller gap of 0.64 mm could be maintained for the bus. No adjustments were made to the aerodynamic coefficients for model blockage, tunnel wall effects, groundplane boundary layer thickness, or wheel-to-ground air gaps. Some additional detail on the test setup can be found in Mason (1975).

Flow Visualization – Flow fields were visualized with vaporized-oil smoke from a remotely controlled wand. The wand and its bracketry blocked an insignificant portion of the test section near the side wall, as can be seen in the top photo of Fig. 3. Flow visualization was conducted at a Reynolds number of 1.0×10^6.

References pp. 89-90.

FOREBODY FLOW FIELD

The forebody of a bus is geometrically simple compared to that of a tractor-trailer. There is only one major blunt surface, and the projected area of that surface is equal to the maximum cross-sectional area of the body. A tractor-trailer has two major forward facing blunt surfaces because of its two-body configuration. The tractor and trailer are bluff bodies in tandem, with the tractor "shielding" a large portion of the trailer. The forebody of a bus is closer to the ground than that of a tractor-trailer, and the underbody geometry is less complicated. A bus has no engine-cooling flow path in the forebody since most are rear-engined, but truck tractors do contain an engine and an associated throughflow of cooling air.

Trailers and Buses – The forebody flow field of a sharp-edged trailer was compared to those of two buses with blunt forebodies. The models are shown in Fig. 4. One of the bus forebodies was detailed; the other was of simplified configuration – flat with sharp edges. The latter was used to establish maximum leading edge flow separation. The flow fields of all three forebodies were similar. There was a large stagnation area on the front face with massive separation from the leading edges. The corresponding drag coefficients were 0.92 for the trailer, 0.71 for the simplified bus, and 0.66 for the detailed bus. The reasons for the larger coefficient of the trailer will be explored later.

Fig. 4. Sharp-edged trailer and two buses with blunt forebodies, detailed and simplified.

Tractor-Trailers – *Zero Gap Length.* The trailer flow field was brought closer to reality by placing the tractor directly ahead of it with no gap. The C.O.E. tractor had well-rounded leading edges ($r \cong 0.1$ d) and a width equal to that of the trailer. The effective diameter of the tractor was 80 percent of the truck diameter; its length was 40 percent.

A large reduction in drag coefficient was measured – from 0.92 to 0.72. Smoke flow visualization revealed that the flow separated from the tractor roof, and did not reattach, as sketched in Fig. 5. There was no separation from the tractor side-edges. A large separation bubble with recirculating flow formed within the concave corner of the tractor roof and trailer face – the familiar flow field of a forward facing step. The step height-to-length ratio was 0.8, and the separated shear surface intercepted the trailer near its leading edges. Some flow separation was present from those edges.

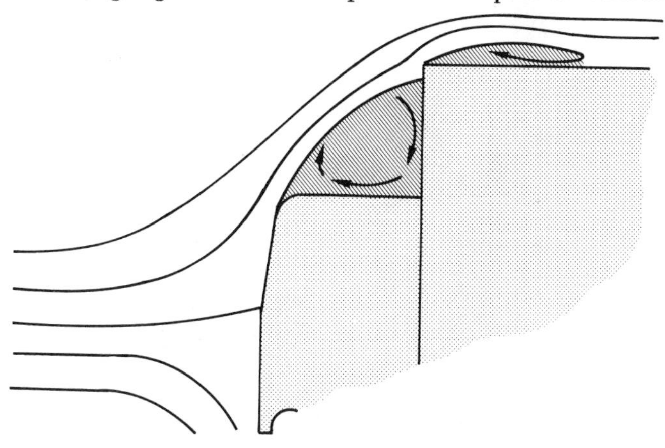

Fig. 5. Forebody flow field of C.O.E. tractor-trailer at zero gap, based on flow visualization.

The zero-gap drag coefficient is probably a function of the degree of matching of the separated shear surface to the leading edges of the trailer. This matching is a function of the height-to-length ratio of the step and the shape of the tractor. When the step height-to-length ratio was decreased from 0.8 to 0.5 by increasing the length of the tractor to 0.6 d, the drag coefficient decreased by 0.05.

Increasing the Gap. To create a realistic tractor-trailer a gap was provided. The gap was varied from zero up to a maximum of 60 percent of the vehicle's effective diameter. Smoke was used to study the flow field as the gap increased, see Fig. 6. The separation from the tractor roof associated with zero gap quickly diminished, and disappeared. A second stagnation streamline impacted upon the exposed face of the trailer, and was well-established at a gap of 20 percent. Some of the flow over the tractor roof turned and became a high velocity downflow through the gap. There was no evidence of lateral flow – all flow was downward. The flow up the trailer face separated from the sharp leading edge of the trailer roof; the flow to either side

References pp. 89-90.

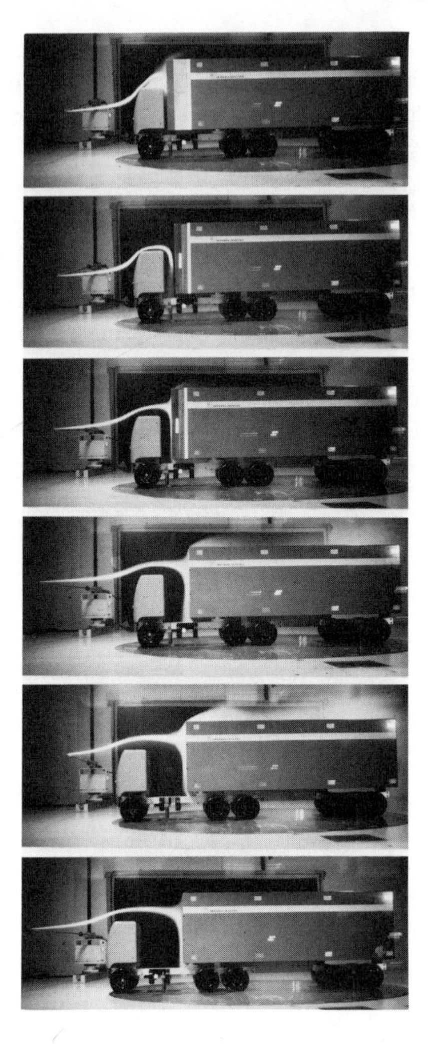

Fig. 6. Development of forebody flow field with increasing gap; maximum gap is 0.6 d.

separated from the upper side-edges of the trailer. The large separation bubbles were between 0.75 and 1.5 diameters in length. As the gap increased, the stagnation point moved down and exposed larger portions of the trailer face. The flow visualization results presented in Fig. 6, and the corresponding drag measurements which are discussed next, should be qualified somewhat by the fact that the leading edges of the tractor were well-rounded. The roof edge of many C.O.E. tractors is sharp, and the gap at which the forward-facing-step separation disappears might be larger than that observed in Fig. 6; in addition, some roof-edge separation would probably remain throughout the range of gaps.

The tractor had a maximum beneficial effect on the drag of the combination at zero gap. Drag is plotted as a function of gap in Fig. 7. The shielding effect decreased as the gap increased and "exposed" increasing amounts of the trailer face to the approaching flow. The drag coefficient increased with increasing gap, and reached a maximum of 0.93 at g/d = 40 percent. This was equivalent to that of the trailer with sharp edges. At gaps larger than 40 percent, the drag decreased slightly, but eventually it has to increase again as the flow fields of the tractor and trailer develop independently. Ultimately, at a gap of infinity a maximum coefficient of 1.3 will be reached. This was determined by measuring the drag of the tractor and trailer individually.

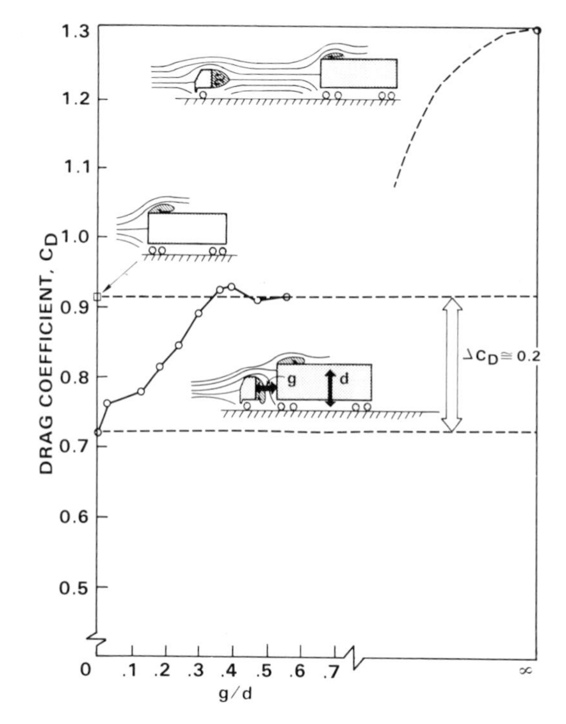

Fig. 7. Drag of C.O.E. non-sleeper tractor-trailer as a function of gap.

Blocking the Gap. The importance of the trailer face stagnation flow as a forebody drag mechanism was demonstrated by blocking the gap with a horizontal plate extending from the tractor roof. Smoke showed that the forward-facing-step flow field associated with zero gap was maintained as the gap increased. This can be seen in Fig. 8. The drag also remained at the zero-gap level, as indicated by the data presented in Fig. 9. The plate prevented increased exposure of the trailer face to the approaching flow, and the shielding effect of the tractor was maintained.

References pp. 89-90.

Fig. 8. Forebody flow fields of C.O.E. sleeper tractor-trailer with open and blocked gap; g/d = 0.25.

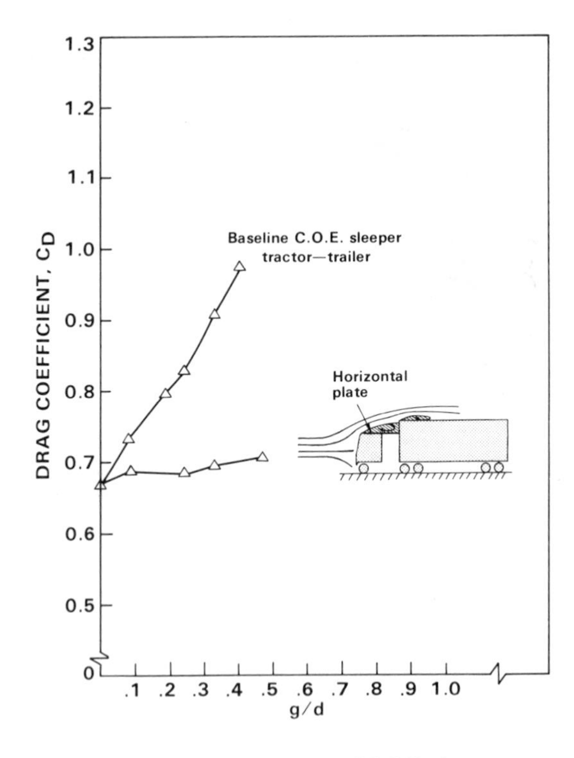

Fig. 9. Effect of blocked gap on drag of C.O.E. sleeper tractor-trailer.

Tractor Size and Shape. Differences in shielding from one tractor size and shape to another were explored over a normal range of gaps (20 to 60 percent of the diameter). The three tractors used covered the range of contemporary design. Two were conventionals and one was a C.O.E. non-sleeper. The effective *geometric* diameters of the two conventionals and the C.O.E. were 72, 74 and 80 percent of the trailer diameter, respectively; the *aerodynamic* diameters (based on $C_D A$) of the three tractors were 57, 61 and 66 percent of the trailer diameter. The conventionals were longer than the C.O.E. – 70 and 80 percent of the effective diameter compared to 40 percent.

Smoke flow visualization photographs of these three tractor-trailers at a gap of 50 percent are shown in Fig. 10. There were local flow field differences from one tractor to another, but these were secondary compared to the similarity of the trailer face stagnation flows. At given gaps within the range where trucks actually operate,

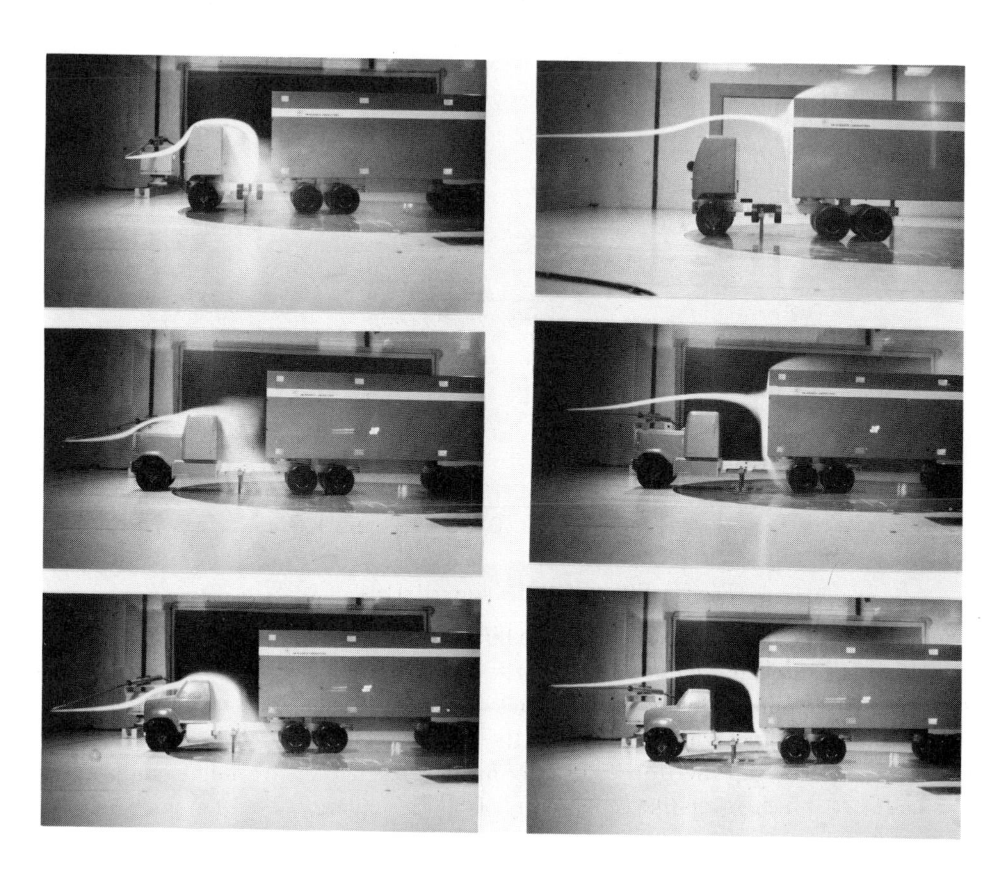

Fig. 10. Visualization of three tractor-trailer flow fields (C.O.E. non-sleeper and two conventionals) at the same gap of 0.5 d. The flow adjacent to the tractor roof is shown on the left; the trailer-face stagnation is shown on the right.

References pp. 89-90.

approximately the *same* drag was obtained for each combination, see Fig. 11. This result was unaffected by the presence or absence of engine-cooling airflow.

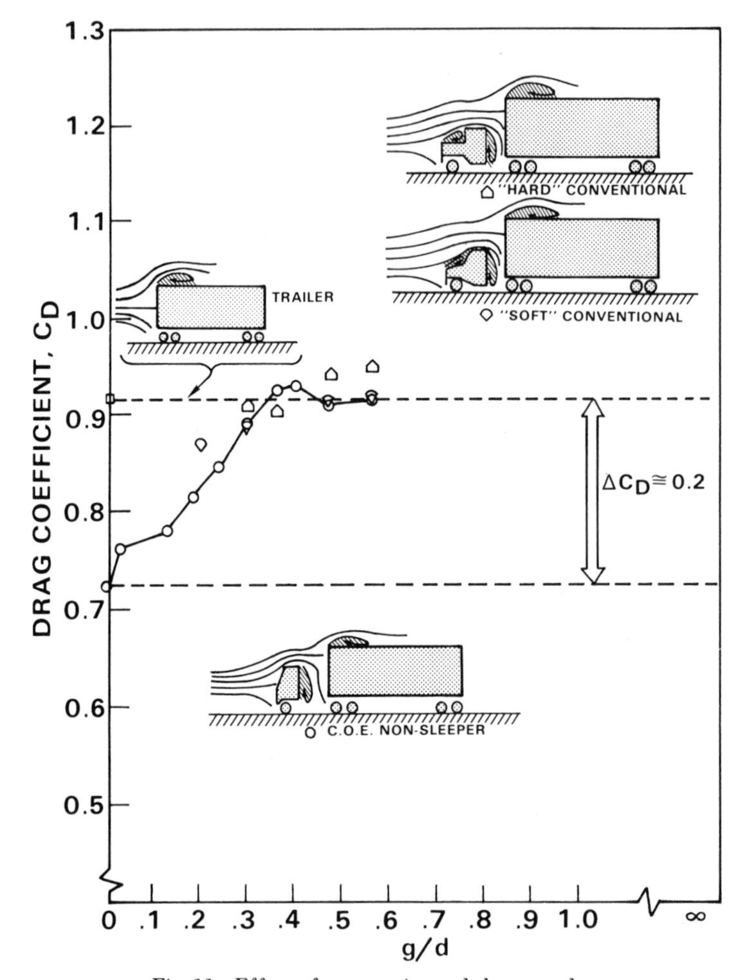

Fig. 11. Effect of tractor size and shape on drag.

Summary — Tractor-trailers and buses exhibit different forebody flows. Because of a gap, the height differential between tractor and trailer creates a flow field with two stagnation streamlines. One of these impacts the tractor, while the second impacts the trailer. Drag increases with gap because the trailer shielding provided by the tractor decreases. At zero gap the flow field only has one stagnation streamline and, except for the forward-facing-step-like separation bubble, is similar to that of a bus.

REDUCTION OF FOREBODY DRAG

Two alternative approaches can be used to reduce forebody drag. One is to reduce flow separation from the edges of any flat, or nearly flat, forebody; the other is to avoid separation in the first place by providing additional forebody length for appropriate shaping. For trucks and buses of fixed length, a "corner rounding" approach will lead to shapes of greater internal volume, i.e. load capacity, than those developed using a "streamlining" approach. From a drag standpoint the advantage of "streamlining" compared to "corner rounding" is not readily apparent, even for bodies away from the ground. Prandtl & Tietjens (1934) showed that the average forebody pressure coefficient for any semi-infinite, rotationally symmetric body is zero in potential flow. Some work referenced by Hoerner (1965), see Fig. 12, suggests that average pressure coefficients near zero, even negative, are indeed possible in *real* flows. These results, however, do not adequately address the drag related significance of "streamlining" compared to "corner rounding" in real flows. For example, if the flat faced, semi-infinite body with $\overline{C}_{p_{fb}} = 0.2$ in Fig. 12 had sufficiently rounded corners to eliminate separation, would its forebody drag be significantly different from that of either the hemispherical or elliptical nose body?

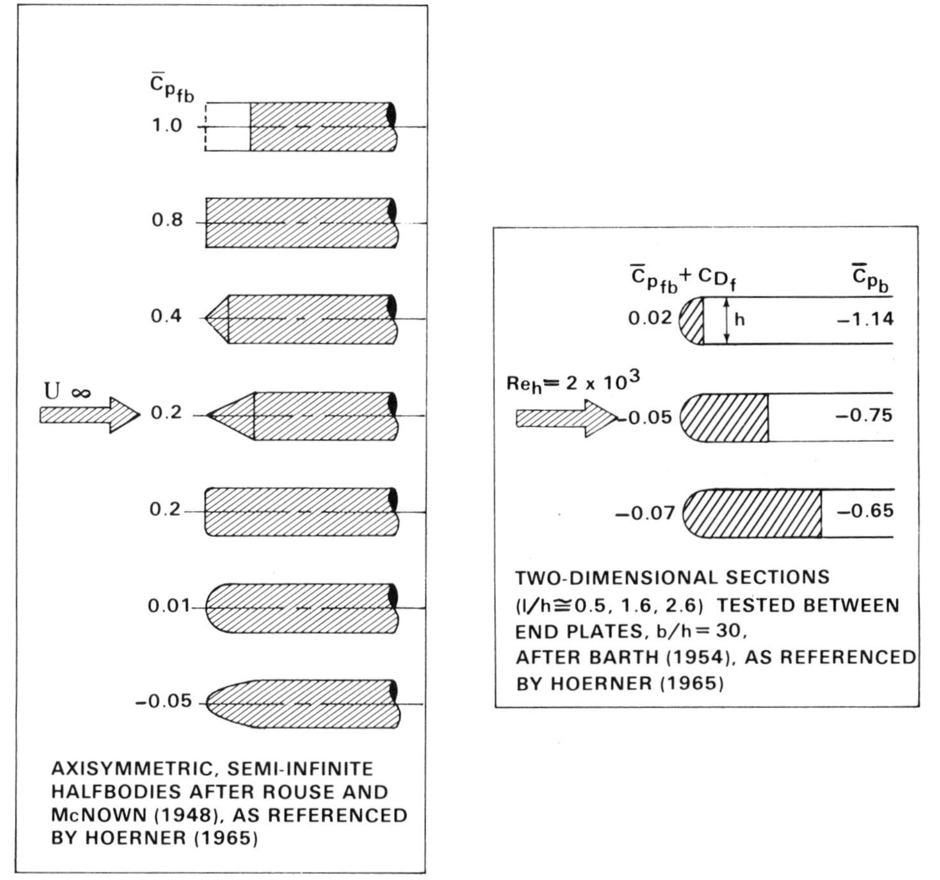

Fig. 12. Effect of shape on average forebody pressure coefficient of axisymmetric, semi-infinite bodies (left), and the effect of length-to-height ratio on the average forebody and base pressure coefficients of two-dimensional sections (right).

References pp. 89-90.

Buses – The "corner rounding" approach is not new or novel. There are many examples in the literature; perhaps one of the earliest and best known is Moeller's (1951) drag reduction of the VW microbus using a combination of "corner rounding" and slight shaping, see Fig. 13, which is referenced in Schlichting's *Boundary Layer Theory* (1960). A similar combined approach was used to reduce the forebody drag of the simplified bus, see Fig. 14, and the result was about the same. The drag coefficient of the simplified bus was reduced by almost 0.3, from 0.71 to 0.44. Smoke and wool-tuft flow visualization, Fig. 15, showed that leading-edge separation was reduced; some separation still existed from the roof.

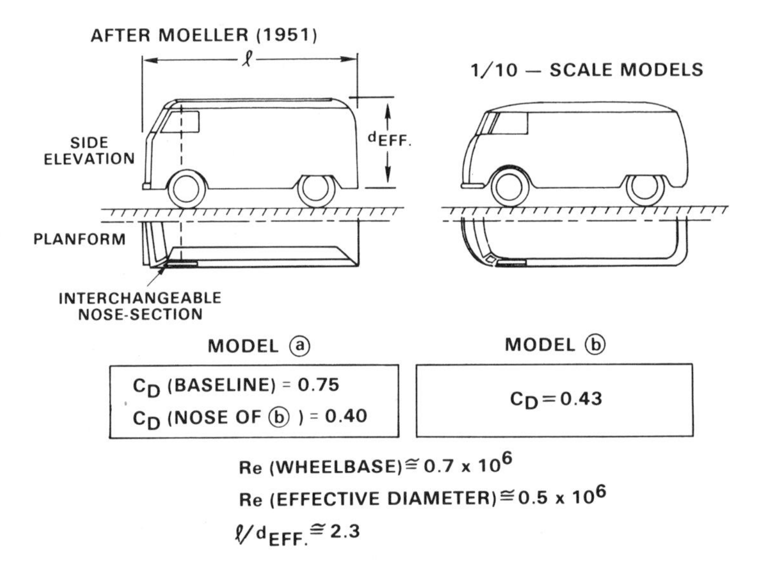

Fig. 13. Drag reduction of VW microbus by Moeller (1951).

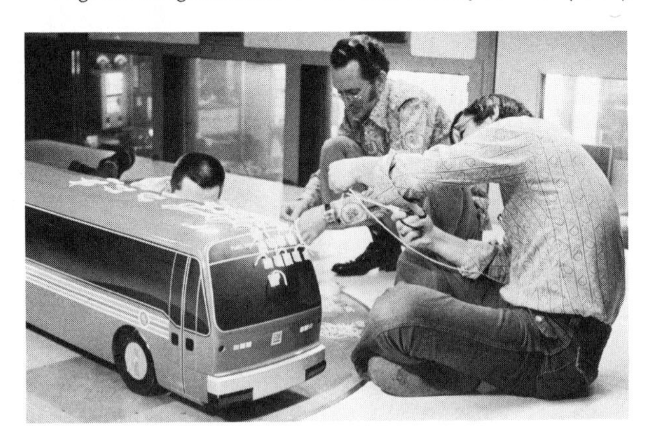

Fig. 14. Practical forebody with corner rounding and slight shaping installed on simplified bus of Fig. 4.

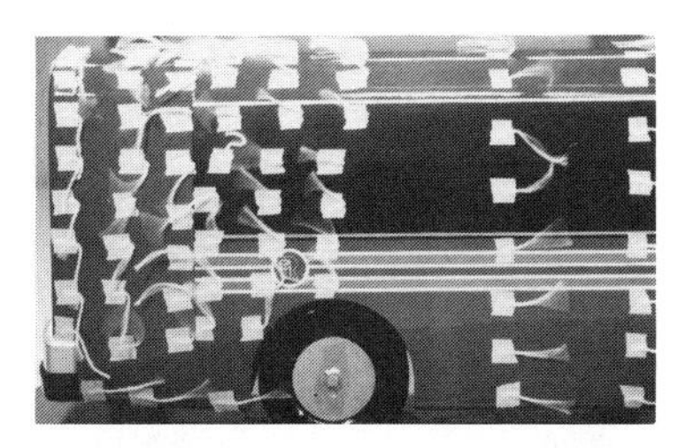

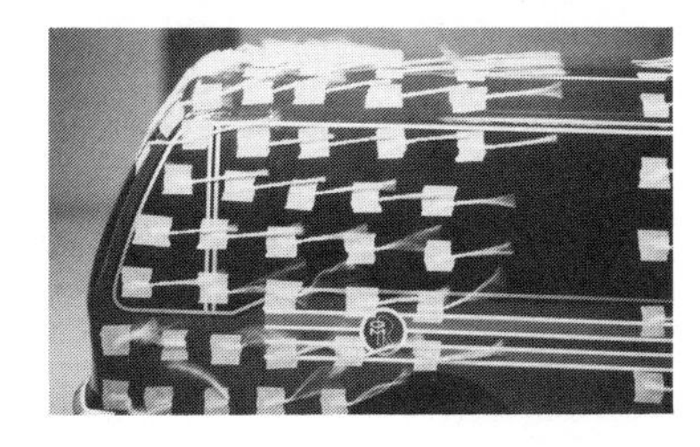

Fig. 15. Wool-tuft visualization of low drag (bottom) and high drag (top) flow fields of simplified bus. Note slight separation from roof of low drag forebody.

The effect of additional streamlined shaping was not investigated. Some basic work done at MIRA by Carr (1967) with rectangular bodies of commercial-vehicle proportions in ground proximity, as sketched in Fig. 16, indicated that there was very little drag advantage associated with "streamlining" once "corner rounding" was used to reduce separation from the flat forebody. The same result might be expected with the simplified bus.

Tractor-Trailers — For tractor-trailers at normal operating gaps, the forebody drag associated with the stagnation streamline on the exposed face of the trailer provides a potential for drag reduction not available on buses. The forebody drag can be reduced by three alternative approaches. The first two are those discussed earlier. The third is to transform the double-stagnation flow field into one that is similar to the single-stagnation field of buses. The low drag of the truck at zero gap is evidence that the single-stagnation field produces lower drag.

Trailer Face Modifications. To provide adequate minimum-radius turning, manufacturers design most trailers with either beveled or radiused *side* edges. Side edge radii of 10 percent of the trailer diameter are fairly common, and when tested, were found to eliminate most of the side-edge separation; drag coefficient reductions between 0.10 and 0.20 were obtained, depending on tractor configuration. Beveled side-edges, where the length of the bevel was the same as the corner radius, provided half as large a drag reduction. For C.O.E. tractors, rounding of the portion of the side edges shielded by the tractor provided no drag reduction.

References pp. 89-90.

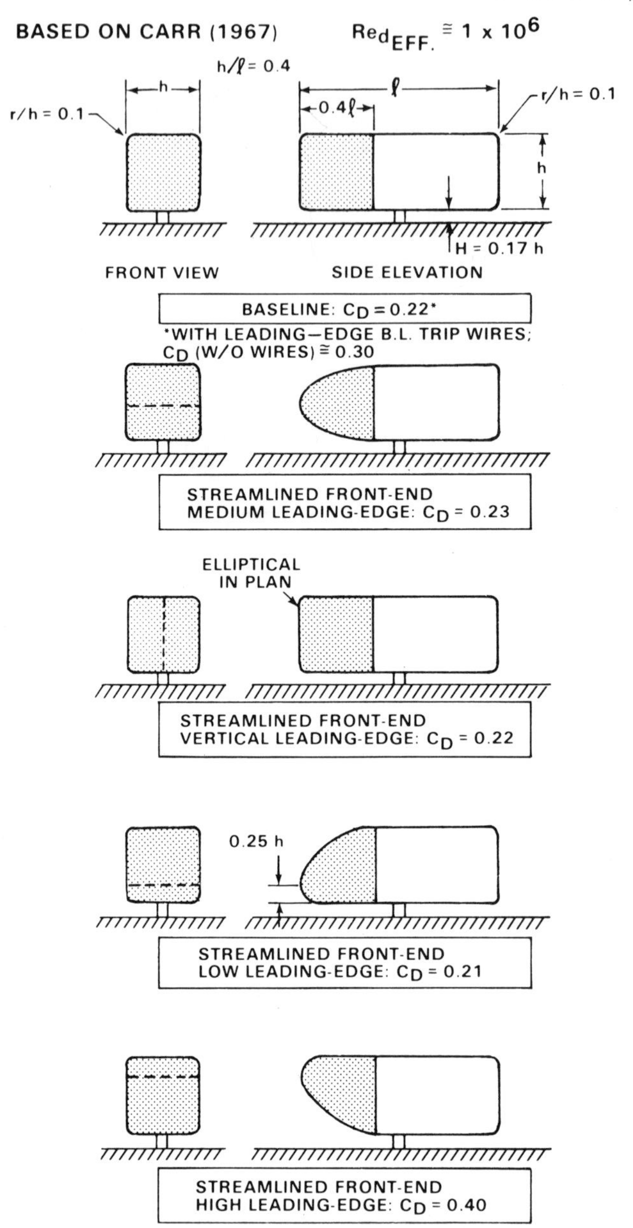

Fig. 16. Effect of forebody streamlining on drag of rectangular body in ground proximity; after Carr (1967).

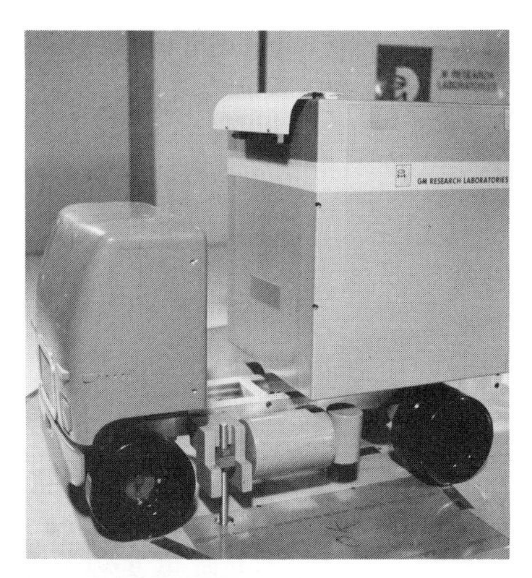

Fig. 17. Drag reducing treatments of trailer-roof leading edge – lip fairing, guide vane, and guide vane with exit area taped; C.O.E. non-sleeper, g/d = 0.3.

The *roof* edge of most trailers is sharp. An effective rounding of this edge with either a vane or a lip fairing, illustrated in Fig. 17, reduced the drag coefficient by 0.07. Out of curiosity the flow between the vane and the trailer was eliminated with tape as shown in Fig. 17. The drag reduction was found to be the same as when the flow path was open. The vane was evidently aerodynamically equivalent to the lip fairing, which would explain the similar drag reductions that were obtained. Kirsch & Bettes (1974) and Steers & Montoya (1974) reported similar reductions for vanes, and Lissaman (1974) reported about the same reduction for a lip fairing.

References pp. 89-90.

The effects of rounding *both* the side and roof edges were not found to be additive. For a particular C.O.E. tractor-trailer, rounding the side edges reduced the coefficient by 0.09; rounding the roof edge reduced the coefficient by 0.07. Simultaneous rounding gave a 0.09 reduction – the same as that obtained with only side-edge rounding.

Would "streamlining" the front end of the trailer beyond "corner rounding" provide significant additional reduction in drag? Drag reducing trailer fairings have been reported by Fitzgerald (1974); however, the advantage of this approach would have to be measured against the drag reduction achievable by rounding the side and roof edges in combination with some decrease of the gap. Such an evaluation has not been made.

Sizing the Tractor. At normal operating gaps the size and shape of contemporary tractors limit their ability to act as a good forebody for the trailer. As was shown earlier in Fig. 11, the drag of most tractor-trailers whose gaps are at, or beyond, 40 percent of the diameter is equal to, or greater than, the drag of the trailer with sharp edges. For minimum drag of the combination it would seem that the tractor should be configured such that the separated streamlines from the top and side surfaces, which are inevitable, are matched to the leading edges of the trailer. Depending on the size of the tractor and the length of the gap, such a design criterion may call for purposely *inducing* flow separation from the leading edges of the tractor. This may be contrary to "normal" aerodynamic thinking, i.e. for low drag one should *reduce* separation, but evidence will be presented which shows that it is effective. The idea of sizing *and* shaping the tractor for minimum drag is exemplified by the sketches of tractor-trailer flow fields in Fig. 18. Good matching is illustrated by Figs. 18(a) and 18(b) for both separated and attached tractor side-flows. On the other hand, if a tractor is streamlined for minimum side-flow separation, geometric mismatching can still produce increased drag, as shown in Fig. 18(c). The worst combination, that of Fig. 18(d), occurs when both tractor separation *and* mismatching exist.

One configuration of particular interest which satisfies the minimum drag criterion is a C.O.E. tractor whose frontal area is the same as that of the trailer, and whose leading edges are sufficiently rounded to avoid separation, see Fig. 18(b). If the gap is not too large, the streamlines which separate from the base edges of the tractor should match reasonably well with the edges of the trailer. Such a tractor-trailer combination could be used to establish a minimum drag target for more practical tractor designs. For this purpose the separation from the leading edges of a trailer was minimized with "corner rounding" (optimized vanes were used, see Fig. 19). The incremental drag reduction was 0.3, from 0.92 to 0.63. This is perhaps a little optimistic for a tractor-trailer since it did not include the drag potentially associated with a gap. It is interesting to note that the *incremental* drag reduction of the trailer due to corner rounding was the same as that reported earlier for the bus.

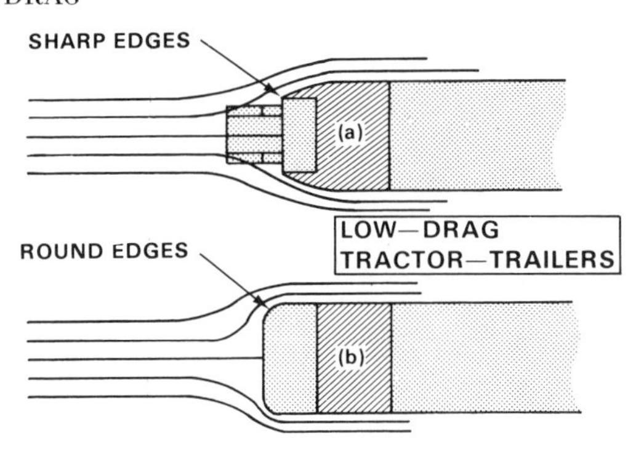

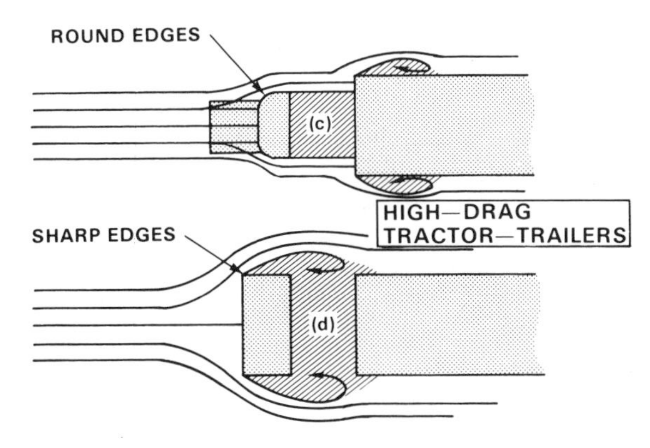

Fig. 18. Sizing and shaping the tractor for minimum combination drag. Planform sketches of low drag flow fields (top) and high drag flow fields (bottom).

Fig. 19. Corner rounding of trailer with optimized guide vanes.

References pp. 89-90.

Rather than redesigning contemporary tractors, add-on devices can be mounted to the tractor roof. Vertical flat plates, as proposed by Saunders (1966), are an example of a class of such devices. Ideally, flat plates at any given longitudinal position on the tractor should be sized such that the separated shear layer from the edges of the plate is matched to the perimeter of the trailer, as sketched in Fig. 20(b). Even when this matching criterion is satisfied, the plate tends to stagnate and separate the flow ahead of itself in forward-facing-step fashion. The details of this separation, for example the vertical location where the shear layer stagnates on the plate and the curvature of the locally concave streamlines, will depend on the tractor shape and the plate location. A perhaps unexpected example of a flat plate drag reducer is a roof-mounted air conditioning unit, Fig. 21, which was found to reduce the drag coefficient by 0.12, from 0.91 to 0.79. When the small flow path between the roof-air unit and the roof was filled in with clay, the incremental reduction increased to 0.28! The same result was obtained with the small flat plate of equivalent area shown in Fig. 21. The accompanying smoke flow visualization photograph, see Fig. 21, indicates the presence of optimum shear layer matching, as called for in the sketch of Fig. 20(b), while the higher drag flow field with the roof-air unit shows that the separated shear layer was intercepted by the trailer face, as in Fig. 20(c).

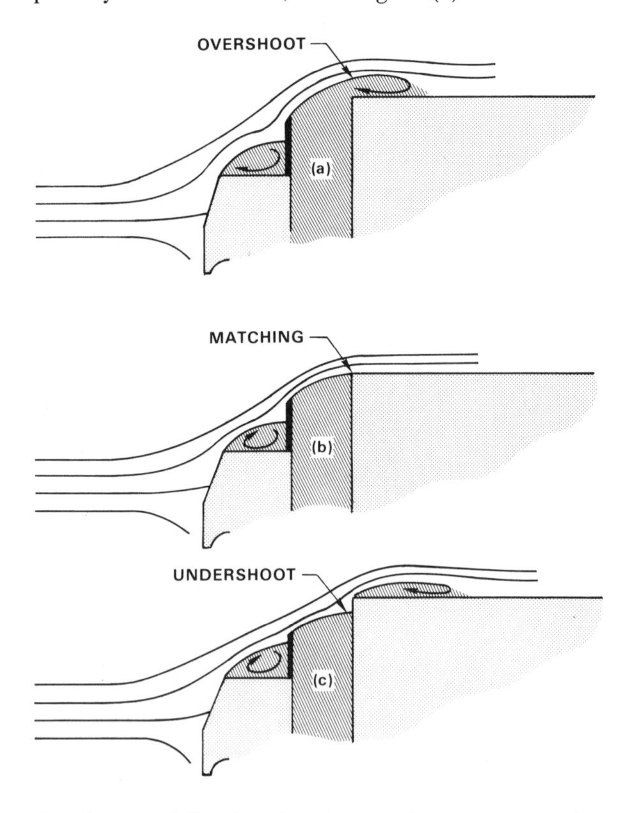

Fig. 20. Sizing of roof-mounted flat plate for minimum drag of tractor-trailer combination.

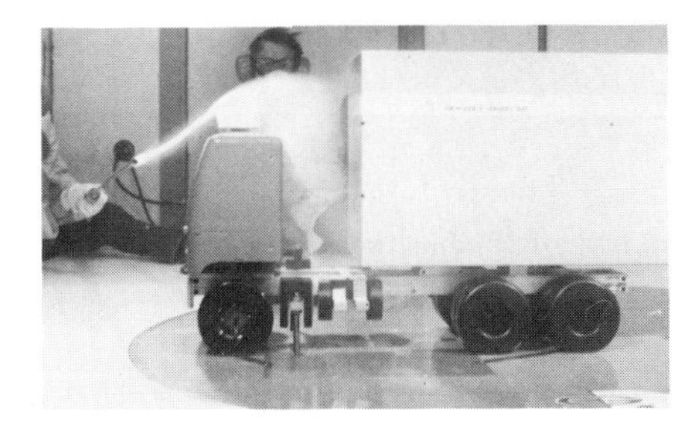
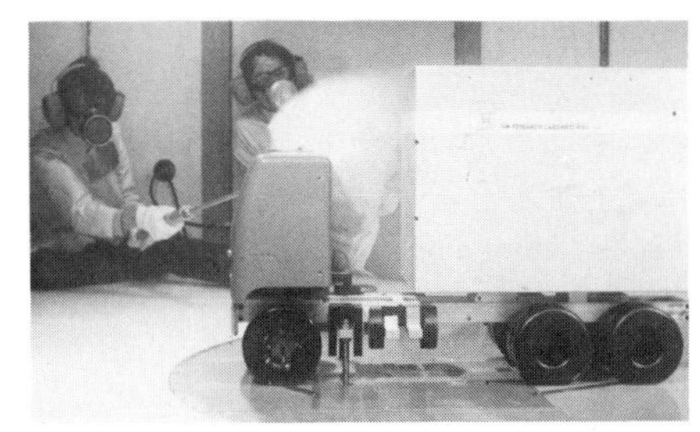

Fig. 21. Roof-mounted air conditioning unit and flat plate of equivalent frontal area; smoke visualization of flow fields at g/d = 0.3.

Add-on roof fairings, see Mason (1975) and Marks, Buckley, & Walston (1976), are an example of another class of roof-mounted devices. Fairings eliminate the separation from the tractor roof that occurs with flat plate devices, and avoid the locally concave streamline curvatures associated with that separation, see Fig. 20 and compare with the sketch shown in Fig. 22. Optimum fairings were designed for the C.O.E. and conventional tractors for gaps of 20 to 60 percent of the diameter using the design procedure published by Mason (1975). Typical fairings are shown in Fig. 23. A low drag coefficient level of about 0.6 was reached for all configurations, Fig. 24, regardless of the gap or tractor size and shape. Visualization of the flow fields, see Fig. 25, showed good matching of the separated shear surface from the fairings to the trailer.

The minimum drag target established by effectively rounding the corners of the trailer was 0.63. Since drag coefficients near 0.6 *were* reached by adding a flat plate or various fairings to the roof of contemporary tractors, i.e. by resizing and reshaping, it appears that the matching criterion for the separated shear surface from the tractor to the trailer *is* a valid approach for low drag, even when the frontal area of the tractor plus add-on device is less than that of the trailer. These results suggest that when the shear surface from the tractor was matched to the trailer, the gap was not a significant source of drag, even as it increased.

References pp. 89-90.

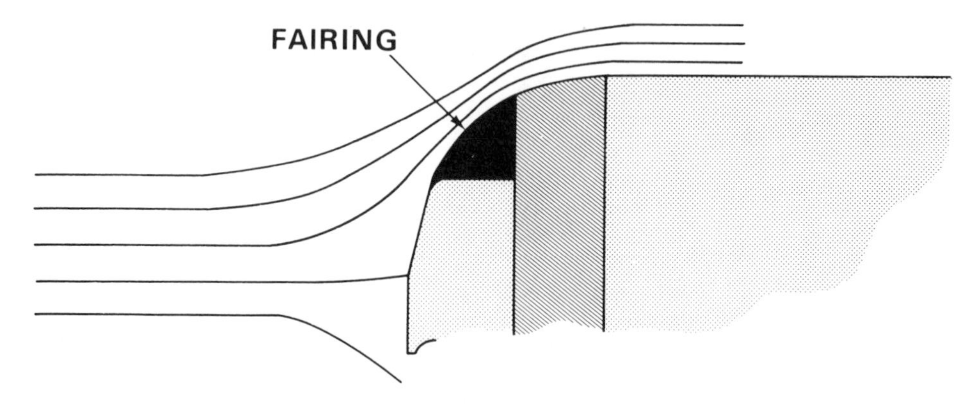

Fig. 22. Roof-mounted fairing sized for minimum drag of combination; separated shear surface matched to trailer leading edges.

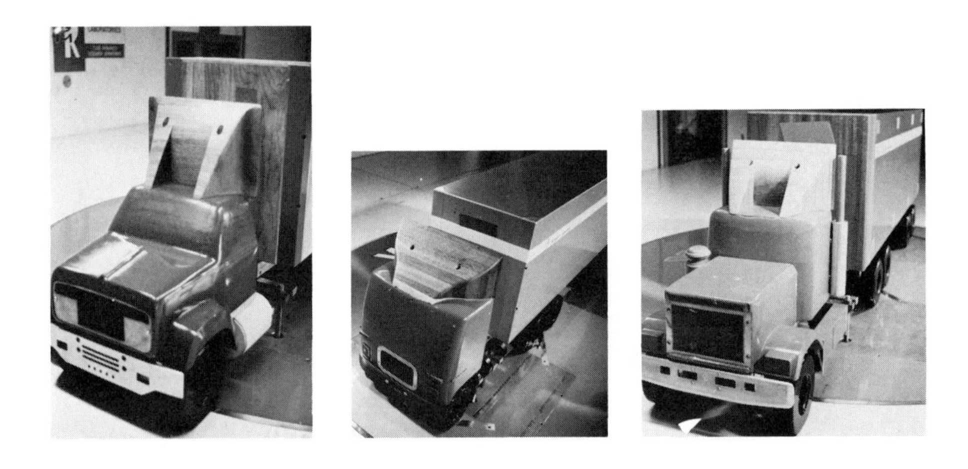

Fig. 23. Optimum fairings for various tractor-trailer combinations.

An interesting consequence of matching for low drag can be inferred from a result reported by Flynn & Kyropoulos (1961). For a C.O.E. tractor-trailer similar to that of Fig. 18(b) they found that the drag of the trailer was *negative,* and 12 percent of the total, i.e. the average pressure on its front face was less than that on the base. This also means that the drag of the tractor was *greater* than that of the combination. The Flynn & Kyropoulos result should not be interpreted to mean that trailer "thrust" is to be exploited, or that the large drag of the tractor might be reduced for an overall drag reduction. It simply suggests that the low pressure in a shear layer enclosed gap is probably an inherent characteristic of matching for low drag, albeit with curious effects on the two drag components.

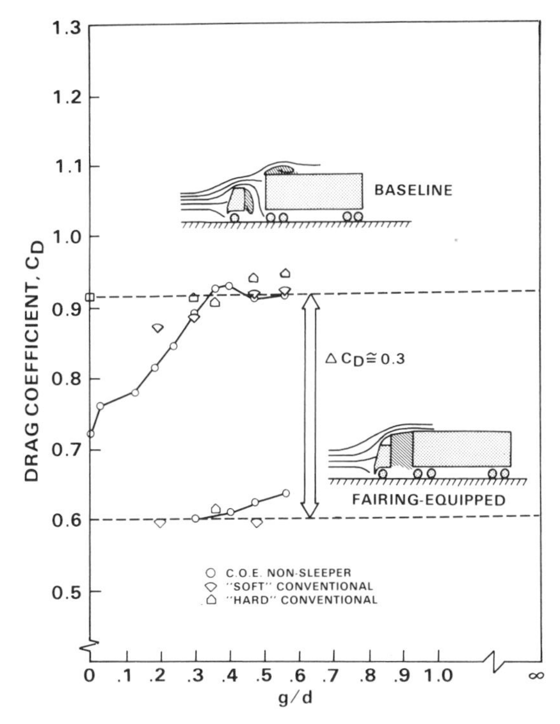

Fig. 24. Effect of fairings on drag of three tractor-trailers over normal gap operating range.

Fig. 25. Flow field visualization of optimum fairing equipped tractor-trailers.

References pp. 89-90.

At this point it is of interest to conjecture about the drag reduction potential available by providing more forebody length for increased "streamlining." As was proposed earlier, it is believed that only a small additional drag reduction would result. The work of Bauer & Servais (1974) tends to support this prognosis. Using a combination fairing and gap enclosure, the drag coefficient of a conventional tractor-trailer was reduced by 0.22, see Fig. 26. A radical reshaping of the tractor only reduced the drag coefficient an additional 0.01 – a relatively insignificant amount.

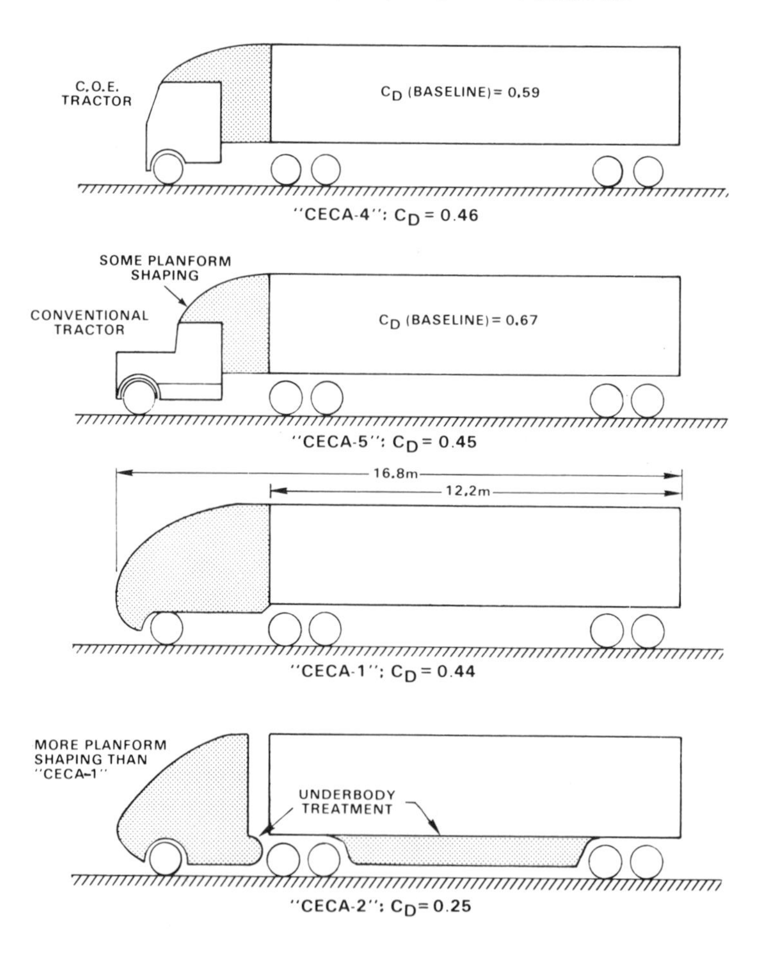

Fig. 26. Drag reduction of C.O.E. and conventional tractor-trailers (1/20-scale) by Bauer & Servais (1974).

Finally, a word of caution is in order regarding assessments of the overall relative merits of flat plates compared to fairings based solely on the preceding material. Although it appears that about the same low drag level can be reached with optimized versions of either class of device, two additional criteria require consideration. One is sensitivity of any *given* device to changes in trailer height and gap length; the other is non-zero yaw drag behavior. Both of these criteria were discussed for fairings by Mason (1975).

Summary – The forebody drag of trucks and buses can be significantly reduced. While tractor-trailer drag *can* be reduced by minimizing flow separation from trailer-face edges, the *lower limits* of drag are reached by transforming the double-stagnation flow field into the single-stagnation field of buses. Such a transformation is effected by using a tractor of proper aerodynamic size. This can be accomplished by tractor redesign, or by fitting add-on devices to the tractor roof, such as flat plates and fairings. The tractor should not be *streamlined* in isolation from the trailer, but *aerodynamically sized* to match it.

For single-stagnation flows it appears that most of the available reduction in forebody drag can be achieved with "corner rounding." That is, once leading edge separation from the perimeter of the maximum cross-section of the forebody has been eliminated, "streamlining" does not appear to provide much additional drag reduction.

Through reductions in forebody drag, apparent minimum drag coefficients of 0.6 and 0.4 were achieved for trucks and buses, respectively.

BASE FLOW FIELD

Since apparent minimum forebody drags have been reached, it is appropriate to examine the characteristics of the base flow field. Smoke was used for flow visualization, and measurements of base pressure distribution were made. For both tractor-trailers and buses a near-wake upwash was observed which carried some of the underbody flow away from the ground, as shown by the photo and sketch in Fig. 27. The upflow was located about one-half diameter from the base. Some of it flowed back toward the base; the remainder was carried downstream. The backflow split at the base. The upper portion did not appear to recirculate; the lower did. The pattern persisted across the width of the base. For both vehicles, a pressure coefficient of -0.15 was found at the top of the base. There was also a characteristic gradient in the vertical distribution., see Fig. 28, that was invariant across the base. Minimum values between -0.2 and -0.3 were found near the bottom.

References pp. 89-90.

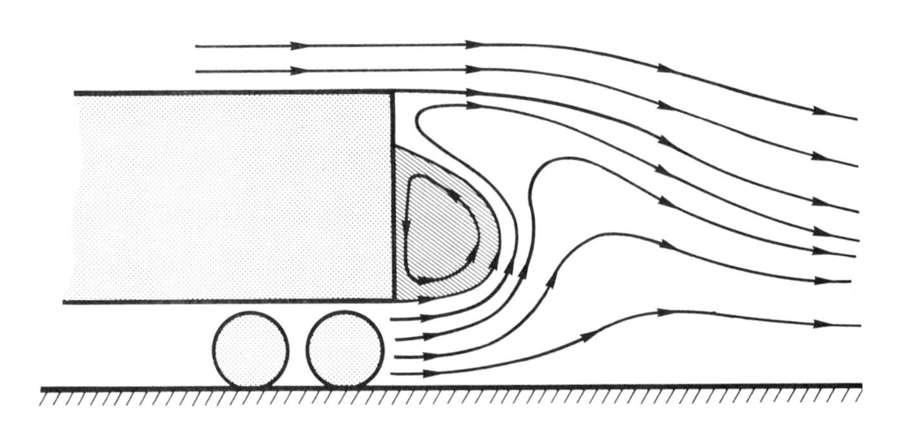

Fig. 27. Photograph and sketch of tractor-trailer's near-wake flow; similar flow field was observed behind bus.

Inflow Conditions – The effect of changes in inflow conditions was investigated by varying the tractor-to-trailer gap, removing the tractor, reducing the forebody drag, removing the wheels, and blocking the underbody flow. These experiments were qualitative in nature since the actual inflow conditions were neither known nor controlled. The presence of the tractor and increases in gap caused the pressure minimum to decrease, see Fig. 29. A similar effect was seen when the forebody flow field was improved, and when wheels were removed from either the truck or the bus. Pressures on the bottom half of the base became lower, while the top pressures didn't change. None of the distributions varied across the base. When the underbody flow was blocked at the rear, a major change in flow pattern occurred. The shear layer

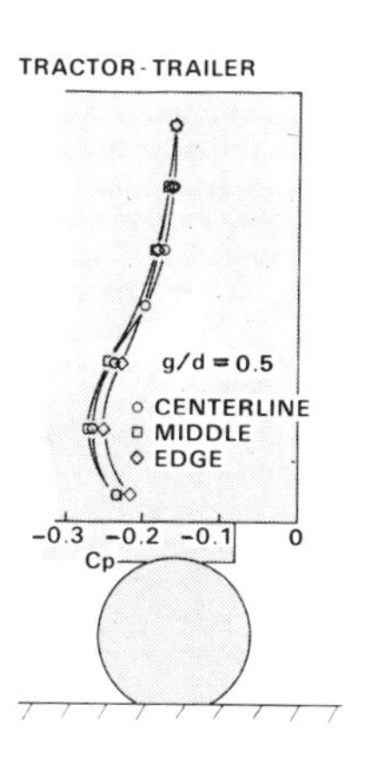

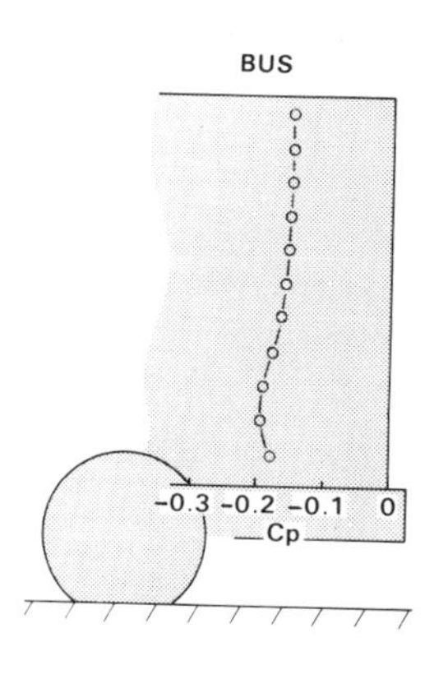

Fig. 28. Tractor-trailer and simplified-bus base pressure distributions. Pressure tap locations can be seen in photographs.

References pp. 89-90.

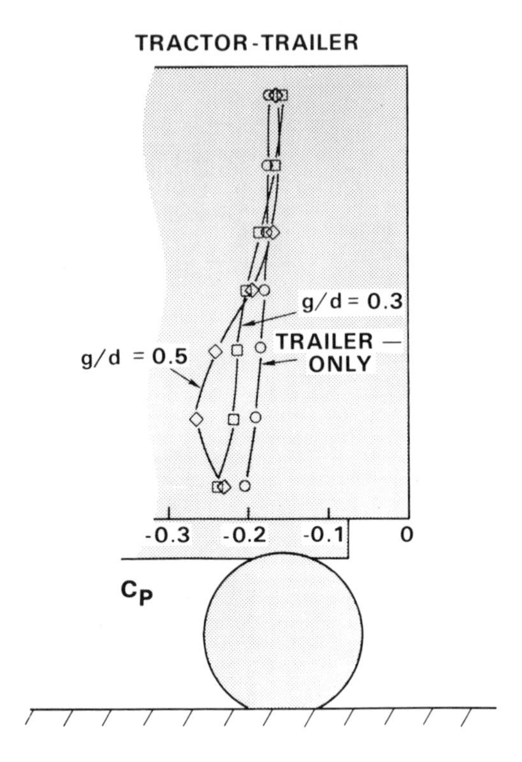

Fig. 29. Effect of changes in inflow conditions on tractor-trailer base pressure distribution.

from the roof edge of both vehicles impinged on the groundplane about one diameter downstream. The pattern is sketched in Fig. 30. In spite of the low aspect ratio of the base, the flow was similar to that of a rearward facing step. Inside the near-wake bubble that was formed there was a single recirculating flow pattern. The base pressures decreased from the baseline values, Fig. 31, and were constant across the base.

Ground Proximity – The trailer was stripped of wheels, axles, and frame and tested at three ground clearances – 0.06, 0.40 (nominal) and 0.71 diameters. The test setup can be seen in Fig. 32. Two configurations were tested at each clearance. One had sharp leading edges; the other had lip fairings. The vertical pressure gradient as well as the improved-forebody-effect typical of both the truck and the bus were obtained at both the nominal and larger-than-nominal clearances, see Fig. 33. However, the flow pattern had top-to-bottom symmetry with two major recirculating regions, rather than being asymmetric with an upwash. Since the vertical base-pressure gradient has been observed with both symmetric and asymmetric near wakes, this suggests that the gradient was not determined by the type of near wake.

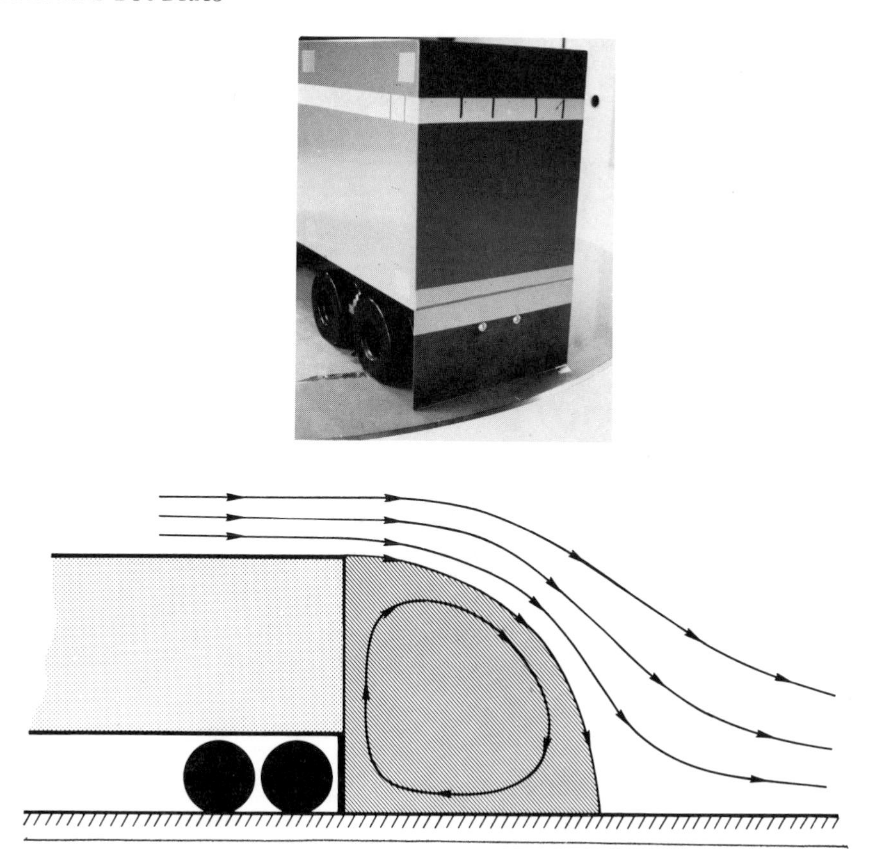

Fig. 30. Near-wake flow pattern associated with blocked underbody. Near-wake pattern of bus with blocked underbody was similar.

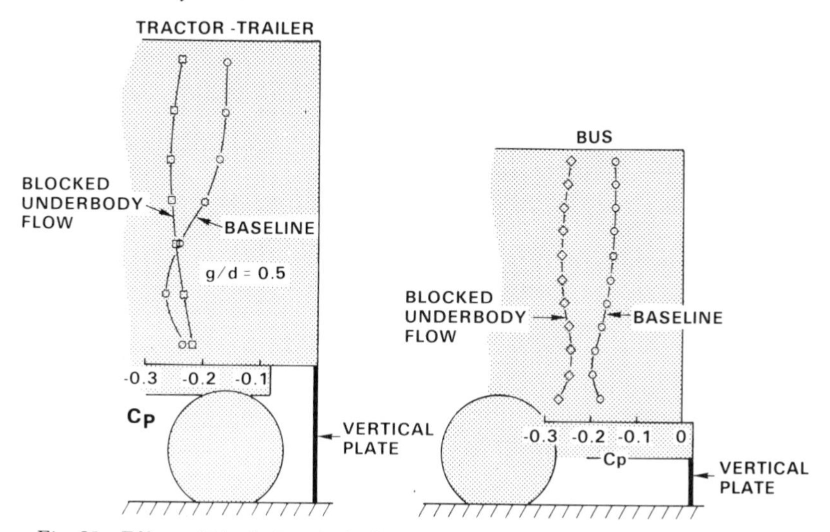

Fig. 31. Effect of blocked underbody on truck and bus base pressure distributions.

References pp. 89-90.

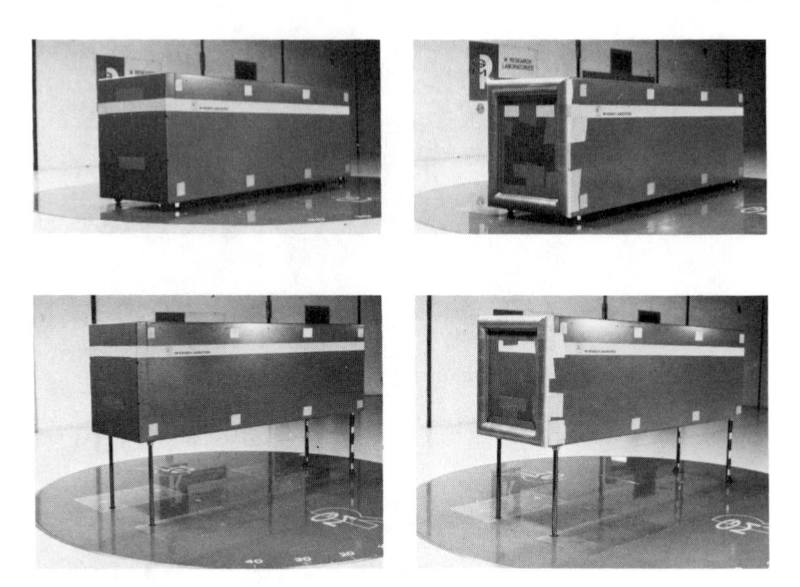

Fig. 32. Trailer stripped of wheels, axles, and frame at two heights above groundplane – 0.06 d and 0.71 d; nominal height of 0.4 d not shown. Sharp-edge (left) and lip-fairing (right) configurations.

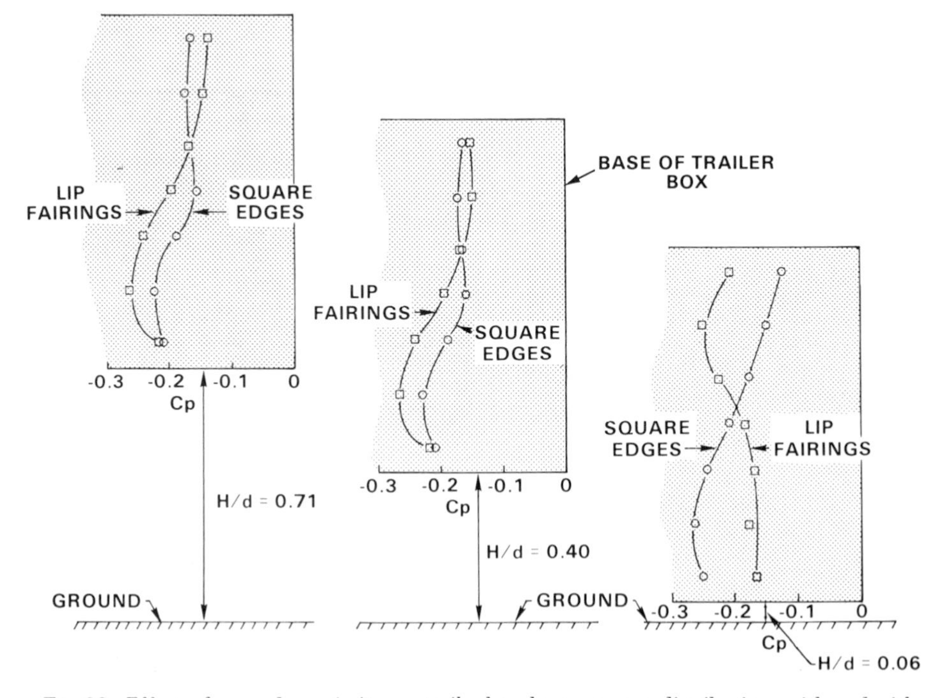

Fig. 33. Effect of ground proximity on trailer-box base pressure distribution; with and without lip fairings.

At the smaller-than-nominal clearance, the asymmetric upwash pattern returned for both configurations. The pressure data presented in Fig. 33 show that with sharp edges the pressure gradient was qualitatively the same as for the larger clearances, but became steeper; however, the gradient was *reversed* by the lip fairings. An explanation is not in hand, but differences in flow pattern were observed, as sketched in Fig. 34. The lip fairings were not optimum; separation was present from *all* edges at the two larger ground clearances. The bottom edge separation, however, was eliminated at the small clearance. For the sharp-edged configuration, separation still occurred from the bottom edge, and partially blocked the underbody flow. In both cases the underbody flow seemed to be confined laterally, smoke introduced at the front exiting from the rear as a coherent stream. With sharp edges the upwash plume was closest to the base – one-half compared to one diameter. The observation of two opposite base-pressure gradients with the same near-wake pattern tends to confirm the conclusion of the preceding paragraph, i.e. the base pressure gradients were not a function of wake type. The effect of the support studs, if any, is unknown.

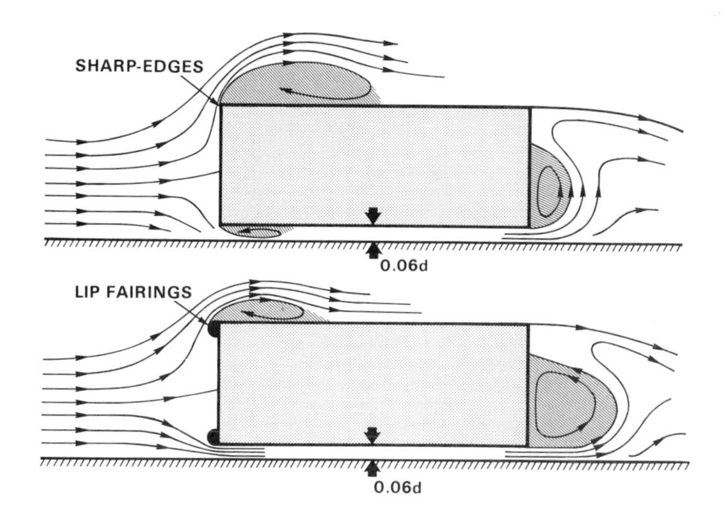

Fig. 34. Sketches of trailer-box flow field at H = 0.06 d, with and without lip fairings (based on smoke flow visualization).

Summary – While most of the configurational changes did affect the base pressure, the qualitative distribution and average level remained the same. In most cases the pressures were lower toward the ground and constant across the base, and the near-wake flow was characterized by an asymmetric upwash. With the trailer box the *same* gradient was observed with a *symmetric* near wake, and the *opposite* gradient with an *asymmetric* wake. It is suspected that the truck and bus gradients were not determined by the asymmetric upwash pattern. The only major overall change in flow pattern, pressure level, and distribution occurred when the underbody flow was blocked.

References pp. 89-90.

REDUCTION OF BASE DRAG

The potential for increasing base pressure was explored. The effect of the trailer base-mounted splitter plates, guide vanes, and cavities shown in Fig. 35 was investigated. Base pressures were not measured, so that the results have to be assessed only on the basis of changes in drag. Both horizontal ($\ell \cong$ d) and vertical splitters ($\ell \cong$ 0.5, 1.0 d) were tested; there were no significant changes in drag. With guide vanes, the results were similar to those of Sherwood (1953) and Kirsch, Garg, & Bettes (1973) – the added drag of the vanes was evidently larger than any base pressure related drag reduction. The dramatic two-dimensional results of Frey (1933) shown in Fig. 36 (also referenced by Hoerner, 1965) do not appear transferable to three-dimensional bodies in ground proximity. The only drag reductions were achieved by the addition of non-ventilated cavities. The best geometry, with a cavity depth of 0.13 d, reduced the drag coefficient by 0.03. If it is assumed that the drag reduction was only due to an increase in base pressure, the average base-pressure coefficient would have had to increase from -0.20 to -0.16 since the base area was only 70 percent of the area used to compute C_D. At the minimum drag levels this would be a 5 percent drag reduction for the truck, and 8 percent for the bus.

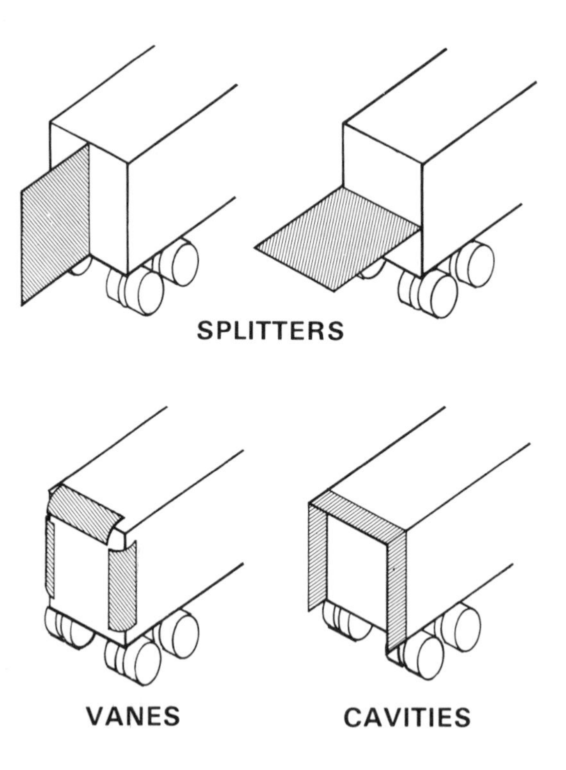

Fig. 35. Base-flow modification devices applied to tractor-trailer.

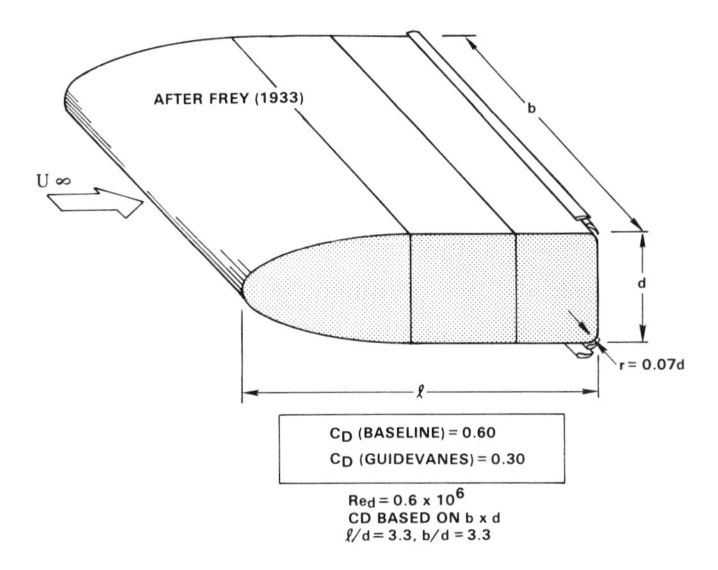

Fig. 36. Reduction of pressure drag by means of staggered guide vanes, Frey (1933). The "two-dimensional" body had a span of 500 mm and was tested between non-metric end plates, 1.3 m long x 1.0 m wide.

SUMMARY OF ZERO-YAW FLOW FIELD CHARACTERISTICS

The material presented in Fig. 37 summarizes the most significant results. The relative drags of all the single body configurations can be seen on the ordinate. The highest drag point, 0.92, is that of the trailer with sharp edges. Moving down the ordinate, the sharp-edged bus has a similar forebody flow field but a lower drag coefficient of 0.71. Continuing down the ordinate, the drag of the bus approaches an apparent limit near 0.4 when forebody treatment is applied. This further reduction of 0.3 in coefficient gives a total reduction of 0.5 relative to the trailer.

When a tractor is added to the trailer, a shielding effect occurs which is maximum at zero gap; its magnitude is a function of the "aerodynamic" size of the tractor. For the two C.O.E. tractors, drag levels about the same as for the sharp-edged bus are reached, a reduction of 0.2 from the sharp-edged trailer. As the gap increases, a stagnation flow develops on the trailer face, the shielding diminishes, and it is totally lost at a gap of 40 percent. This is more-or-less independent of the "aerodynamic" size of the tractor. At gaps larger than 40 percent, the drag must eventually increase again as the two flow fields develop independently. Ultimately, at a gap of infinity, maximum coefficients between 1.2 and 1.3 are reached. The spread is due to differences in tractor drag. Maximum shielding can be maintained by blocking the gap with a horizontal plate. The forward-facing-step flow field associated with zero gap is maintained, preventing the formation of trailer-face stagnation flow.

References pp. 89-90.

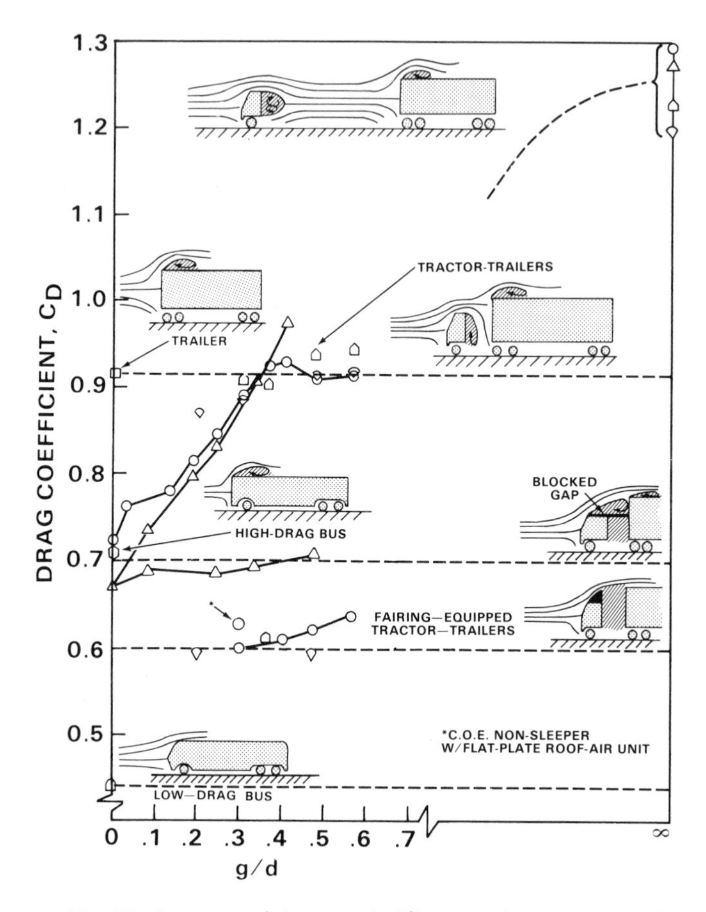

Fig. 37. Summary of the most significant results; zero yaw angle.

For minimum tractor-trailer drag the tractor should be aerodynamically *sized*, and not *streamlined* in isolation from the trailer. The separated shear surface from the tractor should match the leading edges of the trailer. Matching can be achieved by resizing the tractor with add-on devices, such as flat plates and fairings. The apparent limit with this approach is a coefficient of 0.6. This is 0.3 lower than the maximum combination drag, and is the same as that of the trailer with "corner rounding."

The near-wake flow fields of trucks and buses are characterized by an upwash of underbody flow and a vertical base-pressure gradient – the pressures are more negative toward the bottom, and tend to be uniform laterally. The average base pressure coefficient is about -0.2. The only drag coefficient reductions, a maximum of 0.03, were achieved by the addition of non-ventilated cavities.

APPARENT MINIMUM DRAG LIMITS AND UNDERBODY DRAG

The apparent drag coefficient limit of 0.6 reached with the trucks is larger than the limit of 0.4 observed for buses. The low drag configurations of each vehicle have a single stagnation, minimum separation forebody – why, then, is there such a large difference in drag? A similar difference was found between the sharp-edged trailer and the high drag buses, and between the "corner rounded" trailer and the low drag bus. A possible answer to this question was obtained by performing a simplified analysis of truck and bus drag; details are presented in the Appendix. The result of the analysis indicated that the underbody contribution to the drag coefficient of the trucks was 0.22 larger than that of buses, i.e. the difference in the apparent minimum drags observed between trucks and buses was underbody related.

The results of Bauer & Servais (1974) suggest the possibility of a large difference between the underbody drag of trucks and buses. The forebody of a streamlined tractor-trailer was shaped further and the truck's underbody was smoothed and enclosed, see Fig. 26. The combined modifications reduced the drag coefficient by 0.19. The streamlined forebody was probably free of separation before the additional shaping, so it is proposed here that most of the measured drag reduction was due to the underbody. Since the modified underbody typified that of a bus, the 0.19 difference in drag could then be thought of as the difference in underbody drag between low drag trucks and buses. This drag difference is about the same as that between the minimum drag trucks and buses indicated by the preceding analysis. This tends to support the present hypothesis concerning a large truck underbody drag component. It should be noted, however, that Bauer & Servais estimated that most of the reduction came from the slight modifications of the forebody, and not from treatment of the underbody.

LOWER DRAG LEVELS?

In order to assess the possibility of drag levels below 0.6 and 0.4 for trucks and buses, respectively, it is helpful to estimate the drag contributions of the forebody, underbody, side and top surfaces, and base at these low drag levels. The contributions of the side and top surfaces (skin friction) and the base were determined in the Appendix and are given in eqs. (A3) and (A4). If an estimate is made of the underbody contribution for the bus, eq. (A8) can be used to obtain the underbody drag of the truck; this inherently assumes that the average forebody pressure coefficient, $\overline{C}_{p_{fb}}$, is the same for each vehicle. Eqs. (A5) and (A6) for tractor-trailers and buses, respectively, are then left with only one unknown, $\overline{C}_{p_{fb}}$, which can be determined from either equation.

As shown in the Appendix, an upper limit for the underbody contribution to the drag coefficient of the bus is estimated at 0.10. Substitution of this estimated bus

References pp. 89-90.

underbody drag into eq. (A8) yields a total drag contribution of 0.32 from the underbody of low drag trucks. From either eq. (A5) or (A6), the resulting forebody drag contribution for each vehicle is 0.12. Therefore, for trucks, the postulated breakdown of the 0.6 overall drag coefficient is as follows: 0.12 (forebody), 0.32 (underbody), 0.03 (skin friction), and 0.13 (base). Similarly, the overall coefficient of 0.4 for buses is composed of 0.12 (forebody), 0.10 (underbody), 0.03 (skin friction), and 0.15 (base).

As discussed earlier, analytical and experimental evidence suggests the possibility of forebody drags lower than 0.12 for bodies out of ground proximity. Whether such lower drags are possible with ground vehicles is unknown; indications are to the contrary. Modifications of the underbody of tractor-trailers *could* result in lower drag. Bus-like underbodies would not be feasible in the near term, but properly sized and located partial-fairings, deflectors, baffles, etc. may offer some potential. Just how much of the postulated 0.22 increment between trucks and buses is practially realizable is unknown. The fluid-mechanically difficult job of increasing the average base pressure by some means that will produce a net drag reduction, especially in view of functional constraints, makes it perhaps unlikely that much future drag reduction will come through modifications of the base. Non-ventilated cavities might be employed on trucks and buses, while mild "boat-tailing" might be feasible on buses.

The next generation minimum drag limits for trucks and buses are estimated to be 0.45 and 0.35, respectively, based on the following projections. For trucks, practical underbody modifications may lead to a reduction of 0.1, while a practical base modification may achieve 0.05. For buses, a combination of practical underbody and base modifications may produce a total 0.05 reduction.

NON-ZERO YAW DRAG CHARACTERISTICS

The scope of the paper up to this point has been restricted to the zero yaw condition and its associated flow modules. The justification was that *they* establish the basic levels of drag for any given configuration, yet are of a less complex aerodynamic nature. However, it is appropriate to now briefly explore some relevant non-zero drag characteristics, and take a step toward understanding the more complicated non-zero modules. The relative drags of the trucks and buses represented in Fig. 37 are shown at yaw angles of 5 and 10 degrees in Figs. 38 and 39, respectively. The drag of all configurations increased significantly from zero-yaw levels. For trucks, the characteristic increase of drag with gap was still present. The shielding effect of the tractor was again a maximum at zero gap, and was lost completely at gaps larger than 30 or 40 percent. However, as the gap increased between zero and 20 percent, there was a weak, post-initial tendency to maintain shielding effectiveness.

Blocking the gap was not effective at small gaps, and this is probably related to the

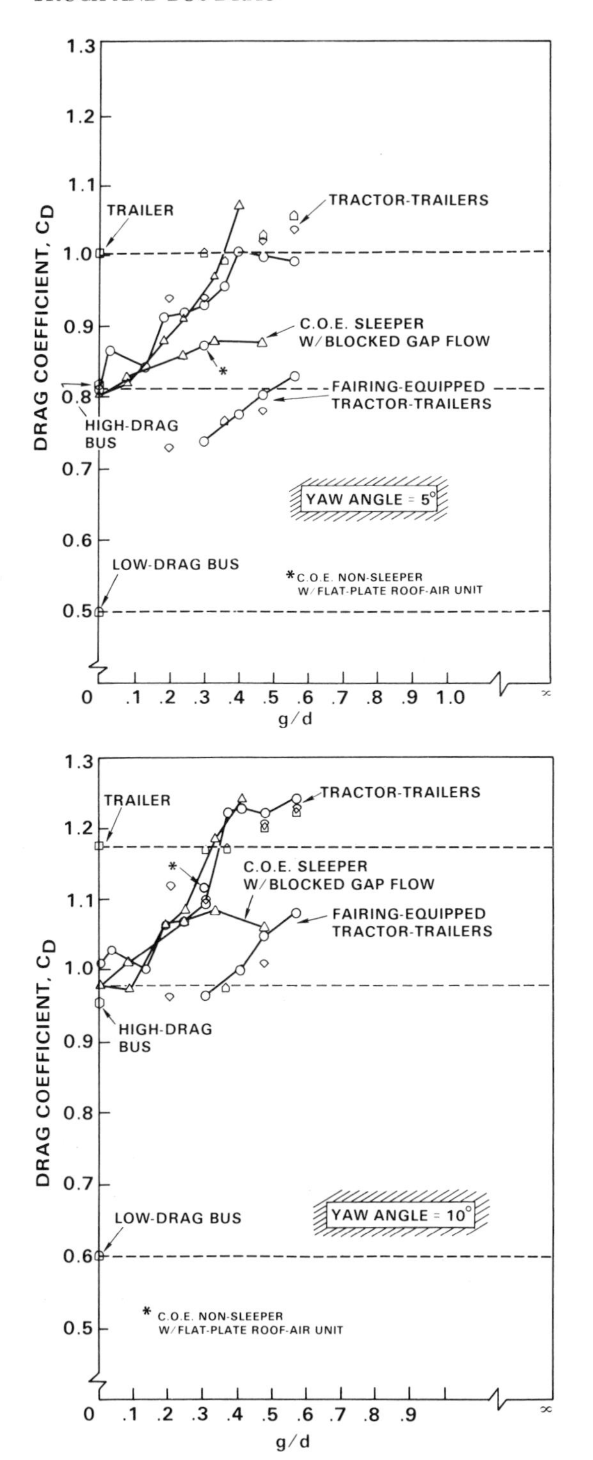

Fig. 38. Relative drags of truck and bus configurations presented in Fig. 37 at 5 degrees yaw angle.

Fig. 39 Relative drags of trucks and buses at 10 degrees yaw angle.

References pp. 89-90.

post-initial tendency for maintained shielding. At gaps larger than 15 or 25 percent the gap blocking effect returned. This may be the point where a significant downflow has developed through the gap for the baseline truck. For the fairing equipped trucks the incremental reduction in drag decreased with increasing yaw angle, the $C_D = C_D(g/d)$ curve developing a positive slope. But even though these fairings were optimized at zero yaw, significant drag reductions were still maintained. In contrast, the flat plate roof-air unit, which had almost the same low drag as the fairing at zero yaw, gave only a small drag reduction at 5 degrees, and *increased* the drag at 10 degrees.

To explore the drag characteristics further, it is more instructive to look at drag as an explicit function of yaw angle. The drags of three tractor-trailers are shown as a function of yaw angle in Fig. 40. The gap ratio was 50 percent; two of the tractors were conventionals, and one was a C.O.E. non-sleeper. The zero-yaw drag levels were the same *and* the yaw characteristics were similar through 10 degrees. The same results are implicitly exhibited by the same data in Figs. 37, 38, and 39.

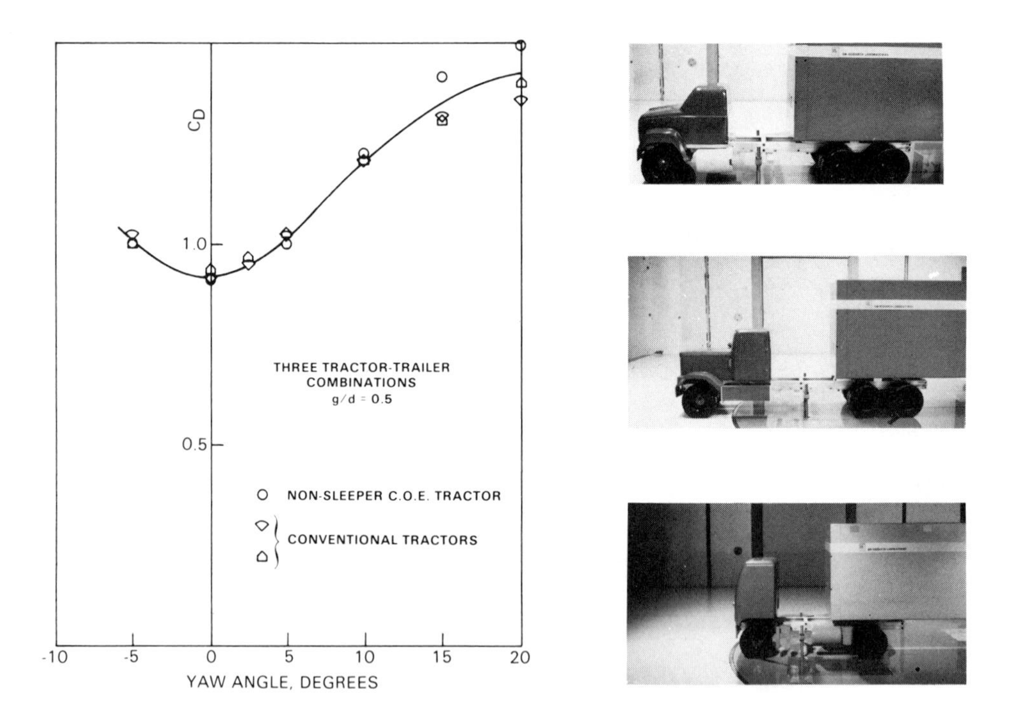

Fig. 40. Drag of three tractor-trailer combinations as a function of yaw angle; g/d = 0.5.

The drag of the *trailer* is compared to the three-truck average in Fig. 41. The zero-yaw drag level and the yaw characteristic up to 10 degrees are the same – this also follows from Figs. 37, 38, and 39. A possible explanation of this correspondence is that the zero-yaw flow module of the trucks, i.e. strong downflow through the gap, persists to 10 degrees. This downflow prevents lateral flows from developing, and the yaw characteristic therefore remains the same as that of a gapless body. However, beyond 10 degrees lateral flows become stronger, and the tractor-trailer yaw characteristics deviate from each other, Fig. 40, and from that of the trailer, Fig. 41. The lateral blocking behavior of the gap downflow *has* been qualitatively observed at yaw angles of 10 degrees or less using smoke for flow visualization. In contradiction, Buckley, Marks, & Walston (1974) indicated a *predominant* lateral gap flow at a yaw angle of 10 degrees. However, their experiments were two-dimensional in planview and therefore precluded the existence of downflow through the gap. Based on the present observations, their result may be misleading with respect to the true nature of the gap flow module, and with respect to non-zero yaw drag-reduction strategies at angles of 10 degrees or less.

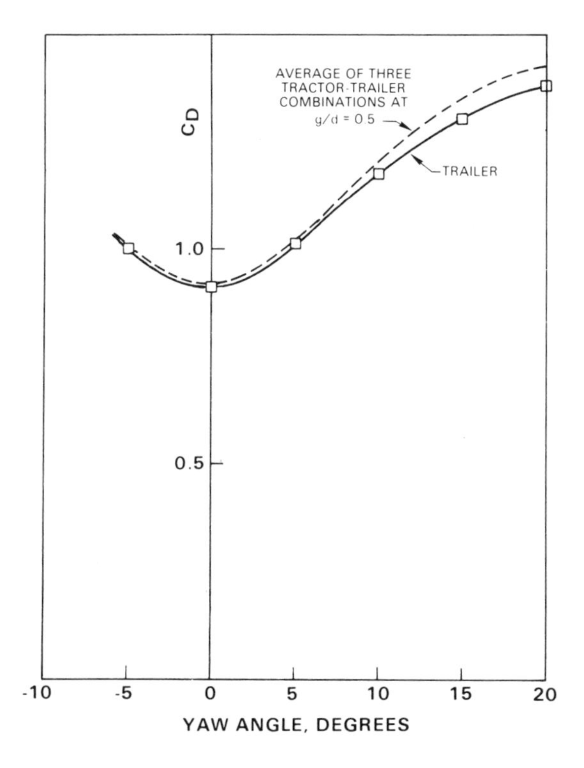

Fig. 41. Comparison between average drag characteristic of tractor-trailers and that of trailer.

References pp. 89-90.

When the corners of the trailer were rounded, the drag level dropped from 0.9 to 0.6. However, the rounded corners *did not* change the yaw characteristic, as shown in Fig. 42. On the other hand, a fairing which reduced zero-yaw truck drag to the same level as the corner-rounded trailer, *did* change the yaw characteristic, as shown in Fig. 43. The fairing was sized and shaped for zero-yaw matching of the separation streamlines from the tractor to the trailer, and the gap flow module was quiescent at that angle. There was therefore no downflow to counter the lateral flows which develop with increasing yaw. Thus, departure of the fairing equipped truck's yaw characteristic from that of a gapless body is probably indicative of an immediate and gradual development of lateral gap flows. Presumably, if the lateral flows were blocked the yaw characteristic would be restored. Some recent results of Cooper (1976) confirm such a restoring effect. Cooper's configuration with a fairing and "gap seal" retains the yaw characteristic of the baseline truck to large yaw angles, see Fig. 44.

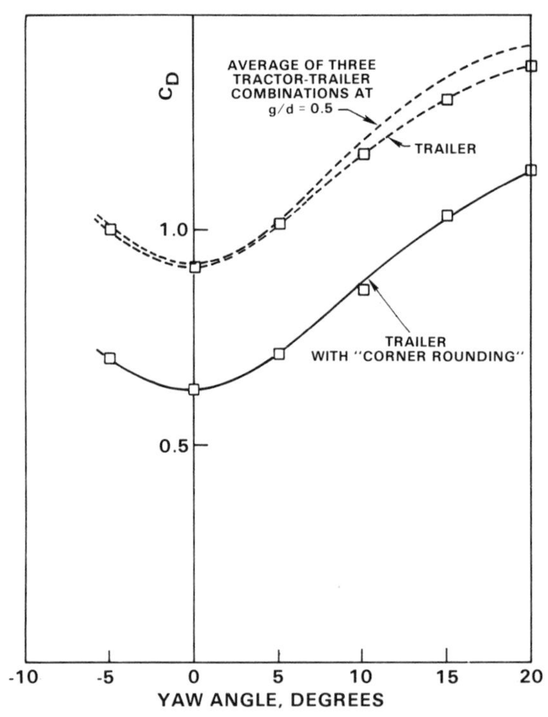

Fig. 42. Effect of "corner rounding" on drag characteristic of trailer.

Finally, the yaw characteristic of the trailer with corner rounding is compared to that of the low drag bus in Fig. 45. The behavior is similar through 5 degrees; however, at larger angles the drag of the trailer increases faster. The reason is unknown, but it may be traceable to the difference in underbody ground clearance.

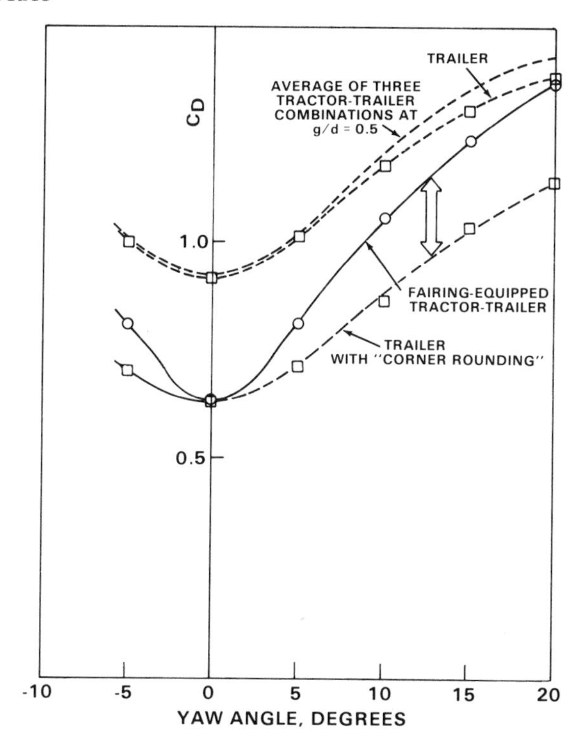

Fig. 43. Drag characteristic of fairing-equipped tractor-trailer compared to those of baseline tractor-trailers and the trailer.

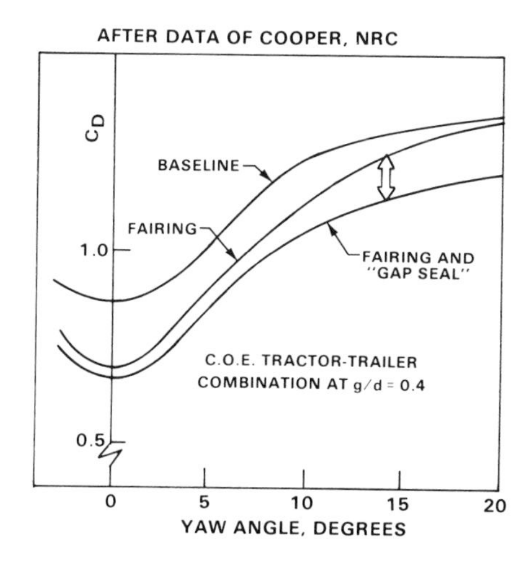

Fig. 44. Effect of vertical "gap seal" on drag characteristic of fairing-equipped tractor-trailer; after Cooper (1976).

References pp. 89-90.

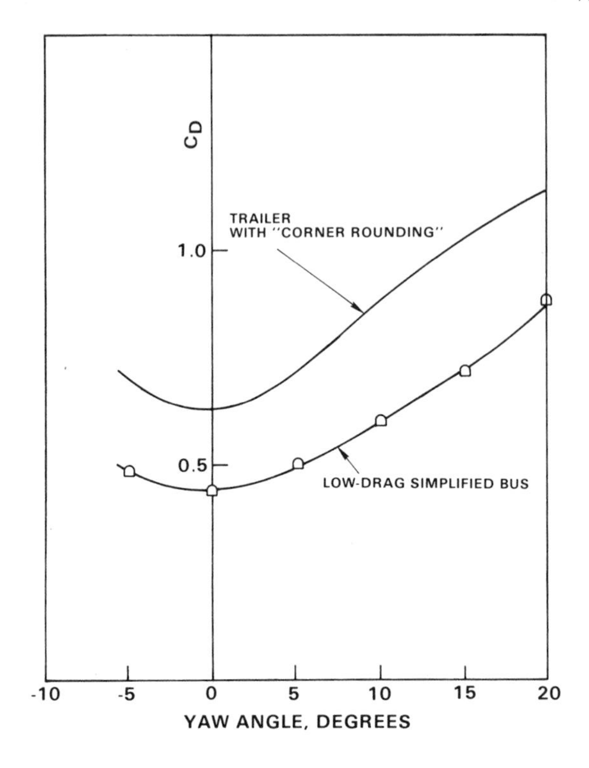

Fig. 45. Drag characteristic of low drag bus compared to that of the trailer with "corner rounding."

Summary – Over the normal operating range of gaps tractor-trailers have the *same* yaw characteristic between zero and 10 degrees as gapless vehicles of similar length-to-diameter ratio, e.g. buses. This appears to be the result of the strong downflow through the gap of tractor-trailers which persists with increasing yaw angle and restricts the establishment of lateral flows. However, with a fairing equipped tractor-trailer the relatively quiescent zero-yaw gap flow allows lateral flows to immediately develop through the gap with increasing yaw angle. This causes a departure in yaw characteristic away from that of gapless vehicles. It appears that when this lateral flow is blocked, the yaw characteristic is restored.

APPENDIX

Drag Analysis – For trucks and buses the total drag can be broken into its components according to the following equation,

$$D = \overline{p}_{fb} A_{fb} + D_{ub} - \overline{p}_{ub} (A_{fb} - A_b) + C_f A_w q_\infty - \overline{p}_b A_b \qquad \text{(A1)}$$

where the variables are defined in Fig. A1 and as follows,

A_{fb}	projected forebody area
A_b	projected base area
A_w	wetted surface area parallel to freestream direction, excluding underbody
$\overline{p}_{fb}$	average forebody pressure
$\overline{p}_b$	average base pressure
$\overline{p}_{ub}$	average pressure on rearward-projected underbody area components for cases where $A_{fb} > A_b$, e.g. tractor-trailers
D_{ub}	underbody drag (wheels and "skin friction")
D_f	skin friction drag on surfaces parallel to freestream direction, excluding underbody
q_∞	freestream dynamic pressure
C_f	skin friction coefficient, $D_f/q_\infty A_w$

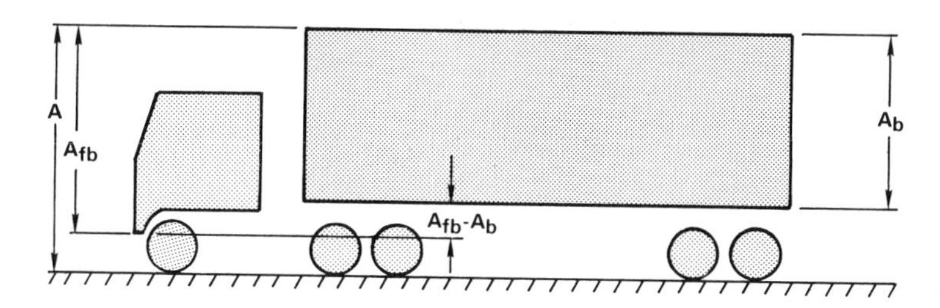

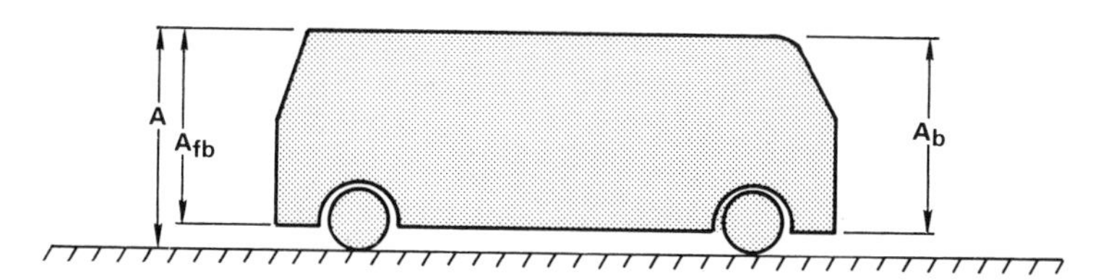

Fig. A1. Schematic sketches of tractor-trailer and bus used to define breakdown of total drag into its components.

References pp. 89-90.

Or, in coefficient form based on the total projected frontal area, A, which is the product of the overall height and width,

$$C_D = \left(\frac{A_{fb}}{A}\right)\overline{C}_{P_{fb}} + C_{D_{ub}} - \left(\frac{A_{fb}-A_b}{A}\right)\overline{C}_{P_{ub}} + \left(\frac{A_w}{A}\right)C_f - \left(\frac{A_b}{A}\right)\overline{C}_{P_b} \qquad \text{(A2)}$$

C_f and A_w/A were estimated at 0.003 and 10, respectively, for both types of vehicles. For a tractor-trailer, $A_{fb}/A = 0.9$, $A_b/A = 0.7$, and $\overline{C}_{P_b} = -0.19$ (see Fig. 28). Eq. (A2) then becomes,

$$C_D = 0.9\,\overline{C}_{P_{fb}} + C_{D_{ub}} - 0.2\,\overline{C}_{P_{ub}} + 0.03 + 0.13 \qquad \text{(A3)}$$

For buses, $A_{fb}/A = 0.9$, $A_b/A = 0.9$, and $\overline{C}_{P_b} = -0.17$ (see Fig. 28). Eq. (A2) becomes,

$$C_D = 0.9\,\overline{C}_{P_{fb}} + C_{D_{ub}} + 0.03 + 0.15 \qquad \text{(A4)}$$

Substituting the measured apparent minimum drag coefficients, 0.6 and 0.4, into eqs. (A3) and (A4) and rearranging, the following two equations for low drag trucks and buses, respectively, are obtained.

$$0.9\,\overline{C}_{P_{fb}} + C_{D_{ub}} - 0.2\,\overline{C}_{P_{ub}} = 0.44 \qquad \text{(A5)}$$

$$0.9\,\overline{C}_{P_{fb}} + C_{D_{ub}} = 0.22 \qquad \text{(A6)}$$

Subtracting eq. (A6) from eq. (A5), gives the following equation for the difference between truck and bus underbody drag,

$$\left(C_{D_{ub}} - 0.2\,\overline{C}_{P_{ub}}\right)_{truck} - \left(C_{D_{ub}}\right)_{bus} = 0.22 + 0.9\left[\left(\overline{C}_{P_{fb}}\right)_{bus} - \left(\overline{C}_{P_{fb}}\right)_{truck}\right] \qquad \text{(A7)}$$

With the observed similarity between the forebody flow fields of these low drag configurations, it is difficult to imagine that the forebody contributions to drag were significantly different, i.e. assume that $\left(\overline{C}_{P_{fb}}\right)_{truck} = \left(\overline{C}_{P_{fb}}\right)_{bus}$. This leads to the hypothesis that the major source of the drag difference was the truck's underbody, and that, according to eq. (A7), the underbody contribution to the drag coefficient of the truck was 0.22 larger than that of the bus, i.e.,

$$\left(C_{D_{ub}} - 0.2\,\overline{C}_{P_{ub}}\right)_{truck} - \left(C_{D_{ub}}\right)_{bus} \cong 0.22 \qquad \text{(A8)}$$

Estimate of Bus Underbody Drag – An estimate of bus underbody drag was made as follows. The frontal projections of the exposed portions of the front and rear wheels were represented as flat plates of corresponding area with $C_D = 1$. Underbody surface drag was estimated using a skin friction coefficient, C_f, of 0.01, which is more than three times that used for the top and side surfaces. To determine the underbody drag *force*, the "wheels" and underbody surface were exposed to freestream dynamic pressure. The resulting contributions of the wheels and underbody surface to the overall drag coefficient were 0.07 and 0.03, respectively, for a total underbody drag, $C_{D_{ub}}$, of 0.10. Since much of the underbody was probably exposed to lower than freestream dynamic pressure, and since the front wheels generated wakes which shielded the wheels at the rear to some extent, the preceding estimate probably should be considered as an upper limit.

REFERENCES

Bauer, P. T., & Servais, R. A. (1974), "An Experimental and Analytical Investigation of Truck Aerodynamics," **Proceedings of the Conference/ Workshop on the Reduction of the Aerodynamic Drag of Trucks,** *California Institute of Technology, October 10-11, National Science Foundation RANN Document Center, Washington, D.C., pp. 55-61.*

Buckley, Jr., F. T., Marks, C. H., & Walston, Jr., W. H. (1974), "An Assessment of Drag Reduction Techniques Based on Observations of Flow Past Two-Dimensional Tractor-Trailer Models," **Proceedings of the Conference/Workshop on the Reduction of the Aerodynamic Drag of Trucks,** *California Institute of Technology, October 10-11, National Science Foundation RANN Document Center, Washington, D.C., pp. 15-31.*

Carr, G. W. (1967), "The Aerodynamics of Basic Shapes for Road Vehicles, Part 1, Simple Rectangular Bodies," Motor Industry Research Association (MIRA) Report No. 1968/2, November.

Cooper, K. R. (1976), "Wind Tunnel Investigations of Eight Commercially Available Devices for the Reduction of Aerodynamic Drag on Trucks," Roads and Transportation Association of Canada National Conference, Quebec City, September.

Fitzgerald, J. M. (1974), "Field Experience Report on Drag Reduction of the Nose Cone," **Proceedings of the Conference/Workshop on the Reduction of the Aerodynamic Drag of Trucks,** *California Institute of Technology, October 10-11, National Science Foundation RANN Document Center, Washington, D.C., pp. 153-159.*

Flynn, H., & Kyropoulos, P. (1962), "Truck Aerodynamics," **SAE Transactions,** *Volume 70, pp. 297-308.*

Frey, K. (1933), "Verminderung des Stroemungswiderstandes von Koepern durch Leitflaechen," **Forschung Ing. Wesen,** *March/April, pp. 67-74.*

Hoerner, S. F. (1965), **Fluid-Dynamic Drag,** *published by the author, Brick Town, N.J., p. 3-12 and p. 3-28.*

Kirsch, J. W., Garg, S. K., & Bettes, W. H. (1973), "Drag Reduction of Bluff Vehicles with Airvanes," SAE 730686, Chicago.

Kirsch, J. W., & Bettes, W. H. (1974), "Feasibility Study of the S^3 Air Vane and Other Truck Drag Reduction Devices," **Proceedings of the Conference/Workshop on the Reduction of the Aerodynamic Drag of Trucks,** *California Institute of Technology, October 10-11, National Science Foundation RANN Document Center, Washington, D.C., pp. 89-120.*

Lissaman, P.B.S. & Lambie, J. H. (1974), "Reduction of Aerodynamic Drag of Large Highway Trucks," **Proceedings of the Conference/Workshop on the Reduction of the Aerodynamic Drag of Trucks,** *California Institute of Technology, October 10-11, National Science Foundation RANN Document Center, Washington, D.C., pp. 139-151.*

Marks, C. H., Buckley, Jr., F. T., & Walston, Jr., W. H. (1976), "An Evaluation of the Aerodynamic Drag Reductions Produced by Various Cab Roof Fairings and a Gap Seal on Tractor-Trailer Trucks," SAE 760105, Detroit.

Mason, Jr., W. T., Beebe, P. S. & Schenkel, F. K. (1973), "An Aerodynamic Test Facility for Scale-Model Automobiles," SAE 730238, Detroit.

Mason, Jr., W. T. (1975), "Wind Tunnel Development of the Dragfoiler – A System for Reducing Tractor-Trailer Aerodynamic Drag," SAE 750705, Seattle.

Moeller, E. (1951), "Luftwiderstandsmessungen am VW-Lieferwagen," **Automobiltechnische Zeitschrift,** *Volume 53, No. 6, pp. 153-156.*

Montoya, L. C., & Steers, L. L. (1974), "Aerodynamic Drag Reduction Tests on a Full-Scale Tractor-Trailer Combination with Several Add-on Devices," **Proceedings of the Conference/Workshop on the Reduction of the Aerodynamic Drag of Trucks,** *California Institute of Technology, October 10-11, National Science Foundation RANN Document Center, Washington, D.C., pp. 65-88.*

Prandtl, L., & Tietjens, O. G. (1934), **Applied Hydro- and Aeromechanics,** *Dover Publications, Inc., New York, 1957, pp. 118-121.*

Saunders, W. S. (1966), "Apparatus for Reducing Linear and Lateral Wind Resistance in a Tractor-Trailer Combination Vehicle," U.S. Patent No. 3,241,876.

Schlichting, H. (1960), **Boundary Layer Theory,** *McGraw-Hill Book Company, Inc., New York, p. 34.*

Sherwood, A. W. (1953), "Wind Tunnel Test of Trailmobile Trailers," University of Maryland Wind Tunnel Report No. 85, June.

DISCUSSION

K. R. Cooper *(National Research Council, Canada)*

Editors' Summary

Mr. Cooper presented and discussed a movie showing an NRC flow visualization study of a 1/50-scale C.O.E. tractor-trailer model in a water tunnel. To visualize the flow, dye was injected from various points on the model. The tests were run at a water speed of 0.5 ft/sec which gave a Reynolds number of 1×10^4 based on effective diameter; this was two orders of magnitude lower than the scale-model work reported by Mason & Beebe. Mr. Cooper's point was that, although the flows were laminar and in many ways *not* characteristic of the higher Reynolds number flows, the general character of the flows *was* well representative *if* you excluded the effects of the laminar separations which sometimes occurred. Flows at zero and 10 degrees yaw angle were shown for the baseline vehicle, and for the same vehicle with a trailer-mounted fairing, and two different tractor-roof mounted devices – a flat plate and a fairing. The effect of a vertical gap seal in combination with the fairing was also shown. Some representative still photographs are shown in Fig. 46.

In the opinion of several discussion evaluators in the audience the results with such extremely low Reynolds number flows should be treated with caution. The flow around the leading edge of the tractor roof is especially critical since its degree of separation has an important effect on the gap flow module and the associated trailer face and tractor base pressure distributions.

T. M. Barrows *(Department of Transportation)*

I was disappointed to see that you didn't have measurements of the base pressure as you were making the various modifications to the rear end of the tractor-trailer. It would be interesting to know whether the average base pressure actually changed in correspondence with the drag changes that you measured for the overall body. Is there some reason why the base pressures were not measured for those configurations?

W. T. Mason, Jr.

Our aerodynamic test facility is currently configured such that forces and moments cannot be acquired simultaneously with surface pressures without risking uncontrolled and unknown tare loads on the model from the Scanivalve™ control and pressure transducer cables. If we had seen significant changes in overall drag when we added the splitters, vanes, and cavities to the rear end of the tractor-trailer, we probably *would* have rerun the model and looked at the corresponding base pressure distributions. As you pointed out, device-related changes in base pressure may have

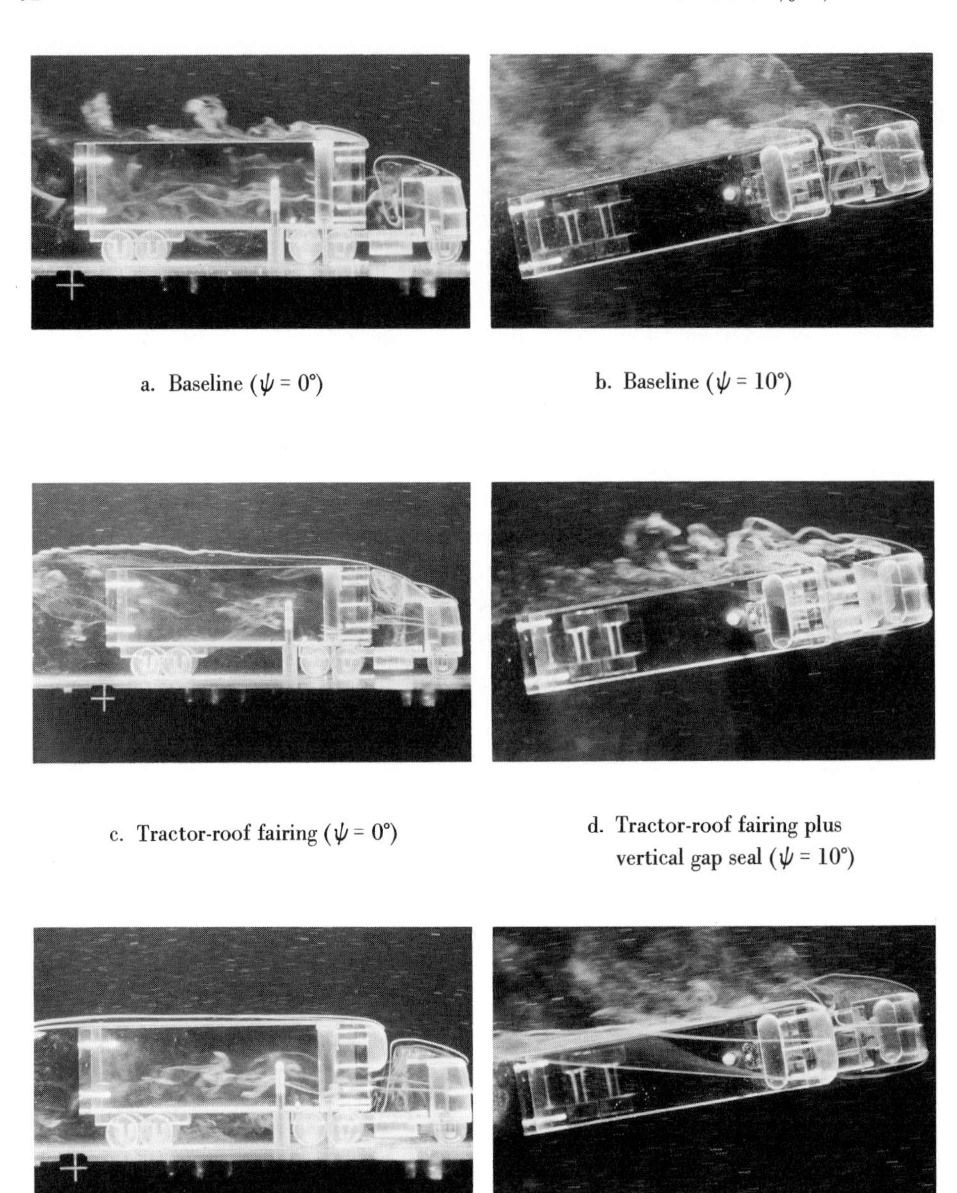

a. Baseline (ψ = 0°)

b. Baseline (ψ = 10°)

c. Tractor-roof fairing (ψ = 0°)

d. Tractor-roof fairing plus vertical gap seal (ψ = 10°)

e. Trailer fairing (ψ = 0°)

f. Trailer fairing (ψ = 10°)

Fig. 46. Flow visualization of 1/50-scale tractor-trailer in Canadian National Research Council Water Tunnel.

been offset by readjustments of the forebody pressure distributions such that no net overall drag change was measured. However, I don't think this was the case for the rear-end configurations that were tested.

P. B. S. Lissaman and W. T. Mason, Jr.

W. H. Bettes *(California Institute of Technology)*

I have two points to make. First, we've had some experience with base-mounted devices similar to those shown by Mason & Beebe which relates to Dr. Barrows' comment. Some unpublished student work at Cal Tech showed that a cavity-type base *does* increase the average base pressure by locally increasing the pressures near the base edges. In some other work* we found that the increases in base pressure achieved with turning vanes were not enough to provide much net drag reduction. However, by combining vanes with a horizontal splitter plate we obtained more overall drag reduction. Greater increases in base pressure were also measured, as a result of the turned flow stagnating against the plate. Editors' note: the base pressure measurements are not presented in SAE 730686; a photo of the vane-splitter configuration is shown in Fig. 21 of that paper.

My second point has to do with emphasizing the importance of the non-zero yaw angle drag characteristics of trucks and buses, and that certain configurations can be extremely yaw sensitive. For example, a flat plate mounted on the roof of a tractor may provide significant drag reduction at zero yaw angle but may actually increase the drag at small yaw angles. Since vehicles nearly always operate in an ambient wind of some sort, the on-road experience with such a flat plate may show an overall increase in fuel consumption.

W. T. Mason, Jr.

Your second point is well made, and we've tried to issue the appropriate cautions in the written version of our paper.

**Kirsch, J. W., Garg, S. K., and Bettes, W. H. (1973), Drag Reduction of Bluff Vehicles with Airvanes, SAE 730686.*

SOME EFFECTS OF FREE-STREAM TURBULENCE AND THE PRESENCE OF THE GROUND ON THE FLOW AROUND BLUFF BODIES

P. W. BEARMAN

Imperial College, London, England

ABSTRACT

Certain aspects of Aeronautical and Building Aerodynamics research are shown to be applicable to the study of the flow around road vehicles. Road vehicles are exposed to the natural wind and the possible effects of free stream turbulence on the mean flow are discussed. A review is given of the effects of turbulence on the drag of bluff bodies with separation from sharp edges, and with separation from continuous surfaces. A road vehicle moves close to a ground plane and the influence that this may have on the flow is examined. Experimental results are presented of the forces acting on an idealised vehicle shape in "ground effect" and they demonstrate the importance of ground clearance. The results also expose differences between wind-tunnel measurements made with a moving wall and a fixed ground plane. Finally a two-dimensional, free-streamline theory is used to predict, qualitatively, the experimental results.

NOTATION

A	plate frontal area
C_D	drag coefficient
C_L	lift coefficient
C_p	pressure coefficient
C_{pb}	base-pressure coefficient
c	body length

References pp. 113-114.

D	depth
H	height
h	ground clearance
L_x	longitudinal scale of turbulence
U	wind speed
$\sqrt{\bar{u}^2}$	r.m.s. horizontal wind-speed fluctuations
V	vehicle speed
W	body width
x	streamwise coordinate
α	angle of incidence
δ^*	boundary-layer displacement thickness
ϵ	ground-clearance/body length; $\equiv$ h/c
κ	circulation

INTRODUCTION

A road vehicle can be considered as a low-aspect-ratio, non-slender body moving close to a ground plane. Road vehicles have a preferred orientation, in contrast to say a building, and thus there is scope for some degree of streamlining. In some "aerodynamically" designed vehicles three quarters of the body may have attached flow, whereas for a building half of the surface or more will often be in a separated flow region. However, since the majority of the drag originates from pressure drag a vehicle will be referred to as a bluff body. With few exceptions bluff-body research has concentrated on understanding the flow about two-dimensional and axisymmetric shapes. The generation of drag and unsteady lift has been studied in great detail but comparatively little is available on the flow about bluff bodies with time-mean lift, such as road vehicles.

The two broad aspects to be covered here are the effect of free stream turbulence on the drag of bluff bodies and the influence of ground proximity on the flow about a bluff body.

THE ROAD VEHICLE AS A BLUFF BODY

In the majority of cases where a flow impinges on a body some separation occurs, and in many instances the flow is dominated by the effects of the separation. In order to gain some insight into complex separated flows it is only natural for the researcher to wish to simplify the situation and we find that much of the work on bluff bodies has been done on various shapes of two-dimensional cylinders. It is well known that the pair of shear layers that spring from the separation points on such cylinders are inherently unstable and interact to form strong vortices at a regular frequency. These strong vortices induce a low pressure in the near wake and it is this low pressure that is responsible for the high drag of bluff bodies. Such flows have been studied in connection with building aerodynamics, and they are clearly important for tall slender structures exposed to the natural wind where the regular shedding of vortices could lead to flow-induced oscillations. However, has this flow any relevance to the problem of road vehicles?

A finite-aspect-ratio bluff body has ends and fluid flows around these ends to relieve the low base pressure. The ends introduce a three-dimensional structure into the separating shear layers which weakens the basic shedding mechanism. Thus as the aspect ratio of the frontal surface, that is the ratio of frontal length to frontal width, is reduced the drag decreases, together with the strength of the shed vortices. It is also observed that the concentration of shed vorticity into discrete vortices becomes an increasingly broader-band process. The shedding of two-dimensional-type vortices is not observed below aspect ratios of 5 or 6. As an example of the effect of aspect ratio on mean drag, some measurements of Bedi (1971) of the drag coefficient of cantilevered rectangular sections are shown in Fig. 1. The results for infinite-aspect-ratio sections and sections with aspect ratios as high as 20 are seen to be quite different. At $H/W = \infty$, as first shown by Nakaguchi, Hashimoto & Muto (1968), rectangular-cylinder drag is highly dependent on cross-sectional shape and when the depth is a little over half the width the C_D may be as high as 3. However, finite-aspect-ratio rectangular sections with depth to width ratios up to 1.2 show little to no dependence on cross-sectional shape.

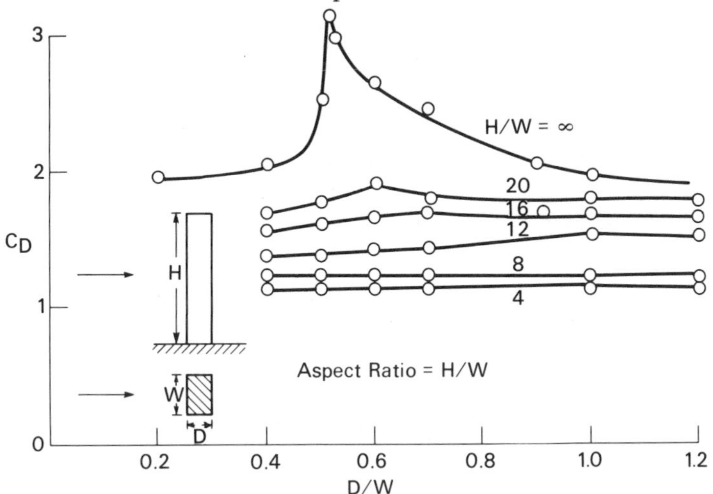

Fig. 1. Drag coefficient measurements of cantilevered rectangular blocks (Bedi (1971)).

References pp. 113-114.

A typical road vehicle has an aspect ratio of order unity so that drag will not be generated by the regular shedding of strong vortices. Therefore devices used to reduce the drag of two-dimensional bluff bodies, such as splitter plates and small quantities of base bleed, which interfere with the interaction between the two shear layers, will not be effective on road vehicles.

THE IMPORTANCE OF APPROACHING FLOW CONDITIONS

The smooth free stream flow (plus a thin boundary layer) approaching a building model placed in a traditionally designed aeronautics wind tunnel does not model correctly the natural wind. One of the major contributions offered by "building aerodynamicists" has been to illustrate the importance of approaching turbulence and shear on the flow about bluff structures (Cermak, 1976). As far as the effect of the wind is concerned, an obvious difference between a building and a road vehicle is that the former is exposed to only the natural wind, whereas the latter has essentially a smooth uniform flow modified to some degree by the wind. The building aerodynamicist is often concerned with the "worst case", say the once in 50 or 100 year wind, or perhaps with some oscillation condition; with a vehicle, if we are considering drag and fuel consumption, we are interested in some suitably weighted integral of all possible wind conditions likely to be experienced during normal driving.

Davenport (1961) and later Harris (1971) have set down general guidelines as to what atmospheric boundary layer characteristics to expect in high-wind conditions, for different types of terrain. Broadly, they tell us that for the same wind speed outside the Earth's boundary layer, the rougher the terrain the lower the mean velocity but the higher the turbulence level. However can this data be applied directly to vehicles? The wind is considered as having a mean velocity based on an averaging time of one hour; all fluctuations with time scales of less than an hour are regarded as turbulence. While an hour is a short time in the life of a building it is clearly a long time in the span of a typical road journey. A vehicle will experience turbulence from different terrains, embankments, cuttings, bridges and other vehicles and this turbulence will not always be of the idealised "stationary" form assumed by "building aerodynamicists". Everytime there is a high wind blowing from a particular direction a building can expect to receive a similar pattern of approaching turbulence in which statistical properties such as turbulence intensity and scale are unlikely to vary with time. A vehicle, however, will receive turbulence modified by local features, and during a journey the intensity and scale will change.

In order to understand the effects of turbulence we must differentiate between "passive" and "active" parts of the turbulence spectrum. The large-scale eddies, large compared to the size of the vehicle, are "passive" and have a quasi-steady effect on the aerodynamics; they change the magnitude and direction of the approaching wind. Smaller scales of turbulence, of the order of the vehicle size and lower, are "active" and can interact with different regions of the flow field and, for example, may affect boundary layer separation and reattachment and thus indirectly have a large influence on the overall flow. Over an expanse of flat terrain the integral scale of the locally generated turbulence, which is a measure of the size of the large eddies, is roughly proportional to the height above the ground. Therefore a vehicle will experience turbulence with a scale size comparable to its own dimensions, and it will be shown in the next section how turbulence of this scale can influence the flow around bluff bodies.

The relative turbulence intensity, defined as r.m.s. wind speed fluctuation divided by mean wind speed relative to a vehicle, can take on any value, depending on the magnitude and direction of the wind. The relative turbulence intensity felt by a vehicle driving more or less into the wind would be $\sqrt{\overline{u}^2}/U/(V/U+1)$, where U is wind speed, V vehicle speed and $\sqrt{\overline{u}^2}$ is the r.m.s. of horizontal wind speed fluctuations. In open terrain the wind turbulence intensity $\sqrt{\overline{u}^2}/U$ is around 0.2, and if the wind speed is as high as a third of the vehicle speed then the turbulence level relative to the vehicle would be 0.05. Higher turbulence levels are likely to be encountered when one vehicle is travelling close in the wake of another, but then this would have to be treated as part of the overall interaction problem.

For many building shapes it is crucial that the shear and turbulence in the approaching flow be correctly simulated in any wind tunnel experiment if there is to be a close tie up between tunnel measurements and the real situation. Although not so important for road-vehicles, these factors could nonetheless influence their drag as well.

THE EFFECT OF FREE STREAM TURBULENCE ON BLUFF BODY FLOW

A body placed in a turbulent stream can be affected in a number of ways, the most obvious of which is the production of fluctuating forces by the approaching fluctuating velocities. The mean flow and mean forces can also be affected because turbulence can promote transition in boundary layers and free shear layers at lower Reynolds numbers than those normally found in smooth flow. A turbulent stream also has some effect on the growth of boundary layers and wakes and can influence positions of flow separation and reattachment.

The most striking examples of the effect of free stream turbulence on bluff-body drag are those of the circular cylinder and sphere. The early data of Fage & Warsap (1929), showing the pronounced effect of free stream turbulence on the drag coefficient of a circular cylinder in the Reynolds number range 5×10^4 to 3×10^5, are reproduced in Fig. 2. The addition of turbulence promotes transition on the cylinder at a lower Reynolds number than in smooth flow. However, the generalization that the introduction of free stream turbulence simulates some higher Reynolds-number regime should be avoided.

References pp. 113-114.

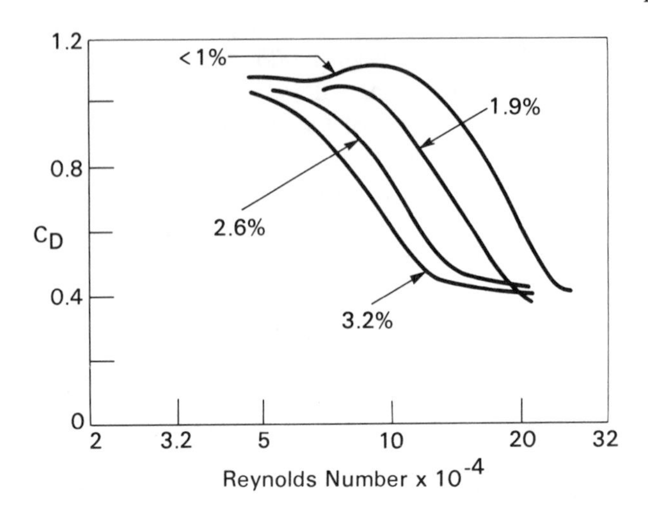

Fig. 2. Effect of turbulence on C_D of a 2-D circular cylinder (Fage & Warsap (1929)).

Will a bluff body with *fixed* separation points be affected by free stream turbulence? Experiments on square plates and circular discs mounted normal to a turbulent flow (Bearman, 1971) show that the presence of turbulence decreases the base pressure. It is argued that this is caused by an extra entrainment process whereby the stream turbulence invigorates the mixing between the shear layers and the near wake. Developing this idea, it is reasoned that the base pressure should be a function of the turbulence parameter $\sqrt{\overline{u}^2}/U)$ $(L_x{}^2/A)$, where $\sqrt{\overline{u}^2}/U$ is the turbulence intensity, L_x the longitudinal scale of turbulence and A the plate frontal area. The variation of base pressure with this parameter is shown in Fig. 3. In a so-called smooth flow, where $(\sqrt{\overline{u}^2}/U)$ (Lx^2/A) was about 5 x 10^{-4}, the base pressure was significantly higher with $C_{pb} = -0.36$.

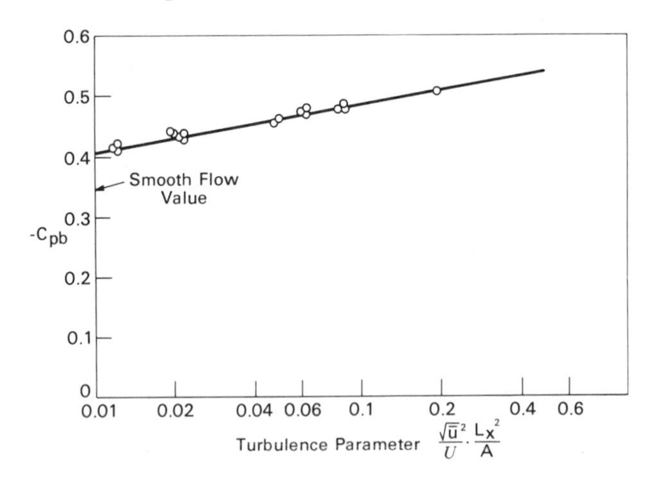

Fig. 3. Base-pressure measurements on flat plates in turbulent flow (Bearman (1971)).

If our attention is now turned to a two-dimensional body with a finite depth we see an *opposite* trend. Vickery (1966) has shown that when a square cross-section cylinder is placed in a turbulent flow the base pressure is increased. His results are shown compared with smooth flow measurements in Fig. 4. The generally accepted explanation for the base-pressure rise, when a side is normal to the flow, is that increased mixing within the shear layers causes them to spread and reattach, or intermittently reattach, on to the side faces. This increased mixing falls in line with the earlier discussion of the flat-plate results, the only difference being that now the shear layers have something to reattach to.

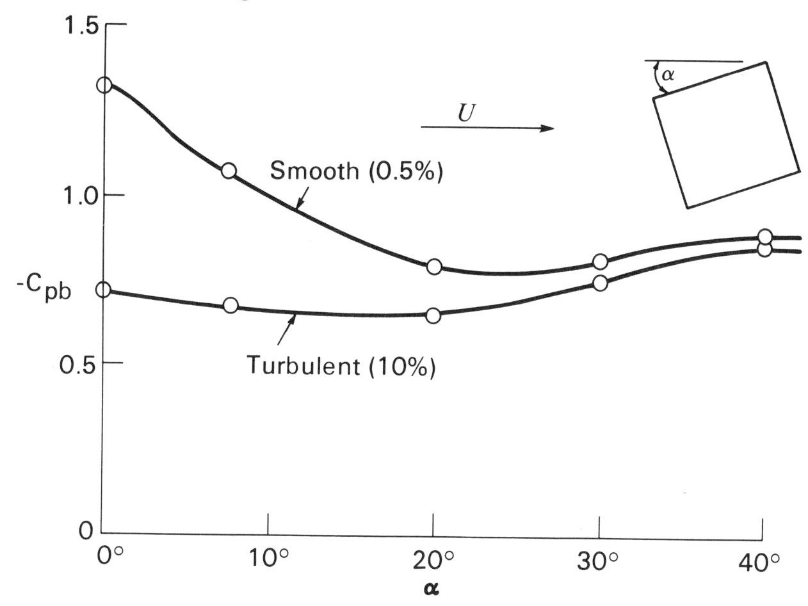

Fig. 4. Variation of base-pressure coefficient on a square-section cylinder in smooth and turbulent flow (Vickery (1966)).

With one face normal to a turbulent flow the base pressure on a square-section cylinder increases with increasing turbulence intensity, however very little dependence on the scale of turbulence is found. Results are shown in Fig. 5, for scales of turbulence varying between 0.4 and 4 times the width of the face of the cylinder, and it can be seen that there is no systematic variation of base pressure with scale. This is in strong contrast to the results shown in Fig. 3 for the square and circular plates, and there is no satisfactory explanation available as to why three-dimensional bluff-body flows should be so much more sensitive to scale. Gartshore (1973) has demonstrated that bluff-body free shear layer development is most affected by turbulence arriving along the mean stagnation streamline. It is this turbulent fluid that finds its way into the boundary layers and hence into the free shear layers. This finding suggests that a relatively small turbulent wake from a vehicle could influence the flow structure around a large following vehicle.

References pp. 113-114.

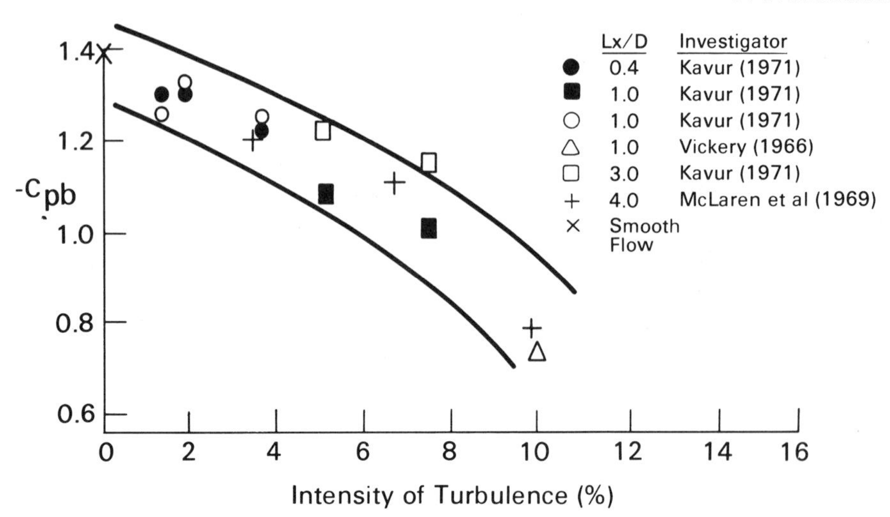

Fig. 5. Variation of base-pressure coefficient with turbulence intensity for a square-section cylinder at zero incidence. Results of Kavur (1971), McLaren et al (1969) and Vickery (1966). The scale of turbulence varies from 0.4 to 4 times the width of the cylinder.

One is prompted to ask whether similar reattachments can be found on low-aspect-ratio bluff bodies. Castro & Robins (1975) have shown that dramatic changes occur in the flow over a surface-mounted cube when the approach flow is varied (Fig. 6). With a thin boundary layer approaching the cube the flow separates from the front face, when a cube face is normal to the wind, and shows no reattachment on any of the other four faces. The pressure distribution on the cube is altered when it is placed in an upstream turbulent shear flow simulating the Earth's boundary layer. The reference pressure and dynamic head used in the definition of the pressure coefficients are the values recorded at a height corresponding to the top of the cube. The flow is strongly influenced by both the shear and turbulence in the simulated Earth boundary layer and the lower part of the cube becomes engulfed in a horse-shoe vortex system. There is a higher growth rate and curvature of the shear layers after separation which leads to reattachment on the top and side faces. The result on the pressure distribution is increased suctions near the leading edges and a pressure on the rear face not much lower than ambient. It is thought that these reattachments are mainly brought about by the action of the turbulence rather than the shear.

THE EFFECT OF GROUND CLEARANCE ON BLUFF-BODY FLOW

One of the most important aerodynamic differences between a building and a road vehicle is that the former is attached to the ground whereas the majority of the latter moves at some small height above the ground. It is easiest to make comparisons with the building if the vehicle is brought to rest and both the air and the ground are considered travelling past the vehicle. In this frame of reference the fluid at the

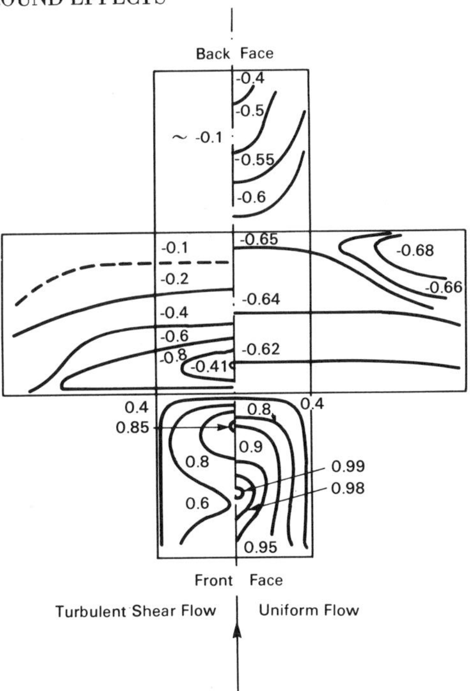

Fig. 6. Pressure-coefficient contours on a cube normal to the incident flow (Castro & Robins (1975)).

ground, approaching and passing under the vehicle, is constrained to move at the free stream speed due to the viscous no-slip condition. Therefore there will be a boundary layer region on the ground, probably extremely thin, developed by the velocity field about and under the vehicle.

Experimenters have used three methods to model the flow about a body moving near to a ground plane. One is to employ an *image*: a pair of identical models are mounted such that the plane of symmetry lies along the position of the ground. This does not necessarily ensure that the assumed position of the ground coincides with a plane of symmetry of the mean flow as the flow may become biased towards one model. Such a technique does not model the viscous effect at the ground, but much more important, it does not eliminate unsteady normal velocities at the "ground". Taking an extreme example, the flow over one half of a full circular cylinder does not model the flow about a half-cylinder moving very close to a ground plane. A better simulation of the half-cylinder flow is achieved by inserting a long splitter plate down the centre line of the wake of the full cylinder to inhibit all flow at the "ground".

The commonest method of representing the ground is to use a *fixed ground board* extending both upstream and downstream from the model. It is impossible to eliminate the influence of the upstream boundary layer and although the boundary layer can be thinned, such that its displacement thickness δ^* is very small, it will still possess non-zero values of $d\delta^*/dx$, $d^2\delta^*/dx^2$, etc. The effect of $d\delta^*/dx$ will be to

References pp. 113-114.

induce a slight incidence on the flow near the ground. The presence of a model will induce pressure gradients on the ground and δ^* may grow quite dramatically.

The effect of the ground can be most closely simulated by using an *endless moving belt.* A sketch of the moving-ground rig at Imperial College is shown in Fig. 7. The moving surface exposed to the flow is approximately 3' x 5' and the rig fits into a closed-return wind tunnel with a working section 4-1/2' wide and 4' high. All remnants of the upstream boundary layer should be removed before the flow meets the belt and this is done by an arrangement of boundary layer bleed and suction applied immediately ahead of the belt. Also, the belt has to be held down by suction otherwise beneath a body such as a road vehicle it would be impossible to maintain the correct ground clearance.

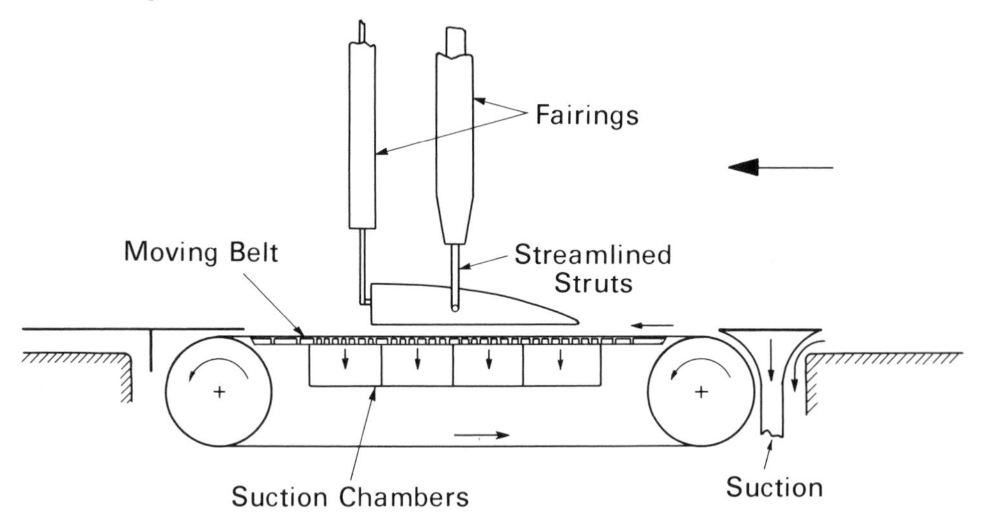

Fig. 7. Moving-floor rig.

The flows about various models moving over a ground have been visualized by Werlé (1963) in a water tunnel. The three techniques, image model, fixed ground and moving belt, were used to simulate the effect of the ground. His photographs clearly illustrate the occurrence of separation on the fixed ground board, ahead and behind the model, and the induced-incidence effect. On two-dimensional aerofoils above a fixed ground the induced incidence can lead to a lower stalling angle for positive incidence and a higher stalling angle for negative incidence compared to the moving-ground case.

There has been both theoretical and experimental work on the ground-effect problem in connection with aircraft landing and take off but the ground clearance generally studied, the ratio of the height above the ground to wing chord, are greater than those encountered with vehicles. For example, Bagley (1960) has calculated the lift of thin two-dimensional aerofoils near a ground by representing the wing and its image system by a suitable distribution of sources and vortices. $dC_L/d\alpha$ is found to increase as the ground is approached; it was also found that the aerofoil thickness causes a force towards the ground. On any finite wing the major effect of the proximity of the ground is a reduction in downwash due to displacement of the trailing vortices, leading to an increase in the effective incidence at a given geometric incidence. These trends on the lift of wings are also observed in experiment and East (1970), for example, showed that the lower the aspect ratio the greater the effect of the ground. East also discusses in detail the effect of the ground-board boundary layer development on his measurements.

The road-vehicle problem, however, involves smaller ground clearances and thicker sections than those normally encountered in aeronautics. Widnall & Barrows (1970) have solved for the flow about both two and three-dimensional flat plate aerofoils above a ground, down to very small clearances, by using the method of matched asymptotic expansions. They use the concept of a local highly-constrained flow through the gap, between the wing and the ground, matched to an outer flow. This approach was also followed by Tuck (1971) who devised an analytic solution to the potential flow around a semi-circular cylinder having its flat surface at a small ground clearance. He also presents an application of the Hess & Smith (1967) method to the calculation of the potential flow around a two-dimensional car-like shape near the ground. Clearly there is a problem about interpreting such results in terms of real bodies where separation will occur. Tuck discusses the possibility of there being some circulation, but is unable to specify its magnitude or sign. Fig. 8, reproduced from Tuck, shows slip velocities arising at the surface of his vehicle-shaped two-dimensional body due to a uniform stream and due to a circulation. The numbers outside the contour represent slip velocities per unit upstream velocity, and those inside are those per unit anticlockwise circulation; all velocities are positive in the anticlockwise direction. These results for $\epsilon = 0.05$, where ϵ is ground clearance divided by vehicle length, indicate that whereas the uniform stream produces almost equal velocities on top and bottom, circulation causes much higher velocities in the gap than over the rest of the body. These higher velocities in the gap suggest a great sensitivity of the gap flow to ϵ, and thus emphasize the need to simulate the underbody region correctly.

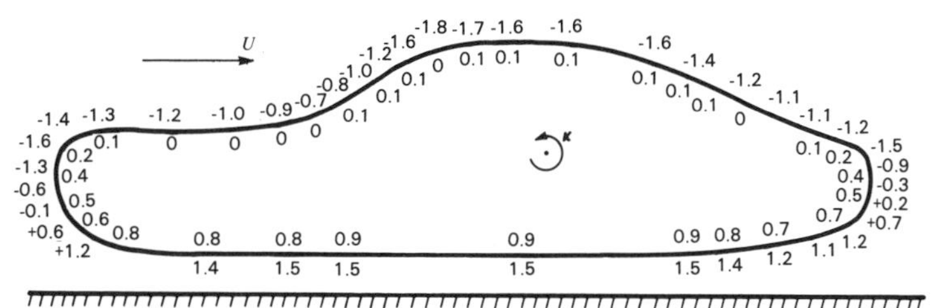

Fig. 8. Potential flow about a two-dimensional automobile-like profile. (Tuck (1971)). Published with the kind permission of the Journal of Fluid Mechanics.

References pp. 113-114.

EXPERIMENTAL INVESTIGATION OF THE EFFECT OF THE GROUND ON AN IDEALIZED VEHICLE SHAPE

The effect of the ground on an idealized vehicle shape has been examined in a wind tunnel at Imperial College, using both moving and stationary ground simulation, by Fackrell (1975). The experiments were performed using the moving floor and wind tunnel mentioned earlier (see Fig. 7) and full details of the experimental technique and apparatus are given by Fackrell. In order to decouple the effects of changing ground clearance and changing flow-separation position on the body it was decided to use a model with fixed, sharp-edged separation at the base. Details of the model are shown in Fig. 9 where it can be seen that it loosely resembled a semi-streamlined vehicle. The nose originally fitted (model 1) gave rise to a small separation bubble on its under-surface, but this was eliminated by fitting a more rounded nose (model 2). Little difference was observed between the overall force measurements for the two models.

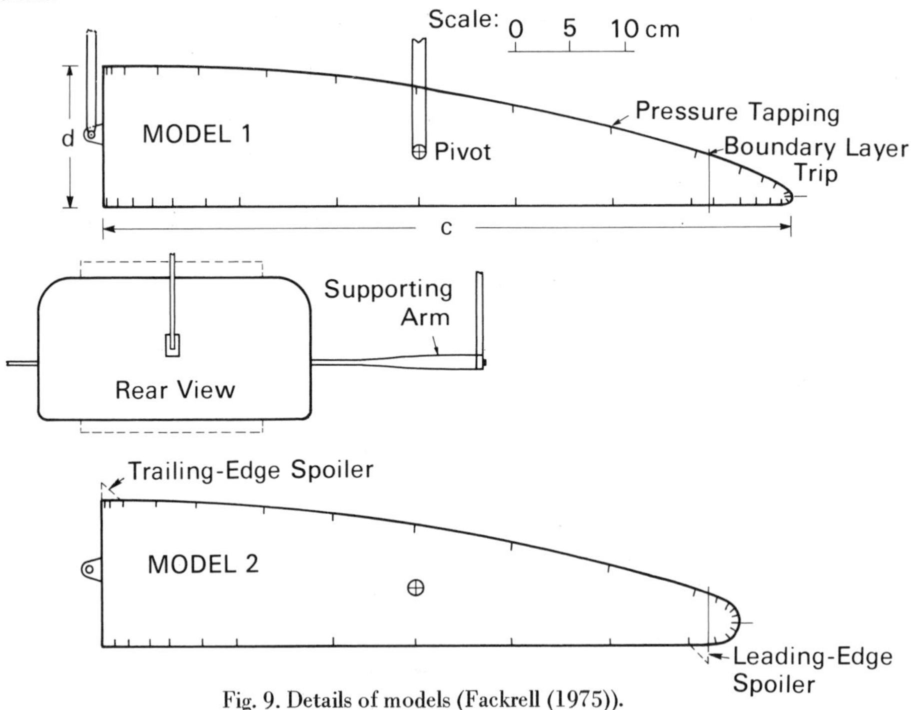

Fig. 9. Details of models (Fackrell (1975)).

The model was fitted with 88 tappings to measure surface pressure. In order to reduce interference with the flow, two remotely operated pressure switches were fitted inside the model and this allowed the pressures to be measured with only four tubes emerging. The bluff body was mounted on arms descending from a three component balance mounted above the roof of the working section. The Reynolds number, based on the length of the model, was approximately 8×10^5 for most of the tests. A boundary layer trip was fitted near the nose of the model. With a fixed ground board the displacement thickness of the boundary layer at the position of the body was estimated to be 0.004c, where c is the length of the model.

Fig. 10 shows the variation of lift coefficient with ground clearance for a 2° incidence range, with the floor moving and stationary. Force coefficients are based on model frontal area and the ground clearance was measured at the zero incidence position; the difference between this height and the minimum height over the 2° incidence range was small. Consistent with potential flow theory the results show an increase in $dC_L/d\alpha$ as the ground is approached. It should be noted that the idealized vehicle shape generates only attached-flow lift, whereas a road vehicle may experience both attached and separated-flow lift. Values of $dC_L/d\alpha$, at zero incidence, for both stationary and moving wall are shown plotted in figure 11 against ground clearance. Stollery & Burns (1969) and Waters (1973) measured the lift on bluff shapes near a fixed ground and they observed an increase in $dC_L/d\alpha$ as the ground was approached, but that this trend reversed very close to the ground. The authors suggested that this could be due to the separation position moving on the model.

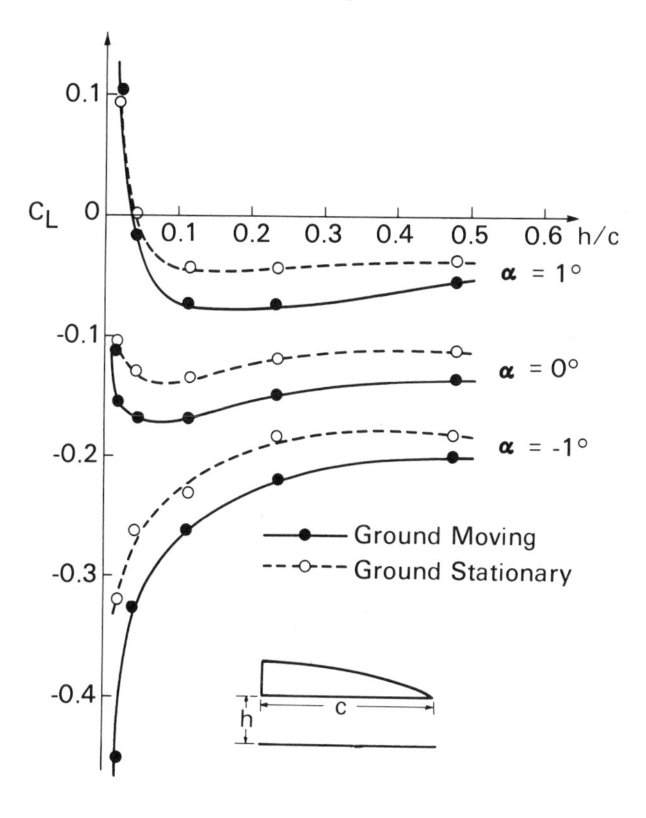

Fig. 10. Variation of lift coefficient with ground clearance - model 1.

References pp. 113-114.

Figs. 10 and 11 show differences in the lift measurements made with the floor stationary and moving. At $\epsilon = h/c$ greater than about 0.1 the values of C_L with the floor moving are less than those with the floor stationary. This is consistent with the suggestion that the growth of δ^* along the fixed ground induces a slight incidence, although it is surprising that at ground clearances as high as $\epsilon = 0.5$ the effect is so pronounced. It is estimated that only about half the C_L change could be due directly to this effect. At values of ϵ less than 0.1, moving the floor induces a higher $dC_L/d\alpha$ (Fig. 11). For a positive incidence of one degree and small ground clearance the induced-incidence and lift-curve-slope effects cancel each other out, whereas for a negative incidence of one degree they act in the same direction (Fig. 10). It is clear that the differences between the curves obtained with and without ground moving cannot be eliminated by a simple horizontal shift by an amount equal to δ^*/c. Thus, it would be false to use the argument that above a stationary flow the effective ground clearance is the physical clearance minus the boundary layer displacement thickness in the absence of the model.

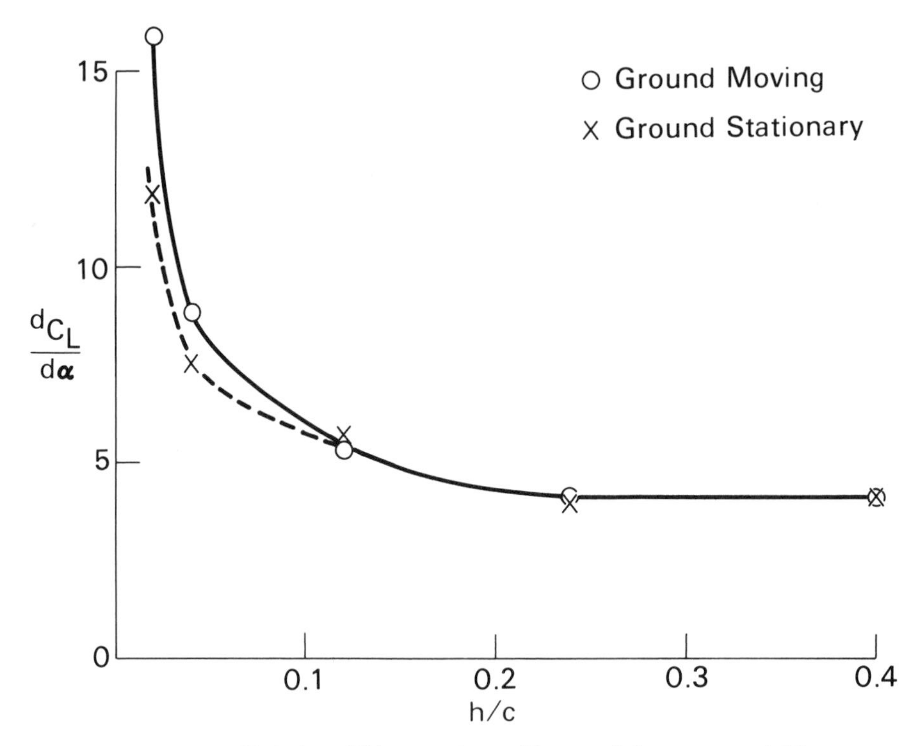

Fig. 11. Variation of lift curve slope with ground clearance - model 1.

Measurements of drag coefficient over the 2° incidence range for both stationary and moving ground are plotted against ground clearance in Fig. 12. Nearly all the drag is base drag and, if pressure is uniform across the base, Fig. 12 reflects the variation of the base pressure coefficient. With the floor moving, the drag increases as the body approaches the floor and reaches a maximum at $\epsilon \approx 0.05$. Except at small ground clearance, $\epsilon < 0.1$, the differences between values of C_D taken with wall moving and wall stationary are not large. This may not have been the case had separation been free to move along the surface of the body. Compared to the variations in C_L the variations in C_D are minor and do not suggest that any appreciable fraction of the drag is lift induced. The drag of low-aspect-ratio bluff bodies is primarily determined by near-wake mechanisms that control the base pressure and these mechanisms may be only weakly dependent, or even independent, of whether the body is developing lift.

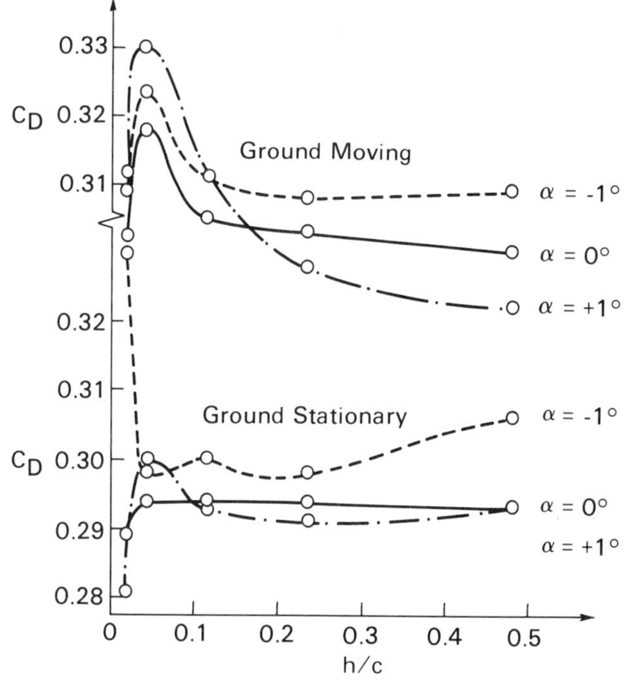

Fig. 12. Variation of drag coefficient with ground clearance - model 1.

Pressure distributions taken down the centre-line of the model with the floor moving are shown in Fig. 13 for $\epsilon < 0.04$ and for angles of incidence of +1°, 0° and -1°. Over this range of incidence C_L changes by about 0.3 and it can be seen that changes in the pressure distribution beneath the body are mainly responsible. This is in agreement with the potential flow predictions of Tuck (1971) who showed that changes in circulation mainly affected the flow under the body. The base of a bluff body is a region of low pressure, and when the body is brought near to a ground plane this low pressure helps to accelerate the fluid through the narrow channel between the underside of the body and the floor. At small ground clearance the ratio of the width to height of this channel suggests that the flow may approach more nearly that of a two-dimensional channel. Widnall & Barrows (1970) discuss how the velocity under flat plate aerofoils, near a ground plane, is set by conditions at the trailing edge. The base pressure feeds upstream to influence most of the undersurface and only a small change in pressure over a large undersurface is needed to substantially alter the lift. Perhaps then, rather than drag having a lift-related component, lift has a drag-related component.

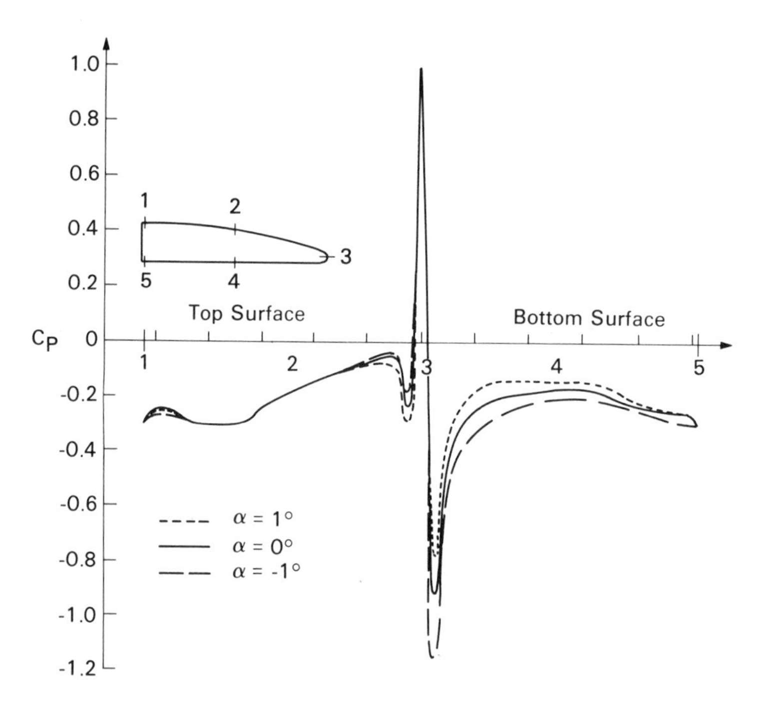

Fig. 13. Centre-line pressure distributions for model 2 at $\epsilon = 0.04$, ground moving.

Fackrell (1975) presents further measured pressure distributions for the model fitted with small spoilers. Two spoiler positions are used, one along the top surface at the trailing edge, and the other just underneath the nose. The pressure distributions are shown in Fig. 14. Both produce a downforce; at $\epsilon = 0.04$ the trailing-edge spoiler produces a negative lift increment of 0.49 and the leading-edge one 0.33. However, it can be seen from the pressure distributions that whereas the leading-edge spoiler only produces a modest decrease in base pressure, the trailing-edge one decreases the base pressure coefficient by 0.25 (nearly 80%). Part of the effectiveness of the trailing-edge spoiler in producing downforce is due to it creating a lower base pressure (and hence higher drag), which feeds along under the vehicle as discussed earlier. The trailing-edge spoiler provides an example of *lift related to drag*.

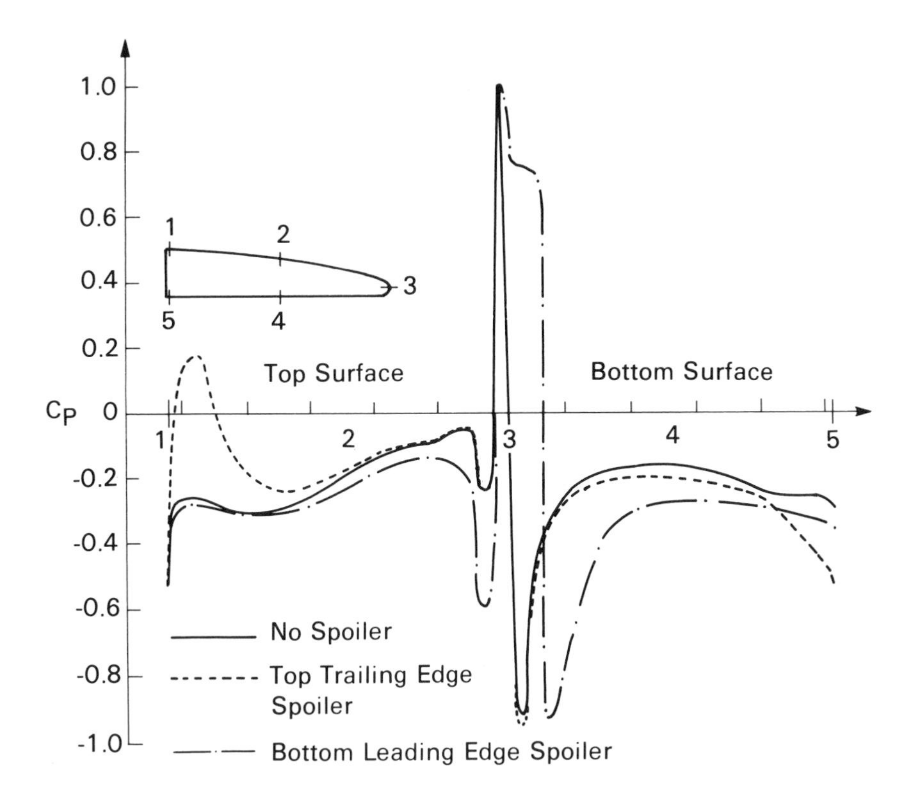

Fig. 14. Centre-line pressure distributions for model 2 at $\epsilon = 0.04$ with spoilers, ground moving and $\alpha = 0°$.

RESULTS OF A FREE STREAMLINE THEORY CALCULATION

Bearman & Fackrell (1975) have presented a numerical free streamline method to calculate the incompressible potential flow external to a two-dimensional bluff body and its wake. The method incorporates the wake-source model of Parkinson & Jandali (1970), and the wake is modelled by placing sources on the rear of the wetted surface of the body. In order to calculate the flow, the base pressure and the separation points must be known. The method has been applied to calculate the flow about the bluff body used in the experimental investigation. In the calculation method the effect of the ground is modelled by an image. The method is two-dimensional, so in order to make some comparison with the three-dimensional body the experimentally determined base pressure coefficient was arbitrarily doubled

References pp. 113-114.

and applied to the two-dimensional calculation method. The results for C_L are shown in Fig. 15, and a comparison with Fig. 10 shows that the trends in the data are well predicted. This work is an extension of that of Tuck (1971) in that it makes due allowance for separation and the effect of the wake. Predicted pressure distributions agree qualitatively with experiment.

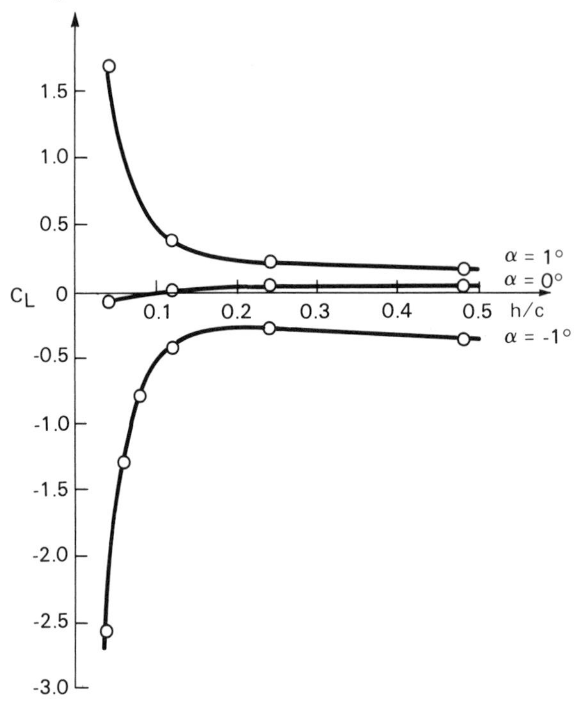

Fig. 15. Variation of lift coefficient with ground clearance, two-dimensional numerical results - model 1.

CONCLUSIONS

Road vehicles are exposed to turbulence, generated both by the natural wind and by other vehicles, and under some conditions this may affect their mean-drag characteristics. It is shown that free stream turbulence can change separation and reattachment positions on bluff bodies by modifying boundary layer, free shear layer and wake development. Two-dimensional bluff bodies and bluff bodies with aspect ratios typical of road vehicles are equally affected. In any comparison between wind-tunnel and on-road data, atmospheric conditions should be considered.

The flow about a road-vehicle-like body in a wind tunnel is best simulated by using a moving ground. Experiments show that using a fixed ground induces a slight incidence in the approaching flow; in addition, very near the ground, it decreases $dC_L/d\alpha$ below the value measured with a moving ground. For the particular model configuration tested the ground condition seemed to most affect lift but this may have been because the model had fixed location of separation. A free-streamline theory demonstrates that flow around bluff bodies near a ground plane can be predicted, qualitatively, if the base pressure and separation position are known. Base pressure influences the lift by accelerating fluid in the gap between the underside of the vehicle and the ground. It is shown that a spoiler under the nose is a more aerodynamically efficient way of generating a force increment towards the ground than a spoiler above the trailing edge. This is because part of the effectiveness of the latter depends on reducing the base pressure.

ACKNOWLEDGEMENT

The experimental work of Fackrell which is quoted in this paper was supported by the Donald Campbell Research Fund and jointly supervised with Dr. J. K. Harvey.

REFERENCES

Bagley, J. A., 1960, Pressure distribution on two-dimensional wings near the ground, R.A.E. Rep. Aero 2625.

Bearman, P. W., 1971, An investigation of the forces on flat plates normal to a turbulent flow, J. Fluid Mech. Vol. 46, pp. 177-198.

Bearman, P. W. & Fackrell, J. E., 1975, Calculation of two-dimensional and axisymmetric bluff body potential flow, J. Fluid Mech. Vol. 72, pp. 229-241.

Bedi, A., 1971, The effect of aspect ratio on the flow around bodies of rectangular section, MSc Thesis, Imperial College, London.

Castro, I. P. & Robins, A. G., 1975, The effect of a thick incident boundary layer on the flow around a surface mounted cube, CEGB Marchwood Engineering Laboratories Report R/M/N795.

Cermak, J. E., 1976, Aerodynamics of Buildings, Annual Review of Fluid Mechanics, Vol. 8, pp. 75-105.

Davenport, A. G., 1961, The application of statistical concepts to the wind loading of structures. Proc. Inst. Civ. Eng. London, Vol. 19, pp. 449-472.

East, L. F., 1970, The measurements of ground effect using a fixed ground board in a wind tunnel, R.A.E. Rep. TR-70123.

Fackrell, J. E., 1975, The simulation and prediction of ground effect in car aerodynamics, Imperial College, Aeronautics Report 75-11.

Fage, A. & Warsap, J. H., 1929, The effects of turbulence and surface roughness on the drag of a circular cylinder, R. & M. No. 1283.

Gartshore, I. S., 1973, The effects of free-stream turbulence on the drag of rectangular two-dimensional prisms, Univ. of W. Ontario, Faculty of Eng. Sci. Report BLWT-4-73.

Harris, R. I., 1971, The nature of the wind, "The Modern Design of Wind-Sensitive Structures," CIRIA, London, pp. 29-55.

Hess, J. L. & Smith, A. M. O., 1967, Calculation of potential flow about arbitrary bodies, Prog. Aero. Sci. Vol. 8, pp. 1-138.

Kavur, C. 1971, An investigation of the effect of turbulence on the flow around cylinders of square cross-section, MSc Thesis, Imperial College, London.

McLaren, F., Sherrat, A. & Morton, A. S., 1969, Effects of free-stream turbulence on the drag

coefficient of bluff sharp-edged cylinders. Nature, Vol. 223, pp. 828-9, and Vol. 224, pp. 908-9.
Nakaguchi, H., Hashimoto, K. & Muto, S., 1968, An experimental study on aerodynamic drag of rectangular cylinder, Journal of the Japan Society of Aeronautical and Space Sciences, Vol. 16, pp. 1-5.
Parkinson, G. V. & Jandali, T., 1970, A wake source model for bluff body potential flow, J. Fluid Mech. Vol. 40, pp. 577-594.
Stollery, J. L. & Burns, W. K., 1969, Forces on bodies in the presence of the ground, 1st Symposium on Road Vehicle Aerodynamics, City University, London, Paper 1.
Tuck, E. O., 1971, Irrotational flow past bodies close to a plane surface. J. Fluid Mech, Vol. 50, pp. 481-491.
Vickery, B. J., 1966, Fluctuating lift and drag on a long cylinder of square cross-section in a smooth and in a turbulent stream. J. Fluid Mech, Vol. 25, pp. 481-494.
Waters, D. M., 1973, Thickness and camber effects of bodies in ground proximity, Advances in Road Vehicle Aerodynamics, BHRA Fluid Engineering.
Werlé, H., 1963 Simulation de l'effet du sol au tunnel hydrodynamique, ONERA Report TP63.
Widnall, S. E. & Barrows, T. M., 1970, An analytic solution for two and three-dimensional wings in ground effect, J. Fluid Mech, Vol. 41, pp. 769-792.

DISCUSSION

Prepared Discussion

F. T. Buckley, Jr. *(University of Maryland)*

My comments relate both to Dr. Bearman's and Mr. Mason's talks, more importantly to Dr. Bearman's. They concern the influence of the turbulence in natural winds on the flow past a class of bluff bodies; in particular past tractor-trailer rigs. They also concern the ability of a wind tunnel to simulate all the effects of the wind on the flow past the vehicle. While turbulence is known to affect critical Reynolds number, the location of flow separation and reattachment, and wake flows, it is significant to note that the turbulence in natural winds has not been simulated in past wind tunnel experiments of surface vehicles. It is of interest to determine the degree to which this error in simulation might affect the data so produced.

We deemed it necessary to perform some on-road tests in a natural wind environment with full-scale vehicles to substantiate some of our wind-tunnel measurements.* We looked at five different truck configurations: a baseline tractor-trailer (B), the baseline vehicle modified with a roof-mounted wind deflector (WD), the vehicle modified with the deflector plus a vortex stabilizing device placed in the gap between the tractor and trailer to improve cross-wind performance of the vehicle (WD + VS), a streamlined roof fairing (SF), and finally, a combination of the streamlined fairing plus a gap seal (SF + GS). These are shown in Figs. 16 and 17. Instrumentation consisted of fifth-wheel measurements of the truck speed, and an

---

**Walston, Jr., W. H., Buckley, Jr., F. T., & Marks, C. H. (1976), Test Procedures for the Evaluation of Aerodynamic Drag on Full-Scale Vehicles in Windy Environments, SAE 760106.*
Buckley, Jr., F. T., Marks, C. H., & Walston, Jr., W. H. (1976), Analysis of Coast-Down Data to Assess the Aerodynamic Drag Reduction on Full-Scale Tractor-Trailer Trucks in Windy Environments, SAE 760850.

anemometer (with one-meter response) mounted on a boom in front of the vehicle to provide the speed and direction of the relative airstream.

Fig. 16. Test vehicle with wind deflector and vortex stabilizer system.

Fig. 17. Test vehicle with tractor roof fairing and full gap seal.

The data were obtained in natural winds at truck speeds ranging from 60 to 45 mph by the coastdown test procedure. The two cases shown in Fig. 18a,b indicate the extremes of the data that were measured. In Fig. 18a we have a steady breeze of 3 mph and a turbulence intensity with respect to the relative airstream of 0.6 percent, while in Fig. 18b we have gusty winds with an average velocity of 16 mph and a turbulence intensity of 4.1 percent. Most of the data were obtained in the range of 2-6 percent turbulence intensity, with a few cases obtained at less than 1 percent. The data were sampled at a rate of 0.9 sec per sample, digitized, and reduced with a statistical-numerical procedure to evaluate the drag coefficients. An average of 14 runs was performed for each of the five configurations.

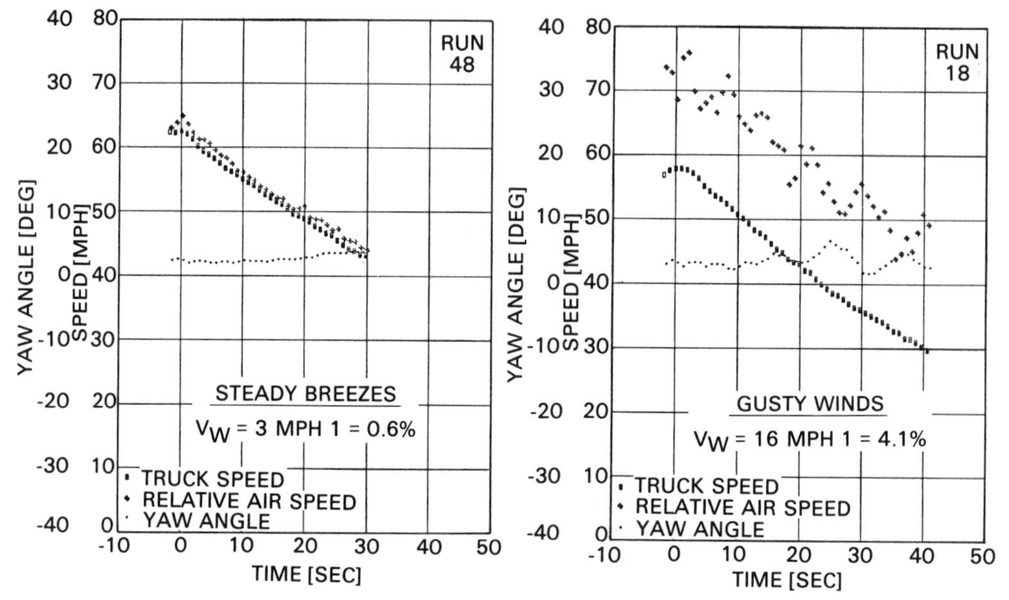

Fig. 18. Typical data records.

F. T. Buckley, Jr.

The calculated results are shown in Fig. 19, where the solid lines represent the means of the data points. There is some scatter in the data which, I am sure, is partly a result of error in measurement. But more importantly, the scatter demonstrates that we are dealing with unsteady flow fields. Therefore, one would not expect to obtain a single, repeatable, measurement of drag at a given yaw angle. The absolute values of the drag coefficients have a systematic error, which I would estimate to be about 5-7%, owing primarily to an uncertainty about the correct magnitude of the rolling resistance of the vehicle. However, we feel that since the same rolling resistance model was used for all the data reduction, the error in the differences between drag coefficients should be much smaller than that in the absolute level.

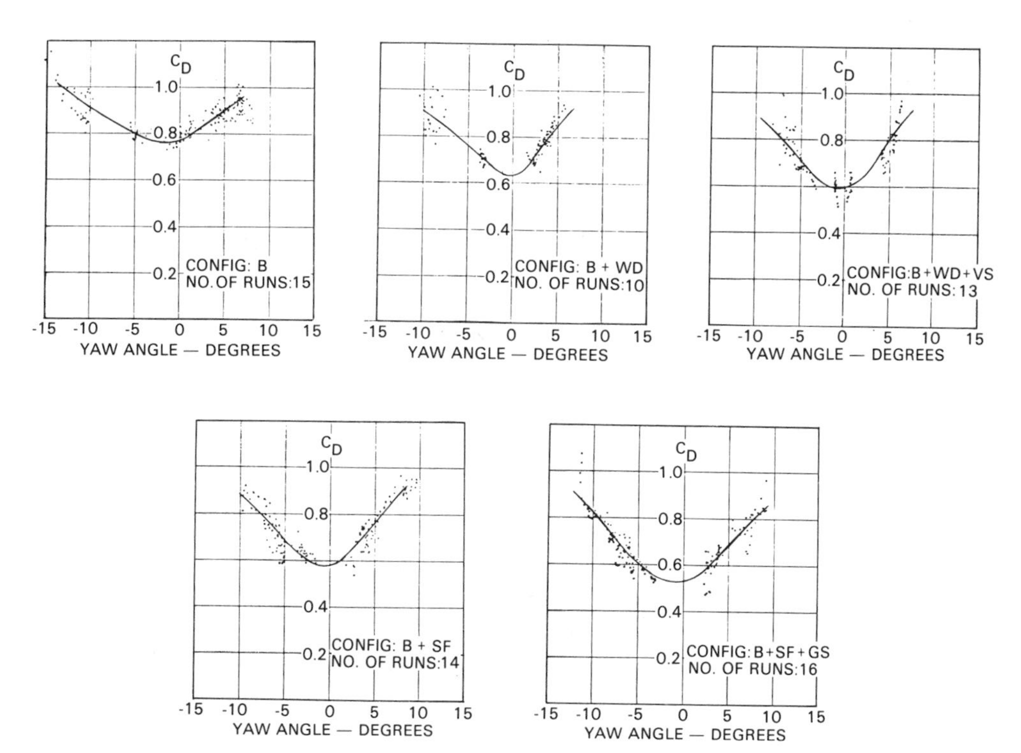

Fig. 19. Drag coefficient variations with yaw angle for full-scale vehicle.

Fig. 20 compares measurements of drag coefficient reduction vs. yaw angle obtained in a wind tunnel*, with those measured in on-road tests. It is apparent that the shapes of the curves are similar. However, there are some important differences in the results which I feel are the result of a turbulence effect.

**Marks, C. H., Buckley, Jr., F. T., & Walston, Jr., W. H. (1976) An Evaluation of the Aerodynamic Drag Reduction Produced by Various Cab Roof Fairings and a Gap Seal on Tractor-Trailer Trucks, SAE 760105.*

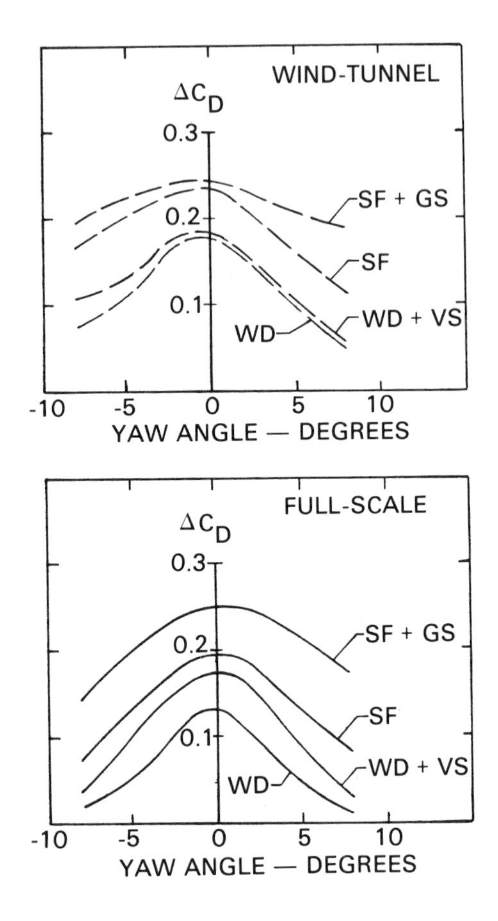

Fig. 20. Comparison between wind tunnel and full-scale drag reductions.

Some further comparisons are presented in Tables 1 and 2 showing the drag coefficient reductions resulting from the addition of a roof-mounted drag reducer, ΔC_{DR}, a device in the gap, ΔC_{DG}, and the total due to a combination, ΔC_{DT}. Table 1 is for zero yaw, while Table 2 contains wind-averaged drag coefficients* based on a vehicle speed of 55 mph and a constant wind speed of 9.5 mph.

It is apparent that better agreement exists in some cases than in others. Considering first the 0° yaw condition, we see that the drag coefficient reductions following the addition of a roof-mounted device were higher in the wind tunnel than in the full-scale tests. Note also that the addition of a device in the gap produced

Buckley, Jr., F. T., & Sekscienski, W. S. (1976), Comparisons of Effectiveness of Commercially Available Devices for the Reduction of Aerodynamic Drag on Tractor-Trailers, SAE Transactions, Vol. 84, Sec. 3.

TABLE 1.
Drag Reduction at 0° Yaw.

Config	"Wind-Tunnel"				Full-Scale			
	C_D	ΔC_{D_R}	ΔC_{D_G}	$\Delta C_{D_{TOT}}$	C_D	ΔC_{D_R}	ΔC_{D_G}	$\Delta C_{D_{TOT}}$
B	.760	–	–	–	.770	–	–	–
B+WD	.580	.180		.180	.635	.135		.135
B+WD+VS	.575		.005	.185	.595		.040	.175
B+SF	.525	.235		.235	.575	.195		.195
B+SF+GS	.525		.000	.235	.525		.050	.245

TABLE 2.
Average Drag Reduction

Config	"Wind-Tunnel"				Full-Scale			
	$\overline{C_D}$	$\Delta \overline{C_{D_R}}$	$\Delta \overline{C_{D_G}}$	$\Delta \overline{C_{D_{TOT}}}$	$\overline{C_D}$	$\Delta \overline{C_{D_R}}$	$\Delta \overline{C_{D_G}}$	$\Delta \overline{C_{D_{TOT}}}$
B	.895	–	–	–	.915	–	–	–
B+WD	.800	.095		.095	.870	.045		.045
B+WD+VS	.795		.005	.100	.835		.035	.080
B+SF	.740	.155		.155	.800	.115		.115
B+SF+GS	.690		.050	.205	.725		.075	.190

insignificant drag reductions in the wind tunnel, but significant ones in the full-scale tests. Furthermore, the total drag reductions obtained in the last situation (roof device plus gap treatment) agreed more favorably between the two different types of tests. An identical trend was found in a comparison of the wind-averaged drag coefficients, indicating that the cause of the discrepancy extended throughout the yaw angle range.

These results provide a consistent set of data for speculation on a wind turbulence effect on drag. First, the effectiveness of the roof-mounted drag reducers is dependent on their ability to produce a stable separated wake region. It appears reasonable to assume that these wakes would be less stable in the presence of the large-scale turbulence that existed in the full-scale tests than they would be in a non-turbulent wind tunnel airstream. This would explain why lower drag reductions were found in the full-scale tests of the roof-mounted devices tested alone. Second, the devices that were added in the gap were such that they would act to stabilize the wake flows of the roof-mounted devices with respect to lateral eddying motions. This would explain why the addition of these devices produced the significant drag reductions found at 0° yaw in the full-scale tests, and why the wind tunnel and full-scale tests compared more favorably when these devices were used.

These results suggest that further research is needed to explore the importance of the role of freestream turbulence in the flows past surface vehicles. The results of such research could have important implications with respect to the manner in which future wind tunnel tests of such vehicles are conducted, and could affect the development of mathematical models for the general prediction of surface vehicle flow fields.

M. V. Morkovin *(Illinois Institute of Technology)*

I would like to ask a question concerning the importance of a moving ground-plane in wind tunnels. I believe that if one wants to know the major differences between the cases with and without a moving belt, one should be looking for the place where the maximum effect on the ground-plane boundary layer might be. I would be looking at the back part of the underside, and immediately behind a vehicle. For instance, let us consider the flow field behind a truck that Bill Mason showed (Fig. 27) where we have a high opening under the trailer. The streamlines of the underbody flow are diffusing very rapidly there, and this must cause an adverse pressure gradient on the floor behind the truck. In a standard wind tunnel (no moving belt) there is a boundary layer on the floor, and it would be affected by such a pressure gradient. I would expect that this would probably be the region where one has the maximum difference between the standard wind tunnel test and the on-road situation. What is known about this flow region where entrainment is going to be influencing its character and the base drag?

P. W. Bearman

There are some very nice flow visualizations done at ONERA by Werlé who has done comparisons of the fixed ground and the moving ground. These visualizations show ground-plane flow separation ahead of a vehicle, where we perhaps would expect it, but they also seem to show separation behind a vehicle just in the region Dr. Morkovin is talking about. Unfortunately, I do not think that we have anything else to go on. I certainly looked through to see if we could find some other results where Werlé documented these effects, but I got the impression that it is really something that needs more looking at. You can see that the displacement thickness is going to be growing very rapidly in that region – certainly if the flow is curving upwards like that. We can expect that the effect of the ground condition could be extremely important back there, for both the wake development and the flow which goes back onto the vehicle. I think that it's an area which needs more research.

M. V. Morkovin

Wouldn't you agree that this is the place where the difference would be the greatest? If you know what's going on in the back, and the difference is not large – wouldn't you be okay with a stationary groundplane?

P. W. Bearman

Yes, that's fair enough, at least for an idealized-flow view of vehicles that allows

P. W. Bearman

flow separation only at the rear. However, if you look at a real vehicle, there can be other separations on the forebody. For example, the ground-plane boundary layer can induce an incidence effect on the flow approaching a vehicle, possibly creating some forebody separations and eliminating others. So when we look at more practical vehicles, there may be other effects of a moving ground in addition to those in the base region. But I agree, for the model that I used, where we were looking mainly at the drag coming from the base, we have to look at the floor there.

H.-J. Emmelmann *(Volkswagenwerk AG, Germany)*

We prepared a slide showing a comparison of the velocity distributions measured underneath a car in wind tunnel and on-the-road tests (Fig. 21). You can see that there are not large differences in the velocity profiles for a realistic Reynolds number. The wind tunnel boundary layer thickness was approximately 150 mm, and the ground clearance of the 4 m long car was 200 mm at the station midway between the axles.

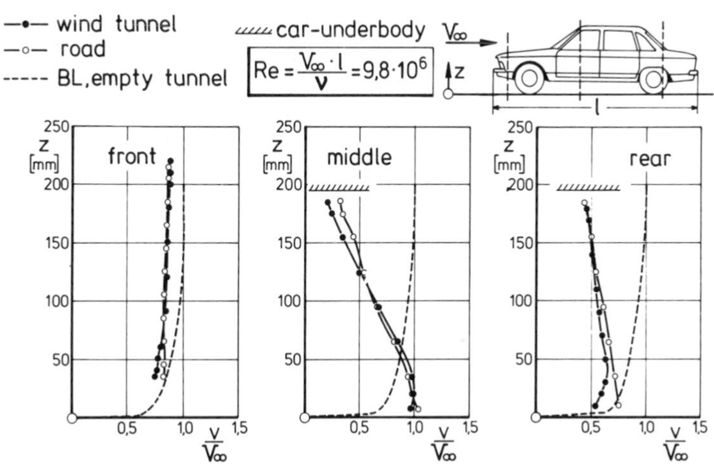

Fig. 21. Velocity distribution underneath a passenger car, comparison of measurements on the road and in the wind tunnel.

P. W. Bearman

It is difficult to comment on this. All I would like to say is that any single set of measurements cannot tell you all about the effect of the moving ground. There certainly are some differences near the floor in your data, and I'm saying that the smaller you make your ground clearance the more the effects of these differences are going to show up; it is not so much on drag, but possibly it may be changing the circulation around the vehicle. There's a different mass flow going through the underbody region than there would be if the floor were moving.

W.-H. Hucho *(Volkswagenwerk AG, Germany)*

I would like to make some additional remarks with respect to our slide. The velocity profiles under the car are also compared to the boundary layer profiles on the floor of the empty wind tunnel test section. The presence of the car changes the ground plane boundary layer in the underbody region completely. It is "compressed" significantly, and is very thin compared to that of the empty tunnel, even without the ground-plane moving. These data were published in an SAE paper*. We also made the same comparison with a sports car having closer ground proximity; the correlation was not quite as good. However, I believe that it still shows that, at least from where we stand now, a moving ground-plane or other kind of sophisticated ground-plane simulation does not promise any better understanding of vehicle aerodynamics, or better drag results. If, in the future, we approach lower drag coefficients and have smooth undersides, as you had on your model, then the picture may change.

P. W. Bearman

The only comment I have is that the moving ground-plane is certainly important on cars with lower ground clearance. We know of one case where a racing car was tested in a tunnel and gave an upforce, but when it was taken on the track it generated so much downforce that they more or less had to scrape it off the track. This must have been due to the change in ground conditions.

C. V. Williams *(Lockheed-Georgia Co.)*

In the last three months I have performed some experiments where we used a blowing boundary layer control system on the tunnel floor to match the velocity distribution at the front bumper so as to have a uniform velocity coming under a car. We found that the drag data for a series of cars, including a 1976 Chevrolet Camaro, compared within 1 to 2 percent, blowing off and blowing on. We have also run an Indianapolis-type race car where the ground clearances are from essentially nothing at the front, to a few inches at the rear, and on which the underbody surface is smooth. We had orifices around its perimeter, and the pressure distribution around the car

Hucho, W.-H., Janssen, L. J. & Schwarz, G. (1975), The Wind Tunnel's Ground-Plane Boundary Layer – Its Interference with the Flow Underneath Cars, SAE 750066, Detroit.

looked pretty much the same whether you had blowing on or off. We have also done some model tests of a truck and we again got 1-2 percent kind of drag number differences, blowing on and off. On the basis of these results, I concur with Dr. Hucho. Where we are now, we couldn't find a fixed ground-plane to be a serious problem.

P. W. Bearman

Just a very short comment on that. In a way it agrees with what I said because, remember, the drag in my case stayed pretty much the same. However, when I used the front-end spoiler and got something nearer to the ground, the differences in the pressure distribution between the fixed-ground and moving-belt cases were enormous.

S. J. Kline *(Stanford University)*

I would like to comment on turbulence and the performance of drag reducing devices for trucks. We have seen that spoilers are used on the top of tractors to make the flow attach at the front edges of trailers. From the flow visualization results that have been shown, what seems to be happening is that the flow separated off the spoiler is somewhat irregular. In addition, when you go over the road you're going to get fluctuations from freestream turbulence. So in fact you're getting fluctuations from two sources. Would it make sense to consider beveling or rounding the face of the trailer, even if you have a spoiler on the tractor, in order to give some leeway in picking up the stagnation point correctly – to get a good flow attachment near the front edges even in the presence of these fluctuations?

P. W. Bearman

I don't think I'm very well qualified to answer that. All I can say is that, yes, I would like you to think about the effect of turbulence when working with any of these devices.

Prepared Discussion

H. M. Nagib* *(Illinois Institute of Technology)*

A smoke-wire technique, we recently developed, was used to study the flow past the bluff body shown in Fig. 22. The technique, which is analogous to the hydrogen-bubble method used in water, is capable of introducing controlled sheets of smoke streaklines, as demonstrated in Fig. 23. The bluff body was placed on a flat plate located in the freestream of a wind tunnel (Case A), and also on the floor of the test section in two different floor boundary layers of increasing velocity gradient and turbulence intensity (Cases B and C, respectively). Thick boundary layers, simulating the atmospheric surface layer, were developed along the 22 ft. floor of a 4 x 6 ft. test section of a wind tunnel with the aid of the counter-jet technique and surface

---

**The reported work was performed in collaboration with D. Koga and R. Slater.*

roughness. A series of grids at the entrance to the test section maintained the turbulence level in the freestream above the boundary layer at 1 percent. The flow past the body was visualized with and without ground clearance using the 1 in. struts (cubes) shown in Fig. 22. All of the photographs shown here depict the condition along the centerplane of the body. The Reynolds number based on b was 10,000.

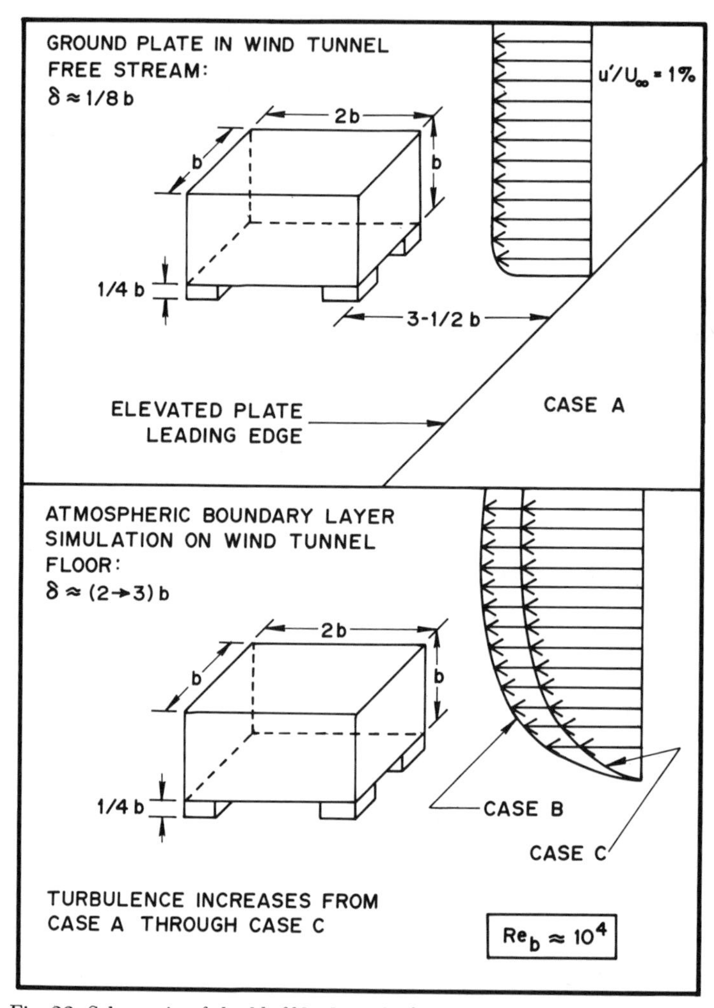

Fig. 22. Schematic of the bluff body and of the three Flow Cases, b = 4 in.

The three realizations of an instantaneous visual record in Fig. 23 demonstrate the continuous changes in the details of the flow field in the presence of only 1 percent turbulence (Case A). In particular, the conditions in the stagnation region appear to vary considerably and the flow through the underbody clearance seems to display some unsteadiness. While in the top two records, as well as in the time-mean record,

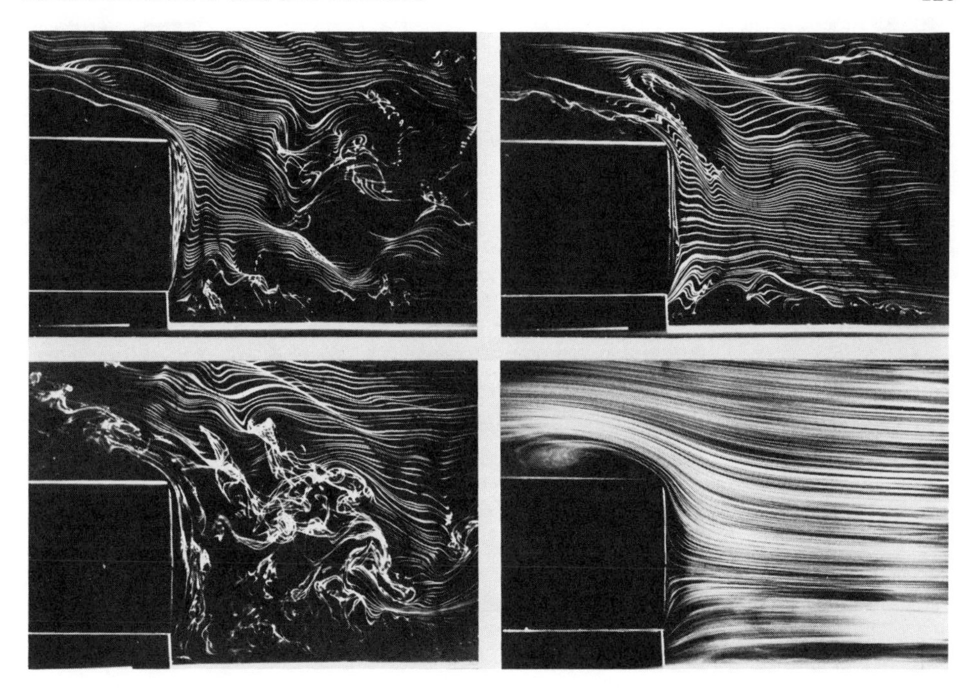

Fig. 23. Three instantaneous and one time-mean visualization records of bluff body with ground clearance in Flow Case A.

H. M. Nagib

some flow is shown passing under the bluff body, the flow through the clearance appears to be momentarily "choked" in the bottom left photograph. The exposure duration for the bottom right picture of this figure, as well as for those of Figures 24 and 25, corresponds to the time required for a particle to travel a distance of about 1 to 2 b.

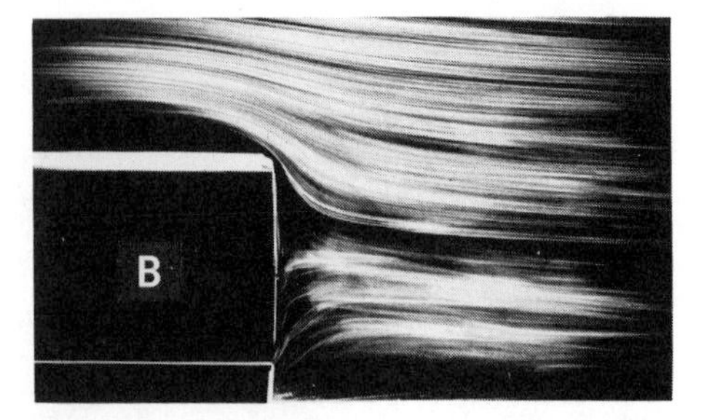

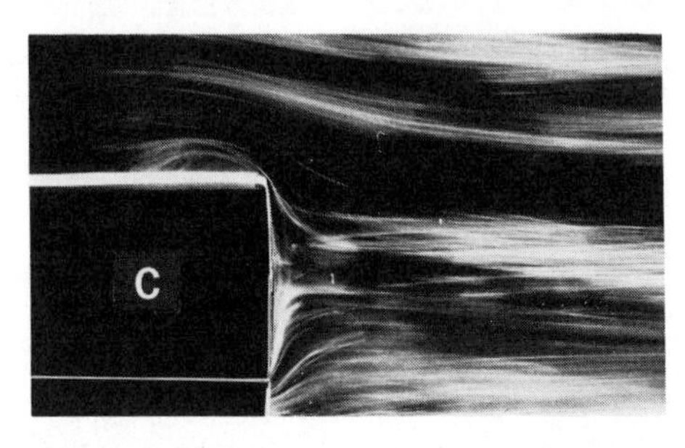

Fig. 24. Visualization of flow past bluff body with ground clearance in Cases A, B and C.

As demonstrated by the photographs, the smoke-wire is a useful tool which may be utilized quantitatively and can handle the turbulence in the flow. Some of the key observations made during our investigations are*: 1) The secondary flow, in the form of a horseshoe vortex at the lower front face of the body, is very unsteady and hence can only be visualized by time-mean records; 2) this secondary flow can be rather complex and composed of several cells, as shown in Fig. 25; 3) the horseshoe vortex disappears in the presence of the particular ground clearance used, producing a flow under the bluff body; without the clearance it grows larger and diffuses as the velocity gradient is increased and the turbulence level is raised, respectively; 4) as displayed in Figs. 24 and 25, increasing the upstream-flow velocity gradient and turbulence for a bluff body, with or without the clearance, leads to an upward shift of the stagnation "point" and to the faster reattachment of the flow separated along the top leading edge; 5) introduction of the ground clearance which allows flow under the body leads to a downward shift of the stagnation region; 6) even with only a relatively short segment of ground-plane upstream of the body, a boundary layer quickly develops, and the rapid change in its displacement thickness leads to some inclination of the average streamlines with respect to the body, as depicted by the smoke streaklines; this inclination is unrepresentative of on-road vehicle aerodynamics; 7) the ground-plane effects may be limited to bodies with small ground clearances, small being measured by the absence of significant flow through the clearance.

I would like to acknowledge stimulating discussions with Peter Bearman, Tom Morel, and Mark Morkovin.

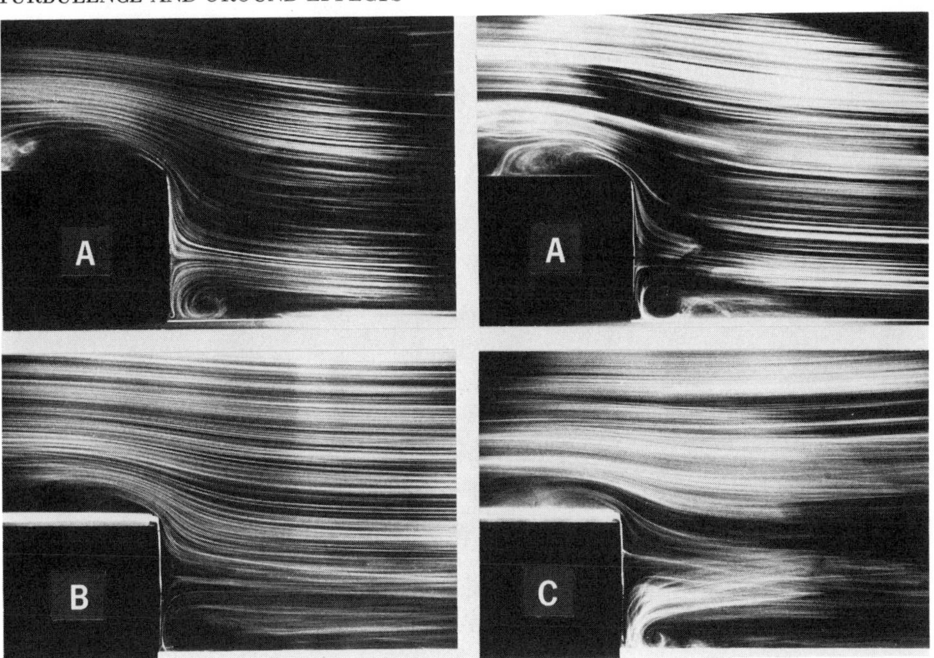

Fig. 25. Visualization of bluff body without ground clearance, shown twice in Flow Case A and once in Cases B and C.

SESSION II

Session Chairperson
S. E. WIDNALL

Massachusetts Institute of Technology
Cambridge, Massachusetts

SOME GENERAL CHARACTERISTICS AND PROPERTIES OF THREE-DIMENSIONAL FLOW FIELDS

E. C. MASKELL

Royal Aircraft Establishment, Farnborough, England

ABSTRACT*

The principal purpose of this paper is to draw attention to physical ideas, derived mainly from basic research in aeronautics, that seem to shed light on the structural forms taken by flow fields generated in the high-Reynolds-number motion of bodies through air. In particular, it is suggested that these ideas lead to the view that any such flow, however complex, can be regarded as a set of much simpler elementary constituent flows which, though they certainly interact with each other, tend to retain much of their individuality for considerable periods of time. Thus, the key to understanding the flow as a whole lies in the recognition of the structure and properties of its basic elements and of the kinds of interaction that can occur between them.

DISCUSSION

S. E. Widnall *(Massachusetts Institute of Technology)*

In preparation for coming to this meeting I told Marten Landahl about an experiment I had once done with a student. Our interest was flow separation on delta wings, but in developing our oil flow visualization technique we tried it out on a two-dimensional wing. The testing was done in a large wind tunnel and at a fairly high angle of attack. We found that instead of getting a single two-dimensional separation bubble on this wing we got a spanwise array of separation cells. It was a very regular

* *Completed manuscript not available at the time of publication.*

pattern, and essentially of the same type that Eric Maskell has just shown us. This two-dimensional wing had a steady three-dimensional separation. It was fascinating.

S. E. Widnall

R. T. Jones *(NASA-Ames Research Center)*

There is an old paper of the 1920's by Prandtl in which he studied two-dimensional flows involving spiral wound-up vortex sheets just like what you showed on your wing. They are described by a family of similarity solutions which change with time. There are two factors involved: $\phi + i\psi$ is dependent on $z^n t^m$.

A.M.O. Smith *(University of California at Los Angeles)*

We have seen that same kind of pattern at the Douglas Company. It was on an airfoil with a chord of about one foot, and at a Reynolds number of several million. It could be eliminated by using a boundary layer trip. At the time I coined the proverb, "A two-dimensional model does not a two-dimensional flow make."

R. T. Jones

You assured us, Mr. Maskell, at the beginning that the Navier-Stokes equations contain all the physics we need to know to determine these flows. I was happy to see that during the rest of the paper you corrected that impression. (Laughter from audience.)

E. C. Maskell

I don't think that's true at all. (Laughter from audience.) I had hoped I was trying to lead you to accept that there are some more-general conclusions that we can draw from the Navier-Stokes equations, but they are certainly consistent with those equations.

R. T. Jones

I don't think anyone would deny that all flows are compatible with the Navier-Stokes equations, but whether those equations contain all the physics needed to determine which of the thousand possible solutions is the real one for any given set of prescribed boundary conditions is another question.

E. C. Maskell

That's fair enough, up to a point. But certainly to get a solution from scratch you've go to start with the external flow, and then find a boundary layer solution that will match it. Indeed it's not very obvious what are the vital conditions necessary in the matching process. But flows *are* reasonably unique – real ones that is.

E. C. Maskell

R. T. Jones

Not in my experience.

S. J. Kline *(Stanford University)*

I agree with R. T. Jones. The question is not whether there is compatibility with the Navier-Stokes equations but whether, since there isn't a unique solution, you can determine which flow pattern will actually occur.

As far as flow structures are concerned, I agree that these have to be determined. However, I'm not sure I agree with you that it all comes out of the boundary layer solution. There are other ways of going at it. We've been doing some separated-flow solutions in the following way. We make guesses about the flow structures, based on a minimal amount of data, and then iterate to closure using zonal models. It works very well, and solves some separated-flow problems that I don't think can be solved by coming at them in the classical manner in terms of very strict boundary layer models.

E. C. Maskell

Although I find it a little difficult to prove, I'm fairly sure that, using the Helmholtz and Euler equations and considering a time-dependent function, if you are given the separation lines, then that is *all* the information you need to define the flow field in the larger scale. You *do* have to be given the separation lines, however.

S. J. Kline

That's exactly the point. By going at it the other way, sometimes you can find where the separation lines are by some iterative closure procedure. In many cases I doubt that they can be found by the classical approach. For example, look at Schlichting's calculation of the sub-critical flow normal to a circular cylinder. To close it, he had to use the measured pressure distribution. If you come at it the other way around, you can *find* those sheets you've been looking for.

E. C. Maskell

I'm not sure what you mean by coming the other way around. I'm not suggesting that one should use my sort of model immediately in any way to calculate solutions. It is my way of drawing inferences about the kind of flow fields that one might have. If one then wanted to calculate the fields, I think one would certainly have to simplify in one way or another. But at least this sort of picture might help to instruct one on the kinds of approximations that are likely to work, or not. Nothing much more than that.

A.M.O. Smith

If I understand you correctly, you said that the separated flow has to come off tangential to the surface at the separation line. This seems contradictory to the finding of Oswatitsch that

$$\tan \theta_S = -3\,\frac{d\tau_w/dx}{dp/dx}$$

E. C. Maskell

You musn't forget that I'm talking here about the infinite Reynolds number situation where the boundary layer gets very thin, and it is in this limit that the separation line is tangential to the wall. However, if we look at the near-wall flow on the boundary layer scale it is still perfectly possible to have a separation at non-zero angle with respect to the wall.

A. Roshko *(California Institute of Technology)*

As a further comment on this, Sychev* in the Soviet Union and Messiter** at the University of Michigan have given the solution for conditions at separation in the

*Sychev, V. V. (1972), On Laminar Separation, Mekh. Zhid. i Gaza, Vol. 3, pp. 47-59.

**Messiter, A. F. (1975), Laminar Separation – A Local Asymptotic Flow Description, AGARD Conference Proc. No. 168, pp. 4-1 to 4-3.

limit $\nu \to 0$. This indeed shows that the separation streamline becomes tangential to the body. Furthermore, this is not in conflict with the Oswatitsch relationship.

In regard to what Eric Maskell has been telling us, I just want to say how delighted I am at this whole expose. I feel this kind of insight is what we need. The complexity really derives from the fact that in most cases the flows are nonstationary. Vortex shedding is a well-known example. But even what were considered to be "more turbulent" flows are now seen to have more organized large structure, possibly of the kind Eric is trying to describe. It may be possible to deal in some simple, average way with the small scale and try to describe in some detail only the large structures, which are the ones that appear to determine all the dynamics, pressures, forces, etc.

E. C. Maskell

I obviously agree with that. I thought it was worth mentioning that if this sort of argument has any validity for the really large-scale motion, then it should continue to have validity for the turbulent boundary layer because it is again an order of magnitude larger than the diffusion scale. So vorticity is going to get out in the same kind of way, and why shouldn't it get out through the same kind of structures? But having said that, the order of magnitude of the turbulent boundary layer is still a lot smaller than that of the things we're interested in. So ultimately what we are thinking about is the following solution. Let's call δ the scale for molecular diffusion, Δ the scale of the turbulent boundary layer, and L the scale of the motion of real interest. We start off thinking about δ/L very much less than one, and then we realize that Δ/L is also very much less than one, but larger than δ/L. So we think of the limit $\delta/L \to 0$ and we are still talking about the right kinds of motion. The only effect that turbulence would have on this would be in changing the position of the separation lines themselves, which of course is a big effect. However, it is not very easy to consider that kind of effect. I don't think that's in prospect.

R. T. Jones

You wouldn't consider the case where Δ/L is comparable to one?

E. C. Maskell

Most of the time when we are talking about a separated flow we are thinking of something on the scale of the body, as opposed to the turbulent boundary layer which is an order of magnitude smaller.

A. T. McDonald *(Purdue University)*

One of the most interesting things I got from your presentation was a suggestion that we try to relate what you've said back to the real-world problem of making measurements in these flows. You said there must be a lot of Strouhal-type phenomena taking place. Measuring them involves coupling with the dynamics of any force-balance system that is used. But even more importantly, very long times are required to establish some fully-separated flow patterns. The characteristic time for a road vehicle of typical length travelling at a typical speed can be as large as a second. If the times to establish a separated flow are five to ten times this, then there are some problems. One is to measure the frequency characteristics of the phenomenon of interest in order to determine the required sampling rates. Another, for road vehicles, is to correlate the response time to a yawed flow, such as might be caused by a wind gust, so that the aerodynamic force characteristics of the vehicle can be extracted. Have you considered any of these problems, or could you make any comments?

E. C. Maskell

No, I have not considered them. What I am trying to say is that I find it very difficult to see how one can ever begin to approach some of these more complex problems without having in mind a very clear idea of the general nature of the flow structure. I've really done no more than introduce that kind of approach. I feel that it is possible to pin down the broad nature of the flow structures, but I accept the fact that very complex flows present quite a problem. But it has to be simpler to understand what is happening if one can at least recognize the individual sort of large-scale flow structures.

S. E. Widnall and E. C. Maskell

MECHANISMS OF TWO AND THREE–DIMENSIONAL BASE DRAG

D. J. MAULL

Cambridge University, Cambridge, England

ABSTRACT

The mechanisms of the production of low pressure on the base of two-dimensional and axisymmetric bluff bodies are discussed. The factors which can influence this pressure are described, with emphasis on the effects of the presence of some three-dimensionality on basically two-dimensional base flows.

The differences between two- and three-dimensional base flows are highlighted, firstly by discussing the effects of tip geometry on a high-aspect-ratio bluff-based body, and then by discussing a very simple three-dimensional bluff body. Some results are given of an investigation into means of influencing the base pressure of this simple body, and the effect of the proximity of the ground is considered.

NOTATION

A	Aspect Ratio
C_d	Drag Coefficient
C_p	Pressure Coefficient
C_{pb}	Base-Pressure Coefficient
$\tilde{C}_{px}$	Modified Pressure Coefficient
ΔC_p	Differential Pressure Coefficient
d	Body Thickness

References p. 153.

h Height Above Ground, Step Height

L Length

n Plate Width

p Pressure

S Displacement Surface

u Velocity

x, y Coordinates

α Angle of Incidence

δ Boundary Layer Thickness

ν Viscosity

δ Density

ω Vorticity

INTRODUCTION

The literature of separated flows is very large, and when it is studied the conclusion that could easily be reached is that the world is two-dimensional. There are innumerable papers on circular cylinders and Karman vortex streets, but very little on practical three-dimensional bluff bodies such as cars, trucks and the upswept fuselages of transport aircraft. Whilst the recent interest in the fluid mechanics of the flow around buildings does shed some light on the latter problems, there are still large areas where very little is known about the flow.

There is no doubt that the three-dimensional flow around a bluff-body base is complicated. Even the pseudo-two-dimensional flow around a circular cylinder with vortex shedding (probably the most researched body in fluid mechanics) still presents many difficulties, particularly when it is realised that the flow is in fact somewhat three-dimensional in that the spanwise correlation of the vortices is not high. In three-dimensional flow vorticity is also shed and large-scale interactions are now possible between the shed vorticity in all *three* directions, sometimes causing a periodic structure to be set up.

There are some common features in two-and three-dimensional base flows, and so some conclusions from the much more researched area of two-dimensional base flows can be carried over into three dimensions; however, some results are not applicable.

MECHANISMS OF BASE FLOWS ON TWO–DIMENSIONAL BODIES

Before discussing the complicated base flow associated with three-dimensional bluff-based bodies it is instructive to think about the simpler two-dimensional and axisymmetric cases. One such case, for example, is the low-speed flow down a step illustrated in Fig. 1. Most analyses consider this a steady flow, with possibly a superposed time-dependent flow.

The mean flow may be divided into three regions: (a) a recirculating region ARB bounded by the step and a dividing streamline AR, (b) a region between the dividing streamline and the displacement surface, S, and (c) the external irrotational flow. In region (a) fluid is entrained by the shear layer originating at A and is reversed back into the region at the reattachment point, R; thus for steady flow, the total mass entrained must equal the mass returned at the reattachment region. The flow in region (b) consists of the original separated boundary layer from A and fluid that it entrains from the external flow. Fig. 1 also shows a typical pressure distribution along the surface behind the step. There is a slight decrease in pressure after the step followed by a rapid rise in pressure near the reattachment point. Most authors, e.g.

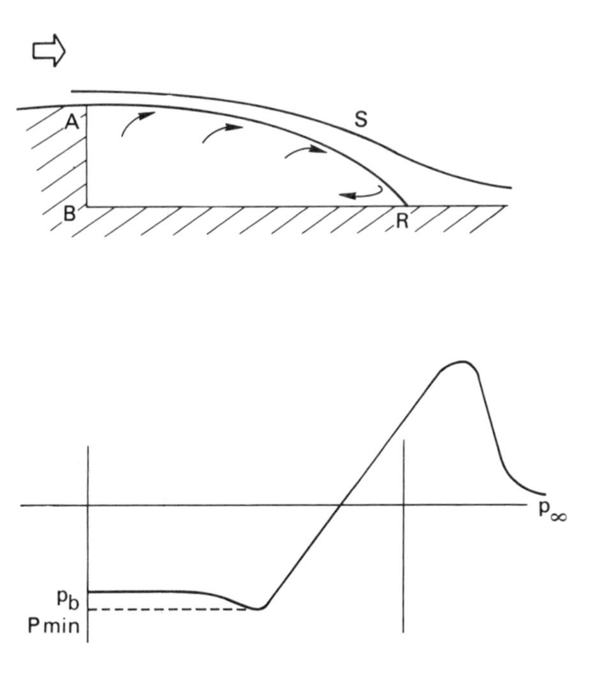

Fig. 1. Typical flow pattern and pressure distribution downstream of a two-dimensional step.

References p. 153.

Roshko & Lau (1965) and Ota & Itasaka (1976) show reattachment occurring before the maximum pressure but Tani, Iuchi & Komoda (1961) show reattachment after this point. In the last case, however, it is probable that the method of determining the dividing streamline AR, using hot wires, was subject to large errors near the reattachment point. It is possible that other methods may also be liable to large experimental error. For instance, Ota & Itasaka in their measurement of their position of the point R found that different methods gave variations of up to 25 mm when the bubble length was about 100 mm.

Any calculation method to determine the base pressure must allow for a matching of the conditions along the dividing streamline with conditions in the external flow, and for an equilibrium of the internal flow, i.e., a balance between mass entrained and mass reversed within region (a). Both these factors are discussed below in an effort to gain some insight into the mechanisms involved.

External Flow Field – The pressure distribution along the dividing streamline can be obtained approximately by assuming that it is given by the flow past a displacement surface S (Fig. 1) and using, say, thin-aerofoil theory. For this, the shape of the surface, S, must be guessed; for instance by choosing a length BC (C is not necessarily the reattachment point) and fitting a polynomial to satisfy the following conditions (see Fig. 2): $x = 0$, $y = h$, $dy/dx = 0 = d^2y/dx^2$; $x = L$, $y = \delta$, $dy/dx = 0 = d^2y/dx^2$; $x < 0$, $y = h$; $x > L$, $y = \delta$. This results in a pressure distribution of the type shown in Fig. 3 for $L = 5(h - \delta)$. It has some similarity with the experimental results of Fig. 1, except of course that the theoretical curve is symmetric due to the shape of S considered.

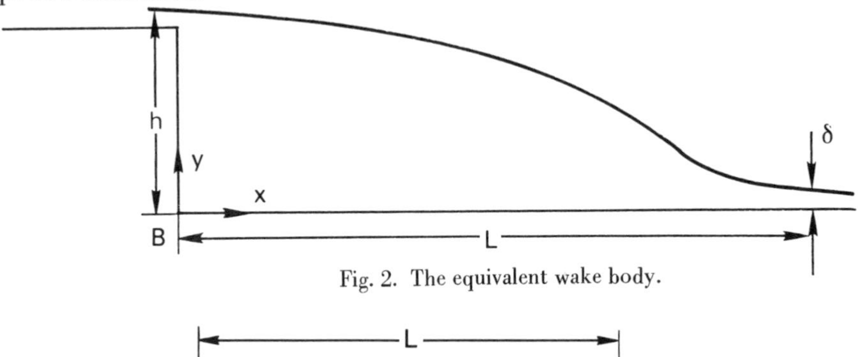

Fig. 2. The equivalent wake body.

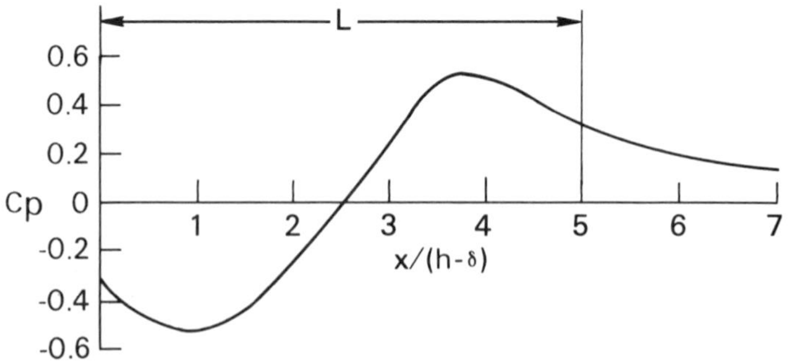

Fig. 3. Theoretical pressure distribution downstream of a two-dimensional step.

The shape of the wake will also determine the pressure distribution on the body (see Pollock, 1972) and conversely, the shape of the forebody will also influence the base pressure. This can be seen in Fig. 4, where base pressures for three simliar two-dimensional bodies with fixed separation points are shown. Therefore, any calculation of the external flow field must include the effect of the forebody. More drastic changes in the forebody shape, particularly those which influence the angle of the separation streamline, will also have large effects; this can be seen in Fig. 5, which is reproduced from Roshko & Lau (1965). Thus models B and C, which have separation streamlines leaving the body at right angles to the free stream direction, have much lower base pressures and longer separated regions than models D and E which have the separation streamline leaving the body in the free stream direction.

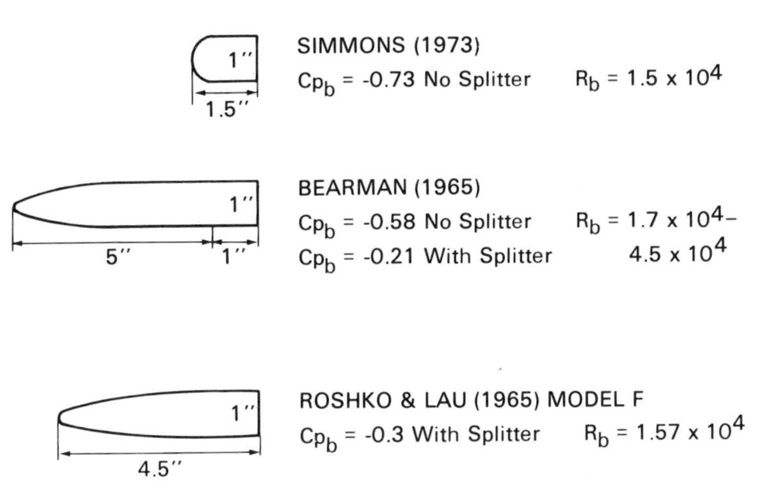

Fig. 4. The effect of forebody on base pressure for three two-dimensional bodies of various length-to-height ratio.

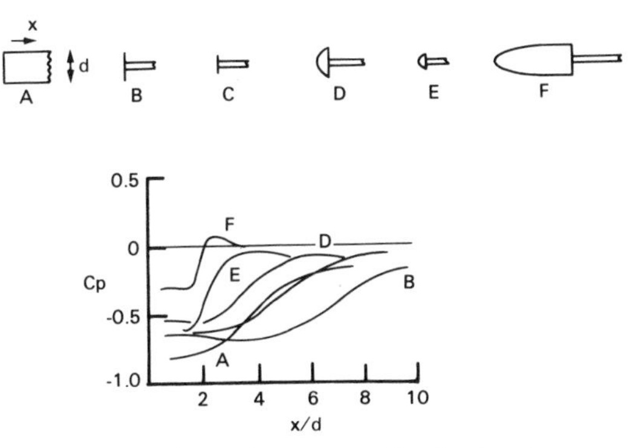

Fig. 5. The effect of separation angle on the pressure distribution downstream of a step (Roshko & Lau, 1965).

References p. 153.

Internal Flow Field – As previously stated, the equilibrium condition to be satisfied is that the mass entrained by the shear layer must be reversed back into the separated region at the reattachment point. Thus, first the development of the separated shear layer must be calculated and, in particular, the stagnation pressure on the dividing-streamline found. Equilibrium is then satisfied by equating the dividing-streamline stagnation pressure (possibly modified empirically for non-isentropic compression) to the reattachment pressure given by a calculation of the external flow field over the equivalent wake body, as described in the previous subsection. Methods based on this model, and some others, are well reviewed by Tanner (1973).

Another method of considering two-dimensional base flow is to investigate the equilibrium of vorticity flux in a control volume, such as that shown in Fig. 6. Let suffix e represent the external, irrotational flow and δ the boundary layer thicknesses at stations 2 and 3. From the vorticity equation for steady laminar flow, vorticity is brought into the control volume by both convection and diffusion in the boundary layers, as well as by diffusion across the solid surfaces 12 and 13 (i.e. the surfaces act as vorticity sources or sinks).

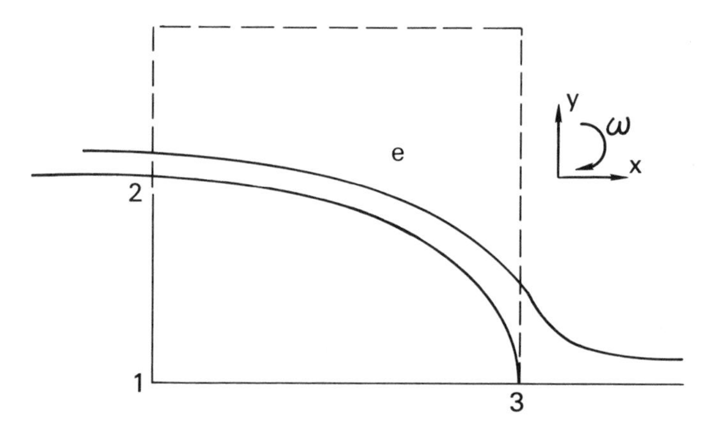

Fig. 6. Control volume for vorticity balance.

As for the diffusion, the vorticity flux per unit length of solid surface into the control volume is $-\nu\, \partial\omega/\partial x$, which may be shown to equal $1/\rho\, \partial p/\partial y$ (Lighthill, 1963). Integrating along 12, the total flux out between 1 and 2 is $-(p_2 - p_1)/\rho$. Similarly, between 1 and 3 the total flux out of the volume is $(p_3 - p_1)/\rho$. In the boundary layer at 2, the vorticity flux by diffusion is, using the boundary layer approximation, $-\nu\, \partial\omega/\partial x = -\nu\, \partial(\partial u/\partial y)/\partial x$. Consequently, the total flux by diffusion in the boundary layer is

$$-\nu \frac{\partial}{\partial x} \int_0^{\delta} \frac{\partial u}{\partial y} dy = -\nu \frac{\partial u_{2e}}{\partial x} ,$$

where u_{2e} is the velocity at the edge of the boundary layer at station 2. Convection in the boundary layer at 2 gives the vorticity flux = $\int_0^\delta u \frac{\partial u}{\partial y} dy = \frac{1}{2} u_{2e}^2$. Similar results are found for station 3. Thus the complete vorticity balance for the control volume in Fig. 6 may be written as

$$\frac{1}{2} u_{2e}^2 - \nu \frac{\partial u_{2e}}{\partial x} = \frac{1}{2} u_{3e}^2 - \nu \frac{\partial u_{3e}}{\partial x} - \frac{1}{\rho}(p_2 - p_1) + \frac{1}{\rho}(p_3 - p_1). \tag{1}$$

If now instead of taking the downstream boundary of the control volume at the reattachment point, 3, it is taken at some point, x, upstream of 3, and if the diffusion terms $\nu\ \partial u/\partial x$ are ignored, which implies high Reynolds numbers, then the total vorticity flux at x is $V_x = \frac{1}{2} u_{xe}^2 = \frac{1}{2} u_{2e}^2 + \frac{1}{\rho}(p_2 - p_x)$. Roshko & Lau (1965), in analysing their experiments on the flow down a step having different forebodies, used a modified pressure coefficient $\tilde{C}_{px} \equiv (p_x - p_2)/(\rho/2\, u_{2e}^2)$ which, together with a proper choice of a scale for the downstream coordinate, led to a collapse of the data shown in Fig. 5. If the above expression for V_x is used to replace $(p_x - p_2)$, one may write $\tilde{C}_{px} = 1 - V_x/(\rho/2\, u_{2e}^2)$, so that $\tilde{C}_{px}$ may be viewed as the ratio of the vorticity absorbed by the solid surface up to position x to the total vorticity flux from the separation point, 2. (Strictly, Roshko & Lau (1965) used p_{min} instead of p_2, but there is little difference experimentally between p_2 and p_{min}). Their results show that $\tilde{C}_{px}$ rises to about 0.35 downstream of reattachment. Using the above expression for $\tilde{C}_{px}$, this may be interpreted to mean that about 35 percent of the initial vorticity flux is absorbed by the solid surface, most of the absorption taking place in the reattachment region.

For most bluff-body flows the base pressure is constant $(p_2 = p_1)$ and, if no splitter plate is present, the term $(p_3 - p_1)/\rho$ is not necessary. The vorticity-equilibrium equation (1) therefore reduces to

$$\frac{1}{2} u_{2e}^2 - \nu \frac{\partial u_{2e}}{\partial x} = \frac{1}{2} u_{3e}^2 - \nu \frac{\partial u_{3e}}{\partial x}. \tag{2}$$

At high Reynolds numbers, where the diffusion terms may be neglected, this equation reduces to $u_{2e} = u_{3e}$. This equality is, in general, not satisfied; to resolve this contradiction, one must include in the vorticity balance the unsteady term $\partial \omega/\partial t$. The unsteadiness represented by this term may be manifested as large scale vorticity shedding, as with a Karman vortex street, and as turbulence resulting in an increased effective kinematic viscosity.

References p. 153.

At low Reynolds numbers, for bodies that are flat-sided near separation, $\partial u_{2e}/\partial x$ is positive, and near the point of confluence of the two free shear layers $\partial u_{3e}/\partial x$ is negative. Since $u_{2e}^2 > u_{3e}^2$, there is therefore the possibility that the viscous terms in Eq. (2) can provide a means of satisfying this equation and that a *steady* separated flow may be established even without a splitter plate.

BASE PRESSURE IN TWO-DIMENSIONS

The low base pressure in two-dimensional flow is associated with the formation of a vortex street, and in particular with the strength of the vortices and the position at which they form (see Bearman, 1965). Thus any method of increasing the base pressure must stop or weaken the vortex shedding, or delay the formation of the vortices.

A sufficiently long splitter plate will stop the vortex shedding, while the effect of a short splitter plate is to delay vortex formation. Base bleed acts similarly to a short splitter plate in that it tends to delay vortex formation. To illustrate this, Fig. 7 shows two results from a computer simulation by Clements (1973) of the effect of base bleed on vortex shedding, showing clearly the delay in vortex formation caused by base bleed. A cavity in the base will also increase the base-pressure coefficient, and Clements shows experimentally an increase from −0.55 with no cavity to −0.475 for a cavity whose depth is equal to the base height. It is not really known how these cavities work, but there is an indication from Clements' computer simulation that the vortex formation is delayed. The calculations, whilst giving a passable prediction of base pressure, did not predict the vortex shedding frequency at all well. It is possible that the cavity walls might act as vorticity sinks, but as already discussed, this would entail a pressure gradient along the sides of the cavity; as far as is known, there are no measurements inside a cavity to confirm this.

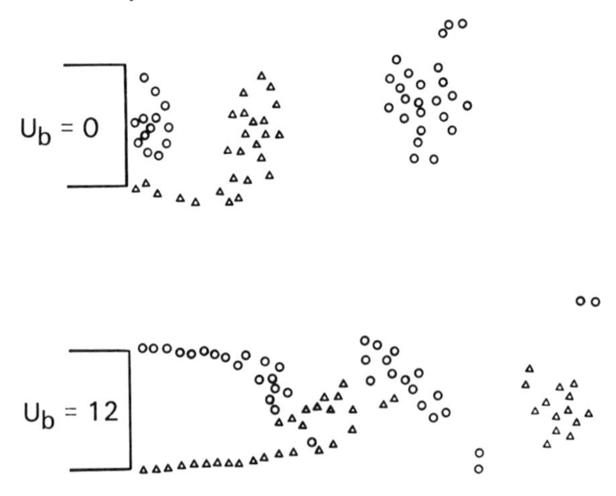

Fig. 7. Computer model of vortex shedding with and without base bleed (Clements, 1973). U_b is the ratio of the base-bleed velocity to the external-flow velocity.

Axisymmetric bluff bodies differ from two-dimensional bodies in one important respect, and that is that there is very little periodic vortex shedding. Various arguments may be put forward to explain this lack of regular vortex shedding; for instance, that there is no possibility of regular interaction between separate shear layers as in the two-dimensional case. The vorticity ideas previously put forward can also be used, but for axisymmetric flow there is now an additional, and rather complicated, term in the vorticity balance, representing the shrinkage of the vortex lines in the area of the confluence of the separated layers, which results in an apparent vorticity "sink" in the control volume.

Three-Dimensional Interference – Since we have seen that base flows in two dimensions are critically dependent upon the equilibrium within a control volume, it is obvious that any three-dimensional disturbance will substantially affect the base pressure. One of the most common instances of three-dimensional effects on nominally two-dimensional bodies is the flow formed when a bluff-based body spanning the width of a wind tunnel penetrates the boundary layers on the tunnel walls. The results for such a configuration are shown in Fig. 8 (Young, 1972). Both the base pressure and the vortex-shedding Strouhal number are seen to vary not only within the wall boundary layers, but for a considerable distance along the body span. In this case the boundary layer thickness was less than two base heights, yet a constant base pressure was only found over the middle five base-heights of the body, which had a span of twenty base-heights.

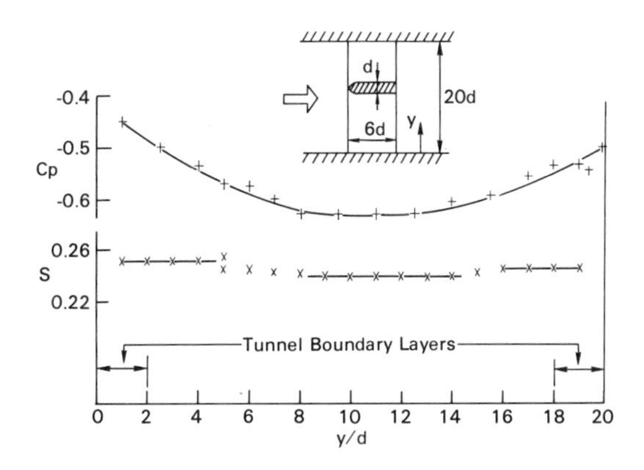

Fig. 8. Base pressure and Strouhal number variations along the span (y) of a bluff-based body of thickness d spanning a tunnel (Young, 1972).

The tip of a finite-aspect-ratio bluff-based body gives an example of a complicated three-dimensional base flow. Young generated such a flow by cutting-off one end of the above two-dimensional body. The resulting body was thus cantilevered, with one

References p. 153.

end attached to the tunnel wall, and with the other end creating a free tip. In the baseline configuration, a, the free tip generates two longitudinal trailing vortices and a rapid acceleration of flow from the tip into the base region, resulting in a low base pressure near the tip. Quite small modifications to the tip region can result in large changes not only in the pressures near the tip, but also in the base pressure along the whole body. For instance, the end plate (model d in Fig. 9) produces almost the same effect as if the body spanned the tunnel (see Fig. 8). On the other hand, if the plate does not extend past the base (model c), the pressure distribution is almost the same as for the unmodified tip. Model f, which is just an extension to stop flow into the base region, produces a large increase in pressure near the tip. Model e which, in addition modifies the shed longitudinal (streamwise) vorticity, produces a further increase in tip base pressure.

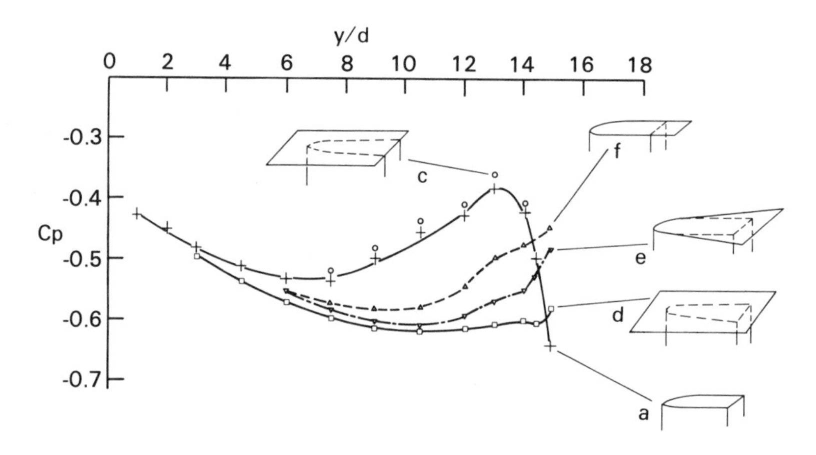

Fig. 9. Effect of end conditions on base pressure on a bluff-based body (Young, 1972).

This interference with the base pressure caused by longitudinal vortices shed from a body can be of great importance. A striking example is shown in Fig. 10, displaying the effect of fitting delta-wing vortex generators to the two-dimensional body which spanned the tunnel, but had end-plates just outboard of the boundary layer on each tunnel wall in order to produce a two-dimensional base flow between the plates. The wings generate streamwise vortices, as shown, and whilst having little effect on the base pressure above the wings, substantially increase the pressure below them. Presumably, the velocities induced by the streamwise vortices serve to keep the shear layers coming off the sides of the body below the wing further apart, thus delaying the formation of the Karman-type vortices and increasing the base pressure. Above the wing the reverse should happen, but apparently, it happens to a lesser extent. Another possibility is that the vortices may also be forcing some fluid into the base region, forming a type of base bleed and thus increasing the base pressure. It is not surprising that the vortex generators also cause the vortex-shedding frequency to change, in a discontinuous fashion, along the span.

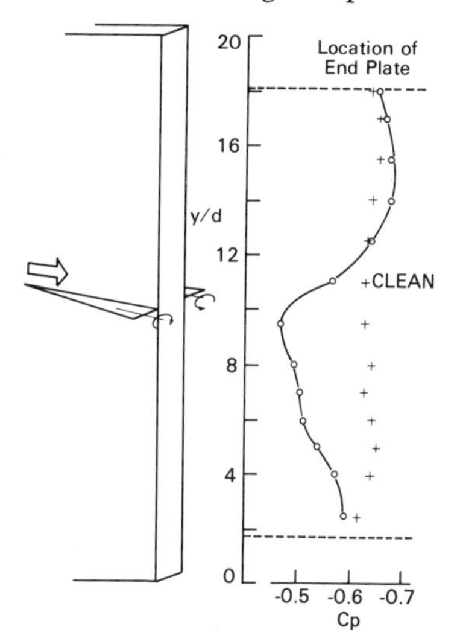

Fig. 10. Effect of vortex generators on the base pressure along the span of a bluff-based body (Young, 1972). o - with vortex generators, + without.

Any introduction of three-dimensionality into a base flow must have an effect on the vortex shedding and consequently upon the base pressure. Many methods have been used, such as a serrated (or segmented) trailing edge, part-span splitter plates and part-span cavities (Pollock, 1972) and other ingenious devices tested and reviewed by Tanner (1975). All these, in one way or another, introduce streamwise vorticity into the flow and break up the vortex shedding.

THREE–DIMENSIONAL BASES

It is perhaps unfortunate that the body of interest to this meeting, namely three-dimensional and bluff, has not been investigated very much. The wake of such a body is undoubtedly complicated. Vorticity shed (in sheets) from the body has components in all three directions, and may roll up into well-identified vortices which may be unsteady. The base pressure of such a body will depend upon the structure of these vortex sheets. In Fig. 11, from Hoole (1968), is shown the base pressure at the centre of the rear of a rectangular flat plate as a function of the plate incidence and aspect ratio. Here is very clearly seen the effect of the generation of lift on the base pressure (i.e., the upper-surface pressure). For all aspect ratios tested, the base pressure decreased as the plates were inclined away from a position perpendicular to the free stream. For instance, for the small aspect ratio of 1/3 the change of incidence from 90° to approximately 45° results in a change of base pressure coefficient, from -0.43 to -0.95.

References p. 153.

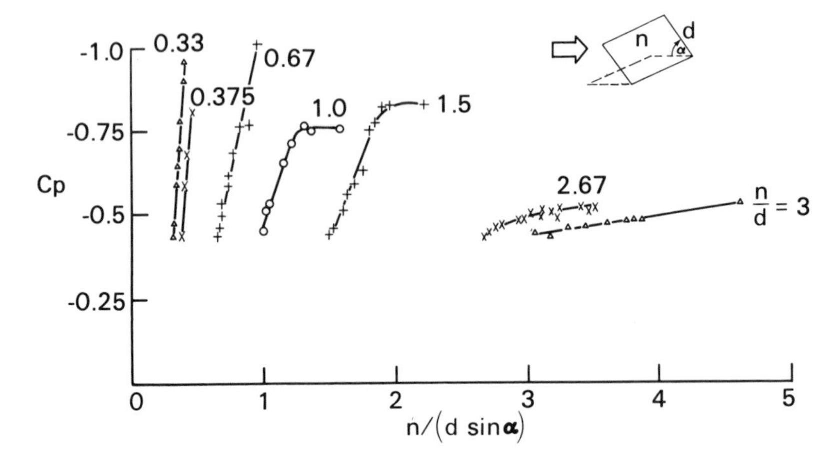

Fig. 11. Pressure at the centre of the rear face of a rectangular flat plate (Hoole, 1968).

Fig. 12 shows some measurements of the drag coefficient for inclined rectangular plates of different aspect ratios, again by Hoole. A point of interest here is the local increase in drag coefficient for the low-aspect-ratio (A = 0.5) plate as the incidence decreases from 50° to 40°; it is due to the leading edge separation reattaching on to the plate. It is worth noting that in most cases a hot wire in the wakes of these plates will pick up discrete frequencies associated with vortex shedding from opposite sides of the plates.

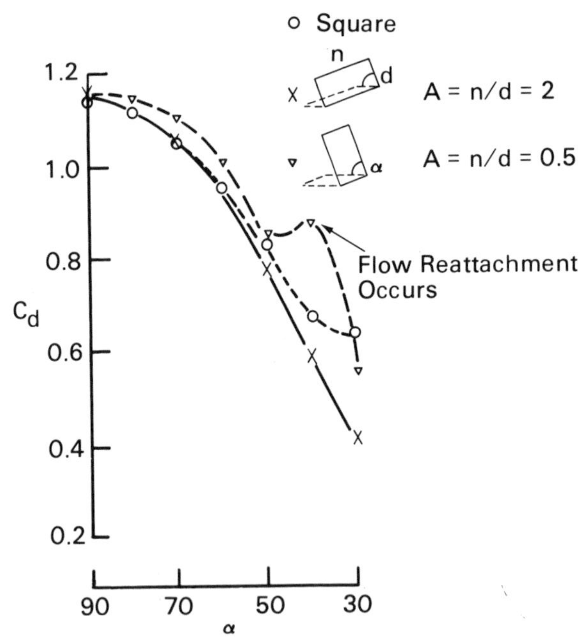

Fig. 12. Drag coefficient of rectangular flat plates as a function of incidence (Hoole, 1968).

Base Pressure on a Rectangular Block – In order to obtain some data on a three-dimensional bluff-based body the rectangular block shown in Fig. 13 was tested by the writer in a low-speed wind tunnel. The body was supported on four cylindrical rods and it could be mounted at various heights (h) above the tunnel floor. The base contained two vertical rows of pressure tappings, one row (A) on the vertical centreline of the base, the other (B) 0.5 inches off centre. The Reynolds number based on model length was 2.6×10^5.

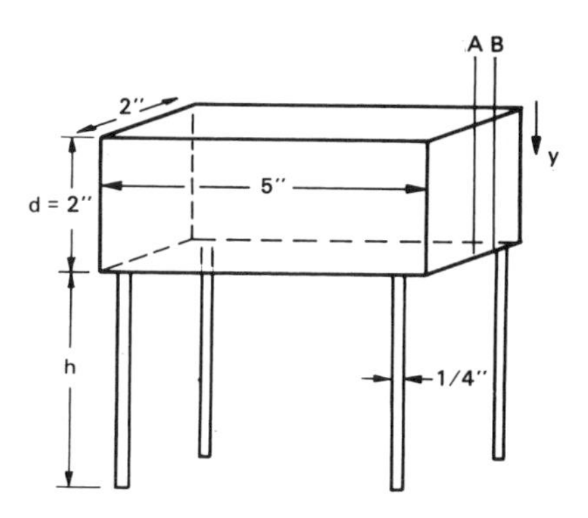

Fig. 13. Sketch of rectangular block.

Effect of Various Devices on the Base Pressure – With the bottom of the block 6 inches from the floor of the tunnel the pressures along lines A and B were measured and are shown in Fig. 14. At this height from the floor no influence of the floor is felt and all further results are presented relative to this set of values. The asymmetry of the C_p distribution with y/d is due to the presence of the supports. In all cases there was little difference between the two lines of tappings.

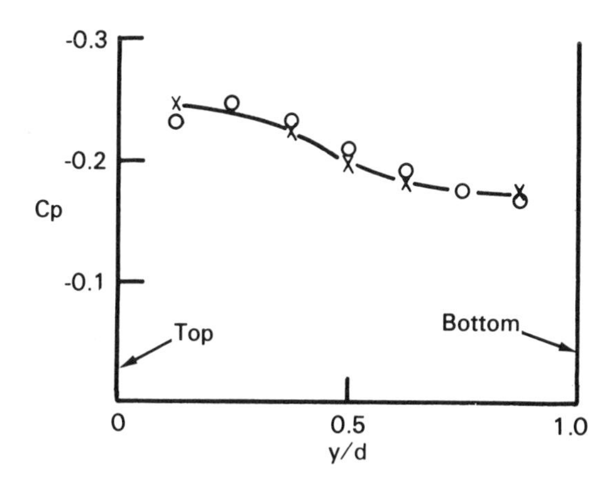

Fig. 14. Pressure on the base of the rectangular block of Fig. 13.

References p. 153.

Since it has been seen that, in almost two-dimensional situations, vortex generators producing longitudinal vortices can have a large effect on the base pressure, the body was tested with vortex generators on the sides to produce vortices as shown in Fig. 15 (seen from the rear of the body looking upstream). The ΔC_p given here is C_p (with generator) – C_p (without). It can be seen that configuration 3 will produce a net reduction in drag compared with the clean model. Presumably, with this arrangement of the generated vortices high pressure air is being forced into the wake from above the model, resulting in an increase of presure in this region. The configuration 1, which is the mirror image of 3, does not produce the same effect because the flow from the underside of the body consists mainly of low-stagnation-pressure air in the wakes of the cylindrical supports.

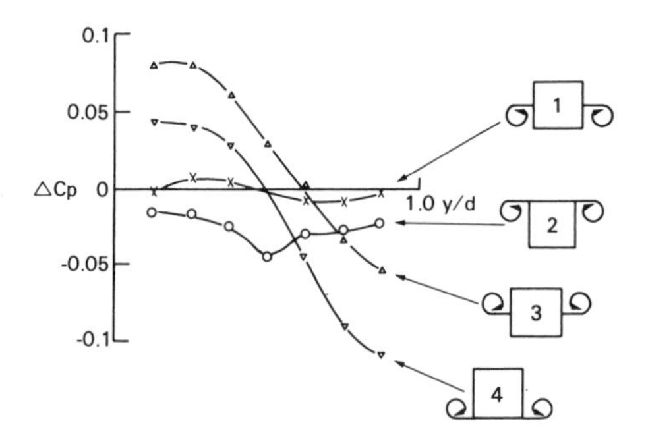

Fig. 15. Change in base-pressure coefficient caused by vortex generators on the rectangular block.

Base cavities are known to have an effect in two-dimensional flow, and Fig. 16 shows some variations in base pressure with this same rectangular body. Configuration A3 gives a useful base-pressure increase, presumably by the same mechanism as at the tip of model f of Fig. 9. Configuration A1 does not give the same effect as A3, again probably due to the presence of the supports, as discussed in relation to configurations 1 and 2 of Fig. 15.

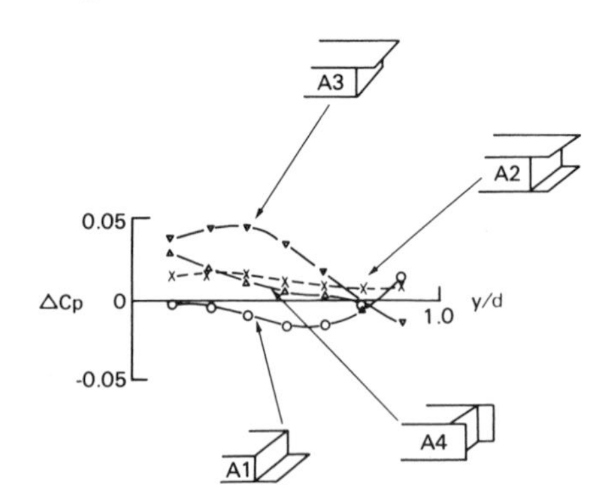

Fig. 16. Change in base-pressure coefficient caused by extensions in the base region of the rectangular block of Fig. 13.

Finally a crude form of boat-tailing, applied on the sides only, is shown in Fig. 17 for the same body. The most convenient manner of incorporating boat-tailing in that test was by increasing the body width with add-on chamfered pieces. The effect of this boat-tailing should be assessed with respect to a new baseline (wider body), the base pressure of which is included in Fig. 17. Again, an increase in base pressure over the central portion of the base can be produced, but not necessarily a decrease in drag, due to the low pressures on the sides of the boat-tailing. The mechanisms of this form of drag reduction will be dealt with in the following paper by W. A. Mair.

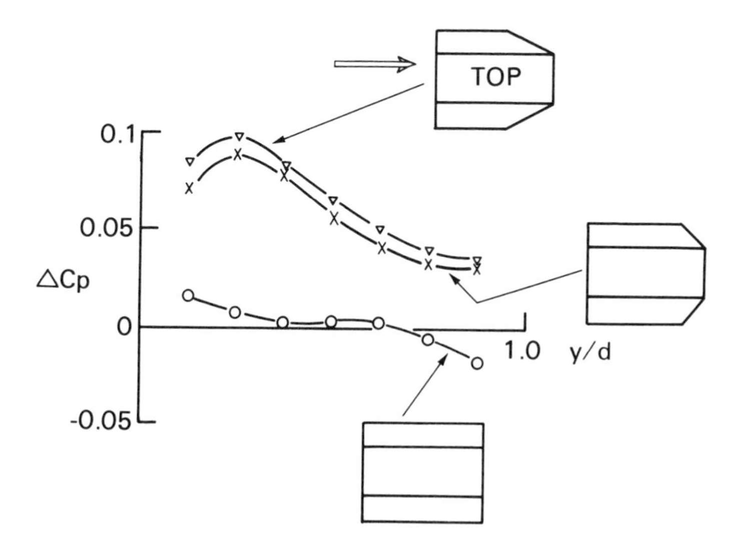

Fig. 17. Change in base-pressure coefficient caused by boat-tailing on the sides of the rectangular block of Fig. 13.

Effect on Base Pressure of Ground Proximity – From two-dimensional tests it is known, e.g. Fig. 18, that the base pressure on a two-dimensional circular cylinder increases as the gap between the cylinder and a wind tunnel floor decreases. In this test, performed by the writer, the boundary layer on the floor of the tunnel had a thickness of about twice the cylinder diameter, and the Reynolds number based on the diameter was 6.5×10^4. A similar drag reduction occurs with the rectangular block described previously. Fig. 19 shows the pressures on the centre line of the base of the block and, as can be seen, there is very little effect on the base pressure until the ground is about 1/4 of the height of the block away. Moving the block nearer the ground produces an increase in pressure at the top of the base and, for some heights, a decrease in pressure near the bottom. This flow is obviously very complex, and it may be that this increase in pressure is caused by a horseshoe vortex from the separated region in front of the body acting in the same way as configuration 3 of Fig. 15.

References p. 153.

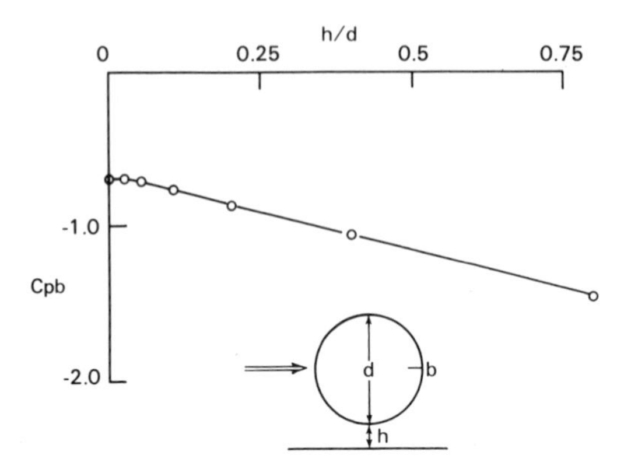

Fig. 18. Base-pressure coefficient on a circular cylinder near the ground.

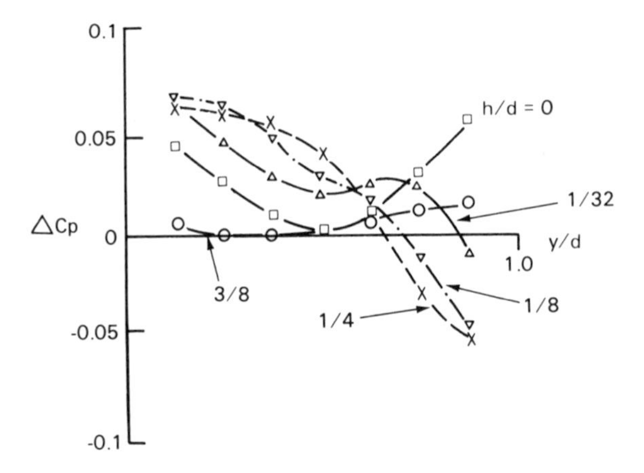

Fig. 19. Change in base-pressure coefficient on the rectangular block of Fig. 13 placed near the ground.

For this series of tests with the block, the boundary layer at the model position was about half the height of the body; the effect of changing the oncoming-boundary-layer parameters is not known. In the context of automobile aerodynamics, this test, which does not simulate the correct dynamics of the problem, may well give misleading results.

CONCLUSIONS

The flow round the base of a three-dimensional bluff body is not well understood, and further work needs to be done before a conceptual model of the flow can be built. Small modifications to the base, and to the forebody, can produce useful changes in the base pressure and a systematic investigation of these needs to be carried out.

The differences between two- and three-dimensional base flows are great, and it is not necessarily true that a drag reducing device which works in two-dimensional flow will achieve a similar result in three dimensions.

REFERENCES

Bearman, P. W. (1965), Investigation of the flow behind a two-dimensional model with a blunt-trailing-edge and fitted with splitter plates. J. Fluid Mech. 21, pt. 2, pp 241-255.

Clements, R. R. (1973), Computer models of separated flows behind two-dimensional bluff bodies. Ph.D. thesis, University of Cambridge.

Hoole, B. J. (1968), The turbulent wake behind an inclined flat plate. Ph.D. thesis, University of Cambridge.

Lighthill, M. J. (1963), In 'Laminar Boundary Layers 'ed. Rosenhead, Oxford Univ. Press, p. 54.

Ota, T. & Itasaka, M. (1976), A separated and reattached flow on a blunt flat plate. Trans. ASME J. Fluids Eng. 98, Series 1, No. 1, pp 79-86.

Pollock, N. (1972), Segmented blunt trailing edges at subsonic and transonic speeds. Aero. Res. Labs. Melbourne, Australia, Rep. ARL/A 137.

Roshko, A. & Lau, J. C. (1965), Some observations on transition and reattachment of a free shear layer in incompressible flow. Proc. 1965 Heat Transfer and Fluid Mech. Inst., Stanford Univ. Press.

Tani, I., Iuchi, M. & Komoda, H. (1961), Experimental investigation of flow separation associated with a step or a groove. Aeronautical Res. Inst. Univ. of Tokyo, Report No. 364.

Tanner, M. (1973), Theoretical prediction of base pressure for steady base flow. Progress in Aerospace Sciences 14, pp 177-225.

Tanner, M. (1975), Reduction of base drag. Progress in Aerospace Sciences 16, No. 4, pp 369-384.

Young, R. A. (1972), Bluff bodies in a shear flow. Ph.D. thesis, University of Cambridge.

DISCUSSION

Prepared Discussion

E. Achenbach *(Institut fuer Reaktorbauelemente, Juelich, Germany)*

An important aspect of bluff body near-wake flows is that they may be unsteady. This unsteadiness is an extra complication in an already quite complex flow situation, as may be seen in the example of a simple three-dimensional body – the sphere. Knowledge of the wake flow behind spheres is not as well developed as, for instance, the wake of 2-D circular cylinders. In particular, there still are uncertainties concerning the periodicity and the configuration of the vortices. We, at our Institute, have observed the vortex shedding mechanism over a wide range of Reynolds numbers, $400 < Re < 3 \times 10^5$ (Achenbach, 1974)*, and I would like to review briefly these results.

**Achenbach, E. (1974), Vortex shedding from spheres, J. Fluid Mech., 62, (1974), pp. 209 - 221.*

E. Achenbach

Vortex Shedding Frequency – At Re greater than 250 - 400, the steady vortex behind a sphere breaks off and forms loops which are periodically released from the sphere. The Strouhal number, S, rapidly increases with increasing Re and reached, in our tests, a value of S = 1.0 at Re ≈ 3000. In the experiments of Moeller (1938)*, a further increase was observed up to S = 2.0 at Re = 8000 (Fig. 20).

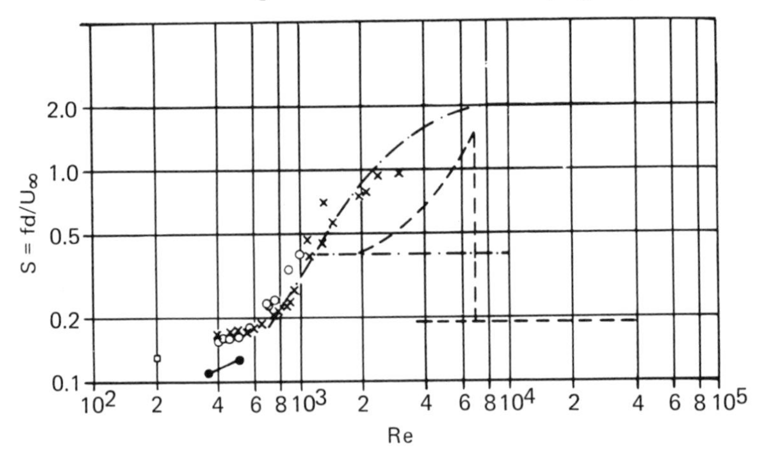

Fig. 20. Strouhal number *vs.* Reynolds number for spheres. -·-·-, Moeller (1938); ----, Cometta (1957)**; •—•, Magarvey & Bishop (1961); □, Marshall & Stanton (1931)†, normal plate. Present results: ○, d = 0·02 m; x, d = 0·04 m.

At around Re = 6000 a change in the separation mechanism seems to occur. The vortex shedding becomes much stronger and its Strouhal number decreases by an order of magnitude (Fig. 21). Beyond Re = 6000 the Strouhal number grows from S = 0.125 at Re = 6000 to S = 0.18 at 3 x 10^4. Beyond Re = 3 x 10^4 the Strouhal

Moeller, W. (1938), Experimentelle Untersuchungen zur Hydromechanik der Kugel, Phys. Z., 39, pp. 57 - 80.

***Cometta, C. (1957), An Investigation of the Unsteady Flow Pattern in the Wakes of Cylinders and Spheres Using a Hot Wire Probe, Div. Engrg. Brown Univ., Tech. Rep. WT-21.*

†Marshall, D. & Stanton, T. E. (1931), On the Eddy System in the Wake of Flat Circular Plates in Three-Dimensional Flow, Proc. Roy. Soc., A 130, pp. 295-301.

number remains constant, except immediately before reaching the critical Reynolds number.

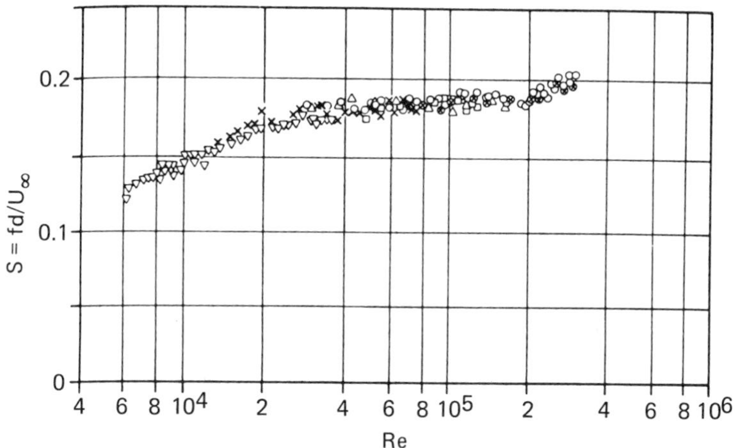

Fig. 21. Strouhal number *vs.* Reynolds number for spheres. ∇, d = 0·020 m; x, d = 0·40 m; □, d = 0·076 m; Δ, d = 0·133 m; ○, d = 0·175 m; ⊕, d = 0·200 m.

The vortex shedding was very periodic in the range 6000 < Re < 3 x 10^5. It was detected by a hot wire flush-mounted with the surface of the sphere. Beyond the critical Re the periodic hot wire signals vanished abruptly, as another change in the shedding mechanism seems to have occured. It should be noted that is setup could also not detect a periodic vortex shedding frequency at Re below 6000.

Vortex Configuration – Above Re = 400 the stationary vortex rings which are observed behind a sphere at lower Re break away periodically and form into vortex loops. The loops are connected to one another like a vortex chain, as demonstrated in Fig. 22. In this schematic representation the arrows indicate the flow direction or the sense of the circulation. Such vortex configurations were found also by Magarvey & Bishop (1961)* in the wake of falling drops (Fig. 23), and by Calvert (1967)** behind inclined disks (Fig. 24).

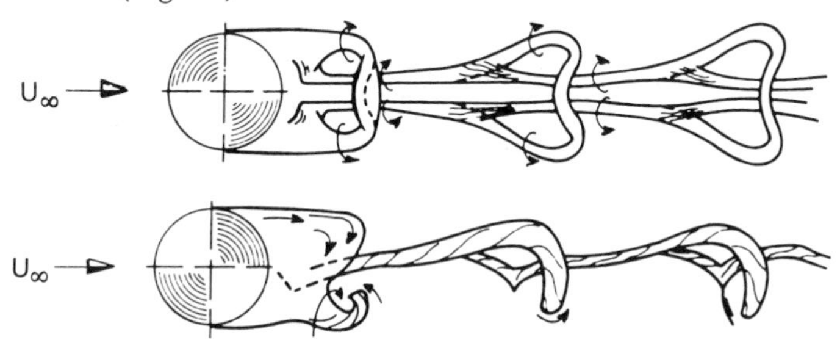

Fig. 22. Schematic representation of the vortex configuration in the wake of spheres at $Re = 10^3$.

**Magarvey, R. H. & Bishop, R. L. (1961), Wakes in Liquid-Liquid Systems, Physics of Fluids, 4, No. 7, pp. 800 - 805.*

***Calvert, J. R. (1967),Experiments on the Flow Past an Inclined Disk, J. Fluid Mech., 29, pp. 691 - 703.*

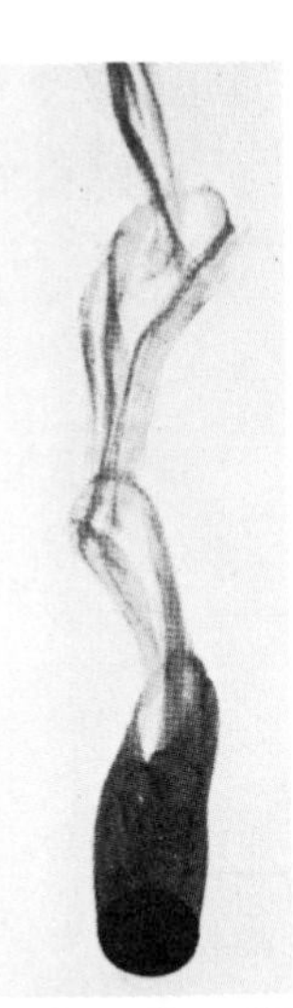

Fig. 23. Vortex chain behind a falling drop, Re = 500, after Margarvey & Bishop (1961).

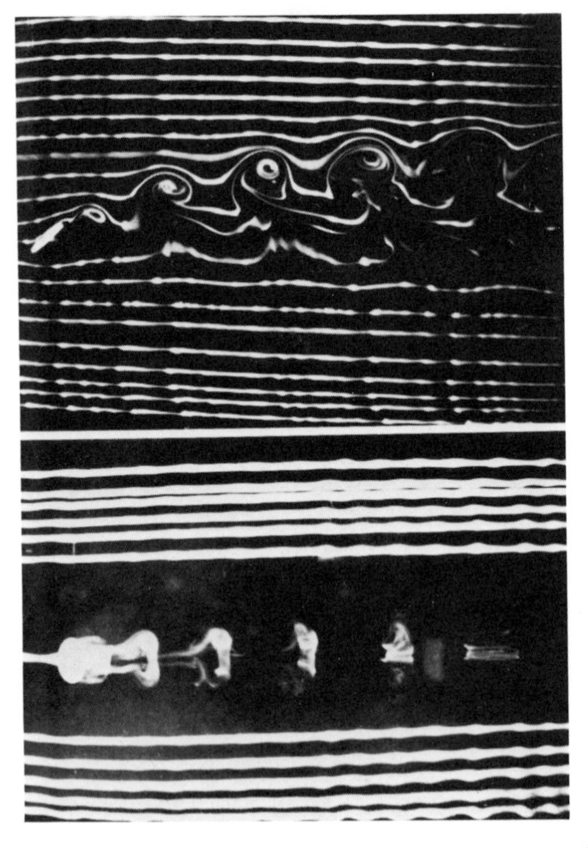

Fig. 24. Vortex chain behind an inclined disk, after Calvert (1967).

In the Reynolds number range $6000 < Re < 3 \times 10^5$, where the Strouhal number increases from S = 0.125 to S = 0.20, a different formation mechanism was found to occur. The shedding mechanism was extremely periodic. For four circumferentially placed hot wires, which were flush-mounted with the surface of the sphere and whose signals were simultaneously recorded, it was concluded that the point of boundary layer separation *rotates* around the sphere at the vortex shedding frequency. Thus, *plane* vortex rings cannot be formed. On the other hand, a helical filament is not possible as it contradicts the circulation theorem

$$\frac{d\Gamma}{dt} = 0$$

where Γ is the circulation of a vortex and t the time.

It is not easy to imagine a vortex configuration which satisfies both the circulation theorem and the evidence of the rotating separation point.

Acknowledgment – Figs. 20, 21, 22 and 24 are reprinted from the Journal of Fluid Mechanics, and Fig. 23 is reprinted from The Physics of Fluids.

W.-H. Hucho *(Volkswagenwerk AG, Germany)*

The vortex generators you showed on the quasi 2-dimensional body, and which you finally applied to the rectangular block, had an angle of attack with respect to the stream direction. Can you tell me this angle?

D. J. Maull

I think the angle is about 10° on the 2-dimensional long body. We did in fact vary the angle. The data is available in the thesis of Young.

D. J. Maull

P. W. Bearman *(Imperial College, London, England)*

I want to ask about possible applications of the computer simulation technique in which you released the point vortices. It seems to be in the spirit of what Eric Maskell was talking about. Can we use it in axisymmetric flows, or perhaps is it possible to even explore more complex cases using that idea? How much further can you go with that notion from 2-dimensional flow?

D. J. Maull

The axisymmetric case is being studied by P.O.A.L. Davies and at NASA–Ames. I think it can be carried a bit further. There is a sort of warning on this, though. Don't push that method too far! It's got faults.

A. Roshko *(California Institute of Technology)*

I thought I heard you say that it's cheaper to calculate a flow than to make a measurement in the wind tunnel.

D. J. Maull

No, I would rather do experiments. The wind tunnel can be looked upon as an analogue computer in that one is testing a model of a real-life situation. I find it easier to get a feeling for the physics of the flow from wind tunnel experiments than from digital computer simulations.

R. T. Jones *(NASA-Ames Research Center)*

Seriously, does anyone feel that there is a real possibility that the quasi-helical periodic wake of a sphere as discussed by Dr. Achenbach could have been discovered with the computer? It's a very strange configuration. That couldn't have been discovered by a computer, could it?

D. J. Maull

I am still not sure about the mechanism of vortex shedding from a sphere. Some results of Calvert have been mentioned. Calvert looked at a disk normal to a stream and found that there was not very strong vortex shedding. However, if he just put a slight angle of attack on the disk then he got strong vortex shedding.

The rectangular plates that I talked about gave strong periodic vortex shedding, with two frequencies which would flip. You can pick up a vortex shedding which seems to tie in with two of the edges of the place, but it will then flip and give vortex shedding from the other edges.

The problem with a sphere is that the method of support may well influence the base flow and the vortex shedding, thus what is needed is a levitation experiment;

that is, a sphere in a tunnel with *no* support. I don't really know how Dr. Achenbach supported his sphere.

D. J. Maull and S. E. Widnall

E. Achenbach

We supported the spheres from the rear, to eliminate any effects of geometric asymmetry. We also changed the diameter of the spheres by a factor of ten for a given support stem diameter and thus varied the supporting conditions substantially. It is, therefore, rather unlikely that small asymmetries occurring naturally in the flow, in spite of the best care of the experimenter, would have the same effect on vortex separation for all sphere sizes. Therefore, I do not believe that the vortex shedding observed in our experiments was a consequence of asymmetrical flow conditions.

DRAG–REDUCING TECHNIQUES FOR AXI–SYMMETRIC BLUFF BODIES

W. A. MAIR

Cambridge University, Cambridge, England

ABSTRACT

The numerous experiments that have been made on drag-reducing devices for *two-dimensional* bluff bodies have been used as a guide to indicate promising lines of investigation for axi-symmetric bodies. For the latter case, experiments on splitter plates, cylindrical extensions, base bleed and ventilated cavities are reviewed. Of these devices, base bleed is the only one that gives any useful reduction of drag. Unfortunately base bleed cannot be effectively applied to road vehicles. The air flow rate available on a typical vehicle from its ventilation system is too small to give any significant effect. If a special air supply giving a larger air flow were to be provided, the intake momentum drag would be more than enough to counteract any drag reduction due to base bleed.

For a blunt-based axi-symmetric body, a boat-tailed afterbody is much more effective in reducing zero-yaw drag than any other device that has been tried. Furthermore, experiments have shown that as the yaw angle of a boat-tailed body is increased from zero, the axial force can decrease slightly up to a yaw angle of about 10 or 15 degrees, although at larger yaw angles it becomes much greater.

The mode of action of a boat-tailed afterbody is explained, and some of the factors leading to a good design are discussed. The possibility of using boundary-layer control in conjunction with a boat-tailed afterbody is considered briefly.

NOTATION

A	Base Area
A_o	Porous area of base (with base bleed)

References pp. 178-179.

b	Width of cruciform splitter plate on a cone (see Fig. 1).
C_D	Drag coefficient referred to maximum cross-sectional area.
ΔC_D	Reduction of drag coefficient.
C_p	Pressure coefficient.
C_{pb}	Base-pressure coefficient.
C_q	Bleed-flow coefficient, $\equiv Q/UA$
C_x	Axial-force coefficient referred to maximum cross-sectional area.
d	Maximum diameter of body of revolution.
d_B	Base height (two-dimensional) or base diameter (axi-symmetric).
d_s	Diameter of a sting-like cylindrical extension.
f	Drag-reduction factor, $\equiv \Delta C_D/0.165$
k	Resistance coefficient at bleed-air outlet.
ℓ	Length of boat-tailed afterbody.
n	Number of air changes per hour, for ventilation.
Q	Volume flow rate of base bleed.
R	Maximum radius of boat-tailed afterbody.
r	Local radius of boat-tailed afterbody.
t	Maximum thickness of two-dimensional aerofoil.
U	Stream velocity.
U_o	Average bleed velocity.
V	Internal volume of vehicle.
X_r	Distance from base to re-attachment on sting or to mean position of bubble closure.
x	Distance downstream from section A in Fig. 5.
β	Boat-tail angle (Fig. 5.)
δ	Boundary layer thickness.
ν	Kinematic viscosity of air.
ρ	Density of air.

INTRODUCTION

Because of the difficulty of theoretical analysis, the study of drag-reducing techniques for bluff bodies has been almost entirely experimental. The bodies that have been studied have usually been two-dimensional, and sometimes axi-symmetric, and much less work has been done on more complex three-dimensional bodies. It is for this reason that attention is concentrated in this paper on axi-symmetric bodies, rather than on general 3-D shapes.

A two-dimensional bluff body in a stream at a low Mach number generates a wake in the form of a Karman street, a regular array of vortices with circulation of alternate sign. This vortex wake is known to be associated with a large drag force on the body, and any device (such as a splitter plate placed in the near-wake) that causes the vortices to form further away from the body gives a reduction of drag.

Some three-dimensional bodies generate wakes in which there is noticeable periodicity, indicating some regular pattern of vortex shedding, but for axi-symmetric bodies any regular vortex shedding is only a minor feature of the flow. Correspondingly, the drag coefficients of axi-symmetric bluff bodies, based on their frontal areas, are usually considerably smaller than those of the related two-dimensional bodies. For example, at a Reynolds number of 10^6 the drag coefficient of a long circular cylinder with its axis normal to the stream is about 0.35, whereas that of a sphere is only about 0.1.

These results lead to two important points. First, a drag-reducing device that is effective on a two-dimensional body because of its action in suppressing or delaying the formation of the vortex street is not likely to be effective on an axi-symmetric body. Second, the maximum reduction of drag coefficient that can be obtained by the use of a device is likely to be less for an axi-symmetric body than for a two-dimensional one, because the drag coefficient without any device present is already lower for the axi-symmetric body.

Nevertheless, since extensive experiments have been made on drag-reducing devices for two-dimensional bodies, it may be profitable to examine whether the results of these experiments suggest any useful forms of drag-reducing devices for axi-symmetric bodies.

The following drag-reducing devices have been found to be effective on *two-dimensional* bodies.

Splitter Plates – Roshko (1954), Bearman (1965), and others have shown that a splitter plate with a length equal to only one base height can in some cases reduce base drag by as much as 50%. One important effect of such a splitter plate is to move the start of the vortex street downstream and away from the base. With a longer splitter plate, the vortex street is suppressed entirely and the base drag may be reduced by 60% or more.

References pp. 178-179.

Base Bleed – Bearman (1967) has shown that outward flow of air from a blunt base can reduce the base drag by as much as 67%. This "bleed" has an effect similar to that of a splitter plate, in that it displaces the start of the vortex street downstream and eventually, at sufficiently large bleed rates, suppresses the vortex street entirely. Bearman has shown that the relationship between vortex-formation position and base pressure is the same with base bleed as with a splitter plate.

Generally, base bleed is most effective in reducing drag when it is distributed over a large proportion of the base area, but Poisson-Quinton & Jousserandot (1957) have shown that a thin plane jet on the centre line can reduce base drag by acting as an "aerodynamic splitter plate".

Ventilated Cavities – Experiments by Nash, Quincey & Callinan (1966) have shown that a thin-walled cavity, with a depth equal to one base height, reduced the base drag of a blunt-based two-dimensional aerofoil at zero incidence by about 23%. When the cavity was "ventilated", by cutting slots in the thin walls at the top and bottom, the reduction of base drag was as much as 60%. Nash *et al.* suggest that the cavity (without ventilation) increases the stability of the wake close behind the body and hence reduces the strength of the vortices in the street. When the cavity is ventilated, there probably is a flow of air from the external stream through the slots into the cavity, and this may have an effect rather like base bleed.

Trailing-Edge Notches – Tanner (1972) has shown, that the drag of a blunt-based aerofoil at zero incidence may be greatly reduced by some form of notched trailing edge. The best arrangement of notches reduced the base drag by 64%. It seems likely that the notches cause the formation of streamwise vortices, and that these prevent the formation of an orderly two-dimensional vortex street.

Boat-Tailing – Experiments on two-dimensional aerofoils with boat-tailed afterbodies and blunt bases have been made by Maull & Hoole (1967). The basic aerofoil, of maximum thickness t, consisted of a semi-elliptical nose of length 3t followed by a parallel portion, also of length 3t. Various boat-tailed afterbodies were added to this basic aerofoil and the pressure drags of these afterbodies (including base drag) were compared with the base drag of the basic aerofoil. The reduction of drag that is obtained depends on the extent of the boat-tailing; in the extreme case of boat-tailing the base height falls to zero and the aerofoil has a sharp trailing edge. When the ratio of base height d_B to maximum thickness t was 0.75, the afterbody drag was reduced by more than 40%, while with $d_B/t = 0.5$ the reduction was 75%.

Of the devices that have been listed here, the trailing-edge notches do not appear to have any useful analogue in the axi-symmetric case. There is also difficulty in conceiving a useful axi-symmetric analogue of a two-dimensional splitter plate, but some attempts have been made to reduce the drag of axi-symmetric bodies by adding extensions at the base, and these will now be considered. Axi-symmetric forms of the other devices that have been mentioned will also be discussed.

BASE EXTENSIONS

In attempts to find axi-symmetric analogues of two-dimensional splitter plates, two different forms of base extensions have been tried on axi-symmetric bodies..

Tanner (1965) attached long cruciform splitter plates, as shown in Fig. 1, to the bases of a series of cones having total included apex angles from 15° to 120°. The base diameter of the cone was 80mm for the 15° apex angle and 98mm for all the others. The cruciform plates were 2mm thick and 250mm long, and their forward edges were 10mm behind the base of the cone. The plates were supported on a central tube of 30mm outside diameter. For the measurements without splitter plates, the cones were supported on a long sting of 25mm diameter. Fig. 1 shows the increase of base-pressure coefficient, ΔC_{pb}, as a function of cone apex angle and b/d, for a Reynolds number Ud/ν of about 2×10^5. There is an appreciable reduction of base drag for the larger apex angles, but the reduction becomes very small as the cone angle approaches zero. At each angle the maximum effect occurs near b/d of unity.

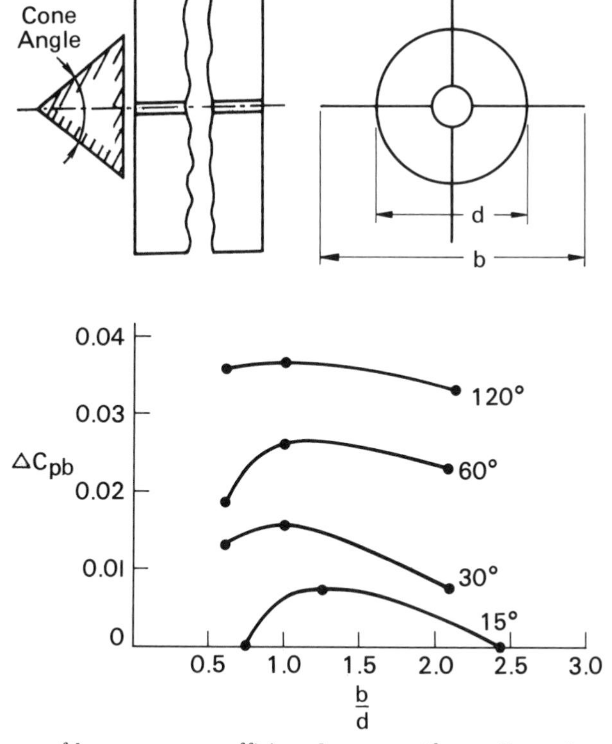

Fig. 1. Increase of base-pressure coefficient due to cruciform splitter plates attached to bases of cones. Total apex angles of cones are marked on curves (Tanner, 1965).

References pp. 178-179.

Calvert (1967) made experiments to determine the effect of long, central, cylindrical extensions of different diameters; they were of the form used as support stings. He remarked that the geometrical axi-symmetric analogue of a thin two-dimensional splitter plate would be a thin central spike; this would not be expected to have any appreciable effect, but cylindrical extensions of various diameters might give results of interest. The basic body consisted of a cylinder of diameter d and length 3.08 d, with a blunt base and a semi-ellipsoidal nose of length 1.33d. The added stings were rather long (8 base diameters), and the ratio of their diameter to the base diameter varied from $d_s/d = 0.17$ to 0.83. The Reynolds number Ud/ν was between 3×10^4 and 9×10^4. Measurements were made of base-pressure coefficient C_{pb} near the outer edge of the base, and of the distance X_r from the base to the re-attachment point on the sting or (for the model with no sting) to the mean position of bubble closure.

Fig. 2 shows that an extension of quite small diameter gives a substantial increase of base pressure, and a slight lengthening of the separation bubble. When d_s/d is about 0.33, the maximum base pressure is obtained; further increase of d_s/d gives a slight reduction of base pressure and a large reduction of bubble length. Although these results show a useful reduction of base drag for $d_s/d \approx 0.3$, it should be noted that all the extensions used were long. It seems likely that an extension with a length less than X_r would have little effect, and it is difficult to see any useful application of the results to drag-reducing devices for road vehicles.

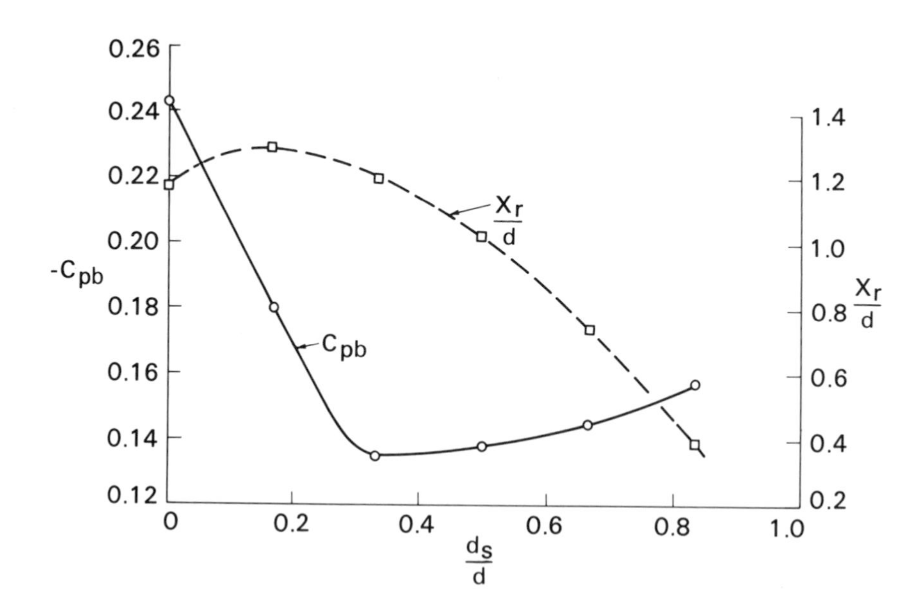

Fig. 2. Base-pressure coefficient and distance from base to reattachment for cylindrical body of diameter d with long sting-like cylindrical extension of diameter d_s (Calvert, 1967).

BASE BLEED

Calvert (1967) investigated the effect of base bleed on a cone with a total apex angle of 60° and a base diameter d of 76.2 mm. The uniformly porous base was made of Vyon sintered polythene, 1.6 mm thick, and the bleed air was pumped into the cone through a long tube, of outside diameter 9.5 mm, extending axially downstream from the centre of the base. The maximum porous area A_o was 3323 mm^2, representing 72.9% of the total base area A. The porous area was reduced progressively by blocking the surface in successive rings from the outside inwards, the minimum value of A_o being 0.283A. The Reynolds number Ud/ν was 6.1×10^4.

Fig. 3 shows the base-pressure coefficient C_{pb} plotted against a bleed-flow coefficient $C_q = Q/UA$, where Q is the volume flow rate of bleed air. As in the two-dimensional case, at a fixed bleed rate the best drag reduction is obtained with a large porous area, giving a low velocity of ejection; however, in the axi-symmetric case the reduction of drag is much smaller. At a given C_q, smaller porous areas increase drag, except at very small flow rates; this is probably caused by entrainment of air from the external stream into the jet of relatively high velocity from the porous base. This increases the curvature of the streamlines leaving the conical surface at the base, and so reduces the base pressure.

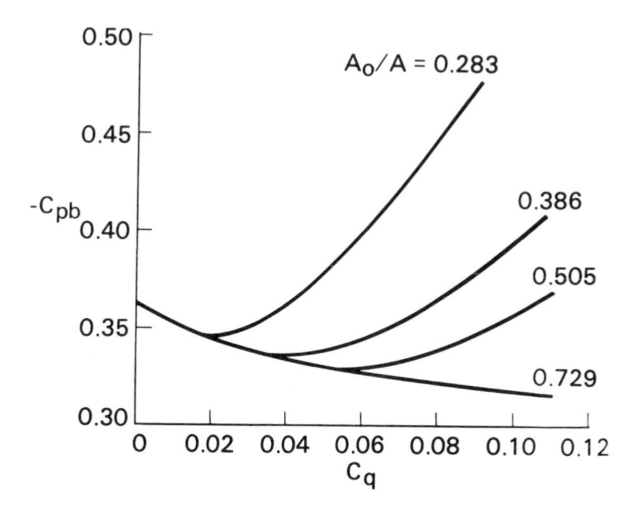

Fig. 3 Base-pressure coefficient on cone of 60° total apex angle with base bleed (Calvert, 1967).

Experiments by Sykes (1969), on a cylindrical body with an ogival nose, have shown a much greater reduction of drag with base bleed. The body diameter d was 100 mm, the length of the cylindrical portion was 5.15 d and the overall length was 6.27 d. The Reynolds number Ud/ν was 1.8×10^5. The bleed air was introduced into the body through a branch pipe of streamline section 0.25 d thick, which entered at the

References pp. 178-179.

side. Although this conduit was about 4 d ahead of the base, it could have had some effect on the flow in the base region.

The results obtained are shown in Fig. 4, where the full lines show base pressures with bleed through an open central hole in a thin-plate covering the base (i.e. through an orifice), while the broken lines shown results obtained with the central hole covered with porous plastic sheet.

The results shown in Fig. 4 are qualitatively similar to those of Fig. 3. Again, the best drag reduction is obtained with a large value of A_o/A, and at this large value the introduction of the porous plastic sheet is beneficial.

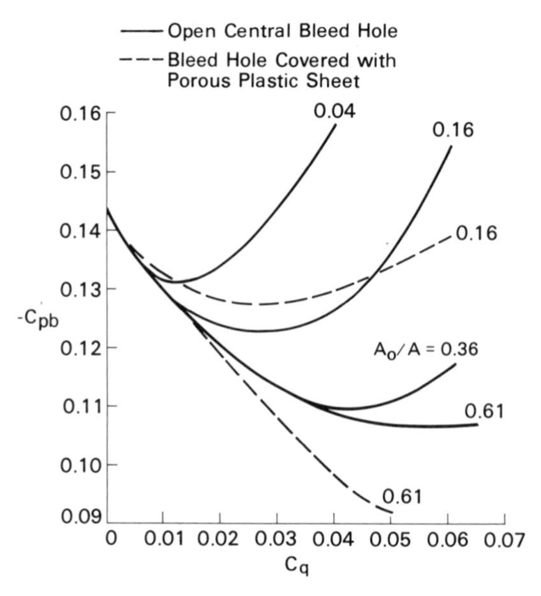

Fig. 4. Base-pressure coefficient on cylindrical body with ogival nose and base bleed (Sykes, 1969).

The effect of the porous plastic sheet in the results of Fig. 4 is difficult to explain. In the absence of this sheet the flow through the bleed orifice would be in the form of a contracting jet, perhaps with some asymmetry caused by the inlet pipe mentioned earlier. Introduction of the porous plastic sheet would tend to eliminate the contraction of the jet, so that its velocity would be lower for a given volume flow rate. This could explain the difference between the two curves in Fig. 4 for $A_o/A = 0.61$, but the difference in opposite sense for $A_o/A = 0.16$ and $C_q < 0.05$ cannot be explained in this way.

To estimate the overall value of base bleed as a drag-reducing device, it is necessary to consider also the intake momentum drag and the power required to pump the bleed air through the outlet resistance. If the pressure drop at the outlet resistance is $\frac{1}{2}\rho U_o^2 k$, where U_o is the average bleed velocity over the area A_o, the ratio of ideal pump power to the power saved by drag reduction is

$$\frac{k\, C_q^3}{(A_o/A)^2 \Delta C_{pb}}$$

where ΔC_{pb} is the increase of base-pressure coefficient due to base bleed. For the values suggested by Fig. 4. this is of order $2k \times 10^{-3}$, and the required pump power is therefore likely to be negligible.

The intake momentum drag is the drag caused by taking in air at the stream velocity U and ejecting it at the bleed velocity U_o. For a volume flow rate Q this is

$$\rho Q(U - U_o) = \rho A U^2 C_q \left(1 - \frac{U_o}{U}\right) .$$

The drag saved by base bleed is $1/2\ \rho A U^2 \Delta C_{pb}$. Hence, the ratio of intake momentum drag to drag saved by base bleed is

$$\frac{2C_q}{\Delta C_{pb}}\left(1 - \frac{U_o}{U}\right) \approx \frac{2C_q}{\Delta C_{pb}}$$

(since $U_o/U = AC_q/A_o$ and is small compared with one). Fig. 4 shows that $2C_q/\Delta C_{pb}$ is likely to be greater than 1, which means that the intake momentum drag is more than enough to counteract the saving of drag due to base bleed. Hence, the only way in which base bleed might be of value in reducing the drag of a vehicle is if the bleed air is taken into the vehicle for another purpose, such as ventilation.

For the purpose of ventilation, the air inside a vehicle of volume V may be required to be changed n times per hour. Then, if all the outlet air from the ventilation system is used for base bleed, $C_q = nV/AU$. For a typical passenger car, with V/A of about 3 metres and traveling at 100 km/h, this yields $C_q \approx 3n \times 10^{-5}$. The value of n required for ventilation is not likely to be greater than 25, so that C_q will only be about 0.00075, too small to have any appreciable effect on drag.

For a commercial truck, V has to be taken as the internal volume of the driver's cab, for that is normally the only space requiring ventilation. The ration V/A will therefore be less than for the passenger car, and again there is no useful reduction of drag to be obtained from using the ventilation outflow for base bleed.

It can be concluded that base bleed shows no promise as a means of reducing the drag of a road vehicle. This is primarily because of the large intake momentum drag, but the need to eject the air over a large proportion of the base area would also

References pp. 178-179.

introduce severe practical difficulties.

VENTILATED CAVITIES.

It has already been mentioned that a cavity at the rear of a blunt-based two-dimensional aerofoil can give a large reduction of drag, especially when the cavity is ventilated by cutting slits in the walls. In contrast, some measurements by Goodyer (1966) showed that a cavity at the base of a body of revolution had no effect on drag when it was not ventilated, while the addition of ventilating slits gave a substantial *increase* of drag. These general results were not Reynolds-number sensitive, as was shown by repeating the experiments with boundary-layer transition fixed at various different positions along the body. The measurements were made on a cylindrical body with an ogival nose and overall length 11.3 d, suspended magnetically in a wind tunnel so that there was no support interference.

In selecting a depth for his cavity Goodyer was guided by the two-dimensional results of Nash, Quincey & Callinan (1966), and chose a depth approximately equal to one body diameter. He made no measurements with cavities of other depths, and it is possible that such cavities would have given different results.

BOAT-TAILING

It has been known for many years that the drag of a blunt-based body of revolution can be reduced significantly by boat-tailing the rear end in the manner shown in Fig. 5. This sketch refers to a basic body shape consisting of a long cylinder with a rounded nose, terminating in a blunt base at section A. The portion AB is a typical boat-tailed afterbody, terminating in a blunt base at B with a smaller area than that of the basic body at A. The addition of a further conical body BC, with semi-apex angle β chosen to match the shape of the boat-tailed afterbody at B, would give the "streamline" tail-piece ABC. Starting with this streamline shape there is, of course, a choice of position for the plane B defining the base of the boat-tailed afterbody. The value of the angle β, defining the slope of the boat-tailed afterbody at its base B, is obviously one very important variable. If this angle is too large the boundary layer separates before reaching B, while if it is too small the afterbody is unnecessarily long for a given diameter ratio d_B/d, where d is the diameter at A

Many of the experiments that have been made on boat-tailed afterbodies have not been relevant to the drag of land vehicles because they have been made at high Mach numbers, often greater than 1. Even when measurements at lower speeds have been included, the lowest Mach number has often been as high as 0.6, so that compressibility may still have had some effect.

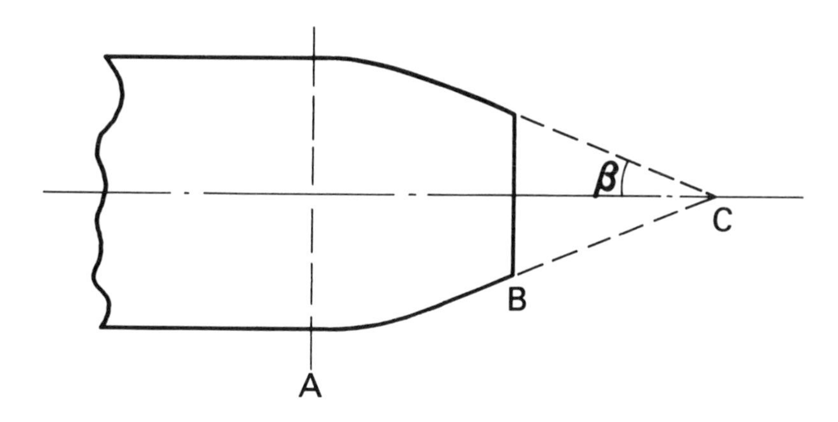

Fig. 5. Boat-tailed afterbody (Mair, 1969).

An undesirable feature of some of the earlier work was a tendency to concentrate too much on base drag, and to pay too little attention to the equally important drag force acting on the curved surface AB (in Fig. 5). There was also a tendency to concentrate on circular-arc or straight conical profiles, the latter often having no rounding at the section A so that there was a discontinuity of slope there. In some of the more recent work, the experiments have been made with one or more jets issuing from the base B, and a single jet has sometimes been simulated by a long cylinder extending downstream from the base, as in the work by Reubush & Putnam (1976). Consideration of the drag component obtained by integration of the measured pressure distribution on the curved surface AB shows that this pressure distribution is seriously affected by the real or simulated jet downstream of B, so that the results are not applicable to a boat-tailed afterbody without a jet.

In some experiments by Mair (1969) at a Mach number of 0.13, attention was concentrated on the form of boat-tailed afterbody that would give the greatest possible reduction of drag. The basic body consisted of a cylindrical portion of diameter d and length 3 d, with a semi-ellipsoidal nose of length 1.3 d. The Reynolds number Ud/ν was 0.46×10^6. A transition wire on the nose ensured that the boundary layer on the cylindrical portion was turbulent, the total thickness δ at the base being about 0.066 d. Some measurements were also made with the boundary layer artifically thickened to give $\delta = 0.2$ d at the base, but the results obtained were little different so all the results given here refer to the thinner boundary layer.

Inviscid flow calculations for the body with various forms of streamline tail attached showed that the pressure coefficient C_p had a nearly constant value of about -0.04 over a length of about one d near the middle of the cylindrical portion. This is only about one-tenth of the value found at the peak suction position on a typical streamline tail just behind A, indicating that the body was effectively "long", so that the pressure distributions at the nose and tail were nearly independent of one

References pp. 178-179.

another. For such a long body, in inviscid flow, the pressure drag of the nose and of the tail would each be zero.

The measured base-pressure coefficient at A on the basic body was -0.165. Thus an "ideal" afterbody, with zero skin friction and with pressure distribution as for inviscid flow, would give a reduction of drag coefficient $\Delta C_D = 0.165$.

The actual experimental values of ΔC_D found for the various afterbodies may be compared with this ideal value by introducing a drag reduction factor $f = \Delta C_D/0.165$. It should be noted that, in a real fluid, f may be considerably less than one even for a conventional streamline tail tapering to a point; for such a tail, early work by Lock & Johansen (1933) found the value of f to be 0.72.

The best boat-tailed afterbody, found in this series of measurements (Mair, 1969) was the one that is drawn to scale in Fig. 5. This is a cone with $\beta = 22°$, joined by a curved fairing of length 0.5 d to the cylindrical body at A. Values of the drag reduction factor f are shown in Fig. 6 for this afterbody, as a function of the length ℓ from the original body-base A to the boat-tail cut-off plane B. The position shown for B in Fig. 5 corresponds to $\ell/d = 0.76$, for which $f = 0.74$. This is actually better than the value found for the streamline tail mentioned earlier, and even with $\ell/d = 0.6$ the value of f is 0.68, not much less than for the streamline tail.

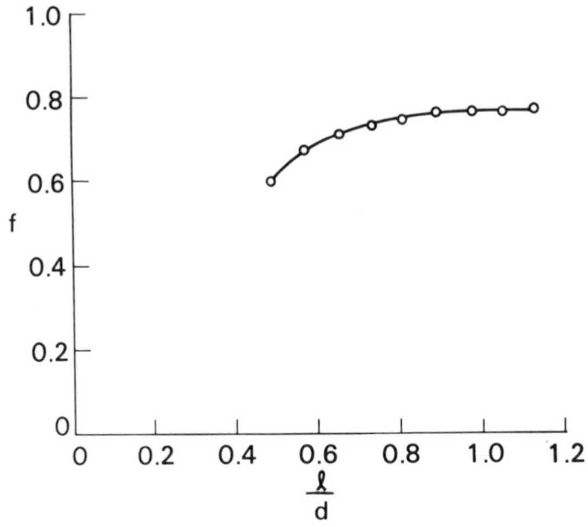

Fig. 6. Variation of drag reduction factor f with length ℓ of afterbody for boat-tailed afterbody shown in Fig. 5, $\beta = 22°$ (Mair, 1969).

For this particular afterbody with $\ell/d = 0.76$ (and for some other related ones) measurements of pressure distribution have been made by Bostock (1972) on a long yawed body at the same Reynolds number. The length of the body was 14 d or greater, and its upstream end was attached to a hinge on the wall of the wind tunnel. This arrangement, which had the body nose very near the wall, did not allow studies of the model at small or zero yaw angles. However, for the angles of yaw considered

($\geq$ 10°) the flow at the tail was not affected by the upstream hinge or by the total body length, provided the latter was greater than about 10 d.

Bostock integrated the measured pressure distributions to obtain axial-force coefficients C_x based on the maximum cross-sectional area. Since the earlier measurements by the present author had shown that increase of boundary-layer thickness made little difference to either the drag or the pressure distribution on the boat-tail, the difference of body length between the two sets of measurements is not likely to have a serious effect on the comparisons.

The axial-force coefficient at zero yaw, which could not be obtained by Bostock, was extracted from the data of Mair (1969). First, for the body at zero yaw, boundary layer calculations by the method of Head (1960), using the measured pressure distribution, have shown that the contribution of skin friction to the drag coefficient of the afterbody is about 0.008 for $\ell/d = 0.76$. For comparison with Bostock's axial-force coefficients C_x for the yawed body (which do not include skin friction) this amount should be subtracted from the measured afterbody drag coefficient at zero yaw. Since Fig. 6 gives $f = 0.74$ for $\ell/d = 0.76$, the value of C_x for zero yaw is $0.165\ (1 - 0.74) - 0.008 = 0.035$. This value, together with those found by Bostock at yaw angles up to 30°, is plotted in Fig. 7. The same diagram also shows results for a boat-tailed afterbody of the same length and the same general form, but with $\beta = 26°$.

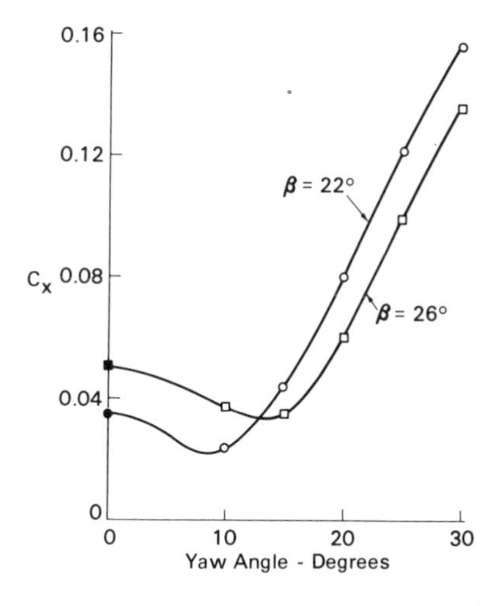

Fig. 7. Axial-force coefficients on boat-tailed afterbodies with $\ell/d = 0.76$ (skin friction not included). Open symbols Bostock (1972), closed symbols Mair (1969).

The striking feature of Fig. 7 is the *reduction* of C_x that occurs with both afterbodies as the yaw angle is increased from zero to about 10° or more. There is good evidence that this trend is genuine and is not due to a spurious comparison

References pp. 178-179.

between two sets of measurements made under conditions that were not identical. For example, the Bostock points for $\beta = 26°$ and yaw angles of 10° and 15° indicate a definite minimum in that region. Similar but rather weaker evidence at these same yaw angles was found for $\beta = 30°$, although C_x was then higher than for $\beta = 26°$ at all yaw angles up to 25°.

FURTHER DISCUSSION OF BOAT-TAILED AFTERBODIES

Because boat-tailed afterbodies are so effective in reducing drag, it may be useful to consider their mode of action in more detail.

Fig. 8 shows the pressure distribution on the afterbody shown in Fig. 5, with $\beta = 22°$, for two different lengths ℓ. The distance x is measured downstream from the section A in Fig. 5. The point C in Fig. 5 is at $x/d = 1.50$, so that the pressure distribution on the complete body with pointed tail may be expected to be almost the same as that shown in Fig. 8 for $\ell/d = 1.37$. The other curve shown in Fig. 8, for $\ell/d = 0.67$, shows that when the body is cut off by a plane such as B in Fig. 5, there is little effect on the pressure distribution upstream of B. Because most of the pressure recovery on the longer body occurs in the range of x/d less than about 0.7, little is to be gained by making the cutoff at x/d larger than that. The cut-off body has in addition a lower skin-friction drag, while the penalty for the lower base pressure, caused by the lost pressure recovery on the cut-off piece, is small as it acts only on a relatively small base area.

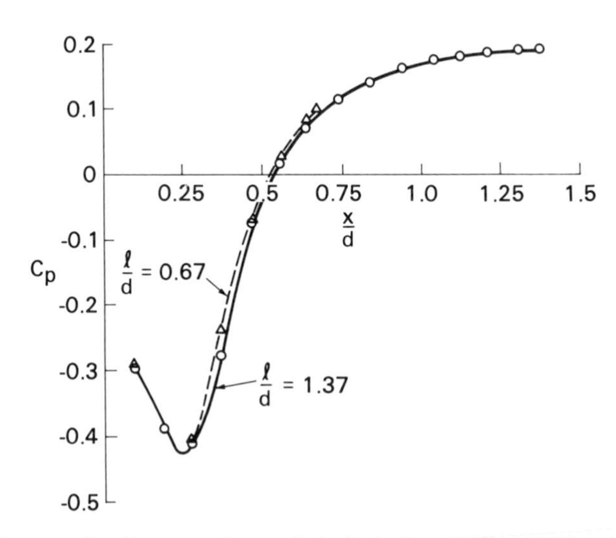

Fig. 8. Pressure distribution on boat-tailed afterbody with $\beta = 22°$ (Mair, 1969).

These results suggest a possible approach to the problem of predicting the drag of a boat-tailed afterbody with a blunt base. The pressure distribution could be calculated for the complete afterbody with a pointed tail, with allowance for the boundary layer displacement effect, and it might be assumed that for an afterbody cut off at $x = \ell$ the pressure distribution for $x < \ell$ would be the same as for the complete body. Unfortunately, calculations of drag made on this basis, using experimental pressure distributions for various afterbody lengths, have shown that no useful estimates of drag can be obtained.

The reason for this disappointing result can be seen by reference to Fig. 9 which shows the total drag and its three components for the boat-tailed afterbody with $\beta = 22°$. Curve A represents the measured total drag, B is the measured base drag, and C is the calculated skin-friction drag. The remaining component, the pressure drag on the curved afterbody, was obtained as the differential D = A-B-C. In the range of ℓ/d that is of most interest, from about 0.6 to 0.8, the positive drag component D due to low average pressure on the curved surface of the afterbody is substantially offset by the base drag B, which is negative in that region. Further, seemingly small changes of pressure distribution can have, when integrated, a substantial effect on the component D. Since, after addition of the negative base drag B, the proportional effect on the total drag A is even greater, the suggested method of calculating drag is of no practical value.

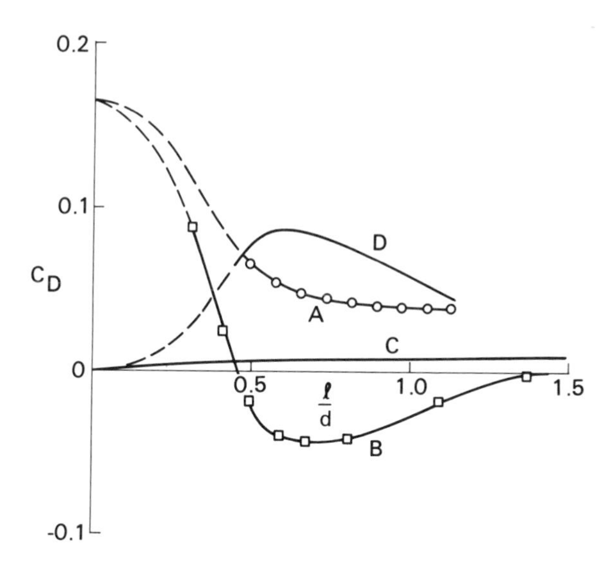

Fig. 9. Components of drag coefficient for boat-tailed afterbody with $\beta = 22°$ (Mair, 1969). A, measured total drag; B, measured base drag; C, calculated skin-friction drag; D, afterbody drag, excluding base, calculated as A-B-C. - - - possible extrapolation to zero length.

The requirements for the design of a good axi-symmetric boat-tailed afterbody may now be considered. First, it is clear that the boundary layer should not separate

References pp. 178-179.

upstream of the base. If separation occurs at some station S, upstream of the base B, there will be little change of pressure between S and B so that the body can be cut off at S without much effect on the drag. The body that is cut off at S will have separation at the base, and, being shorter, will be a better practical design than the original body cut off at B.

For discussion of the other design requirements it is useful to refer to Fig. 10, which shows C_p (r/R) plotted against r/R for a streamline afterbody on a long cylindrical body, as calculated for inviscid flow. For this case the net drag is zero, so that the areas of the two loops A and B are equal. Now suppose that this body is cut off to form a blunt base, say at the section where r/R = 0.72, represented by the point P in Fig. 10. The pressure distribution upstream of P will be nearly the same as for the complete body and the pressure on the base will be the same as at P, so that the pressure drag of the cut-off body will be proportional to the area of the part of the loop A that is above the straight line OP. A body that has a small total area of loop A is also likely to have a small area above the line OP. This means that a good boat-tailed afterbody can be formed by taking a complete streamline afterbody for which each of the loops A and B is as small as possible, and cutting off this body at some section P. The first requirement, that the boundary layer must not separate upstream of P, must also be satisfied, of course.

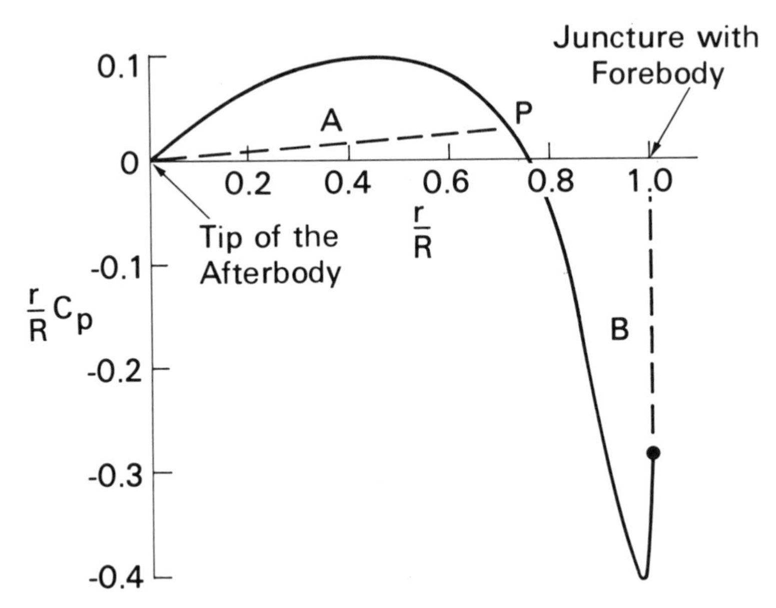

Fig. 10. Radial distribution of $(r/R)C_p$ for streamline afterbody in inviscid flow. r - local radius, R - maximum radius.

Examination of the pressure distributions from which the results shown in Fig. 7 were derived, shows that on the yawed body, the pressure falls as the tail is approached, and then rises, in qualitative agreement with the curves of Fig. 8 for zero yaw. On the yawed body, both the fall of pressure and the subsequent recovery are greater on the windward side than on the lee side and, at stations very close to the base, the pressure becomes nearly uniform all round the body, probably because of the influence of the base. The favourable effect of a moderate angle of yaw, as shown in Fig. 7, indicates that the mean pressure at each section is modified in a favourable sense from the zero-yaw value shown in Fig. 8. As already noted, quite a small change in mean pressure at a section can have a significant effect on drag.

The large increase of axial-force coefficient that is shown in Fig. 7 for yaw angles above 15° is probably caused by a loss of pressure recovery due to separation of the boundary layer.

BOAT-TAILED AFTERBODIES WITH BOUNDARY LAYER CONTROL

The boat-tailed afterbodies that are most useful for practical purposes have a small overall length and a fairly large value of the angle β. Since one of the requirements already noted is that the boundary layer must not separate ahead of the base, it may be useful to consider whether some form of boundary-layer control (BLC) could be usefully employed to delay separation and hence allow greater freedom in design.

The easiest form of BLC to apply to a road vehicle would be a set of vortex generators, although this might be unacceptable for various practical reasons, including safety. Experience with aircraft and other applications suggests, however, that even the best arrangement of vortex generators would have only a marginal beneficial effect, and it seems likely that the best boat-tailed afterbody using these devices would be only slightly better than one without them.

Other possibilities to be considered are the use of BLC by suction or blowing to delay separation. Considering suction first, the best economy of power is obtained by sucking air from the boundary layer to reduce its thickness *before* the layer encounters any severe adverse pressure gradient. The suction flow and the power required will vary from zero, for a boat-tailed afterbody that is satisfactory even without BLC, to quite large values for short afterbodies with large values of the angle β. Thus no useful quantitative statements can be made without studying individual cases in detail.

It is known from studies of BLC in aircraft applications (e.g. Mair, 1966) that suction usually requires less power than slot blowing. Nevertheless, slot blowing has usually been preferred for aircraft because of the ready availability of high-pressure air from the compressors of the propulsion engines. For a road vehicle there would be no such reason to prefer blowing, since a special blower for the BLC would probably be required in any case, and suction would be the obvious choice because of its lower power requirement.

References pp. 178-179.

The use of BLC in conjunction with a boat-tailed afterbody has been suggested here as a possibility, but, at present, there is little reason to think that this possibility is likely to be promising. Boat-tailed afterbodies have been shown to give large reductions of drag without BLC, and only detailed investigations can show whether there might be additional advantages in using BLC to give low drag with even shorter afterbodies.

CONCLUSIONS AND DISCUSSION

Consideration of known methods of reducing the drag of a blunt-based body of revolution has shown that a boat-tailed afterbody is much more effective than any other device that has been tried. Moreover, the boat-tailed afterbody can still give a good reduction of axial force at yaw angles up to about 15°.

A road vehicle must operate near the ground and cannot usually be axi-symmetric, although cylindrical road tankers for carrying liquids are of some interest. Much more experimental work will be needed before it is possible to design with confidence a practicable vehicle with a rear end shaped to take advantage of the boat-tail principle. The "Kamm back" that is already being used may be thought of as a shape with a roof-line as in Fig. 5, but with nearly straight boundaries at the sides and bottom. Such a shape may sometimes give a useful reduction of drag at zero yaw, but in a side-wind there is likely to be separation at the edges of the sloping afterbody and some of the beneficial boat-tail effect may then be lost. More experimental work is needed to develop practicable shapes based on the boat-tail principle, which give a low axial force, even when yawed.

The further complications of shear flow and turbulence in the atmospheric wind must also be considered.

REFERENCES

Bearman, P. W. (1965) Investigation of the flow behind a two-dimensional model with a blunt trailing edge and fitted with splitter plates. J. Fluid Mech. Vol. 21, pp 241 - 255.

Bearman, P. W. (1967) The effect of base bleed on the flow behind a two-dimensional model with a blunt trailing edge. Aero. Quart., Vol. 18, pp 207 - 224.

Bostock, B. R. (1972) Slender bodies of revolution at incidence. Ph. D. Dissertation, University of Cambridge.

Calvert, J. R. (1967) The separated blow behind axially symmetric bodies. Ph. D. Dissertation, University of Cambridge.

Goodyer, M. J. (1966) Some experimental investigations into the drag effects of modifications to the blunt base of a body of revolution. Inst. of Sound and Vibration, University of Southampton, Report No. 150.

Head, M. R. (1960) Entrainment in the turbulent boundary layer. ARC R&M 3152.

Lock, C. N. H. & Johansen, F. C. (1933) Drag and pressure distribution experiments on two pairs of streamline bodies. ARC R&M 1452.

Mair, W. A. (1966) STOL - some possibilities and limitations. J. Roy. Aero. Soc. Vol. 70, pp 825 - 833.

Mair, W. A. (1969) Reduction of base drag by boat-tailed afterbodies in low speed flow. Aero. Quart. Vol. 20, pp 307 - 320.

Maull, D. J. & Hoole, B. J. (1967) The effect of boat-tailing on the flow around a two-dimensional blunt-based aerofoil at zero incidence. J. Roy. Aero. Soc. Vol. 71, pp 854 - 858.

Nash, J. F., Quincey, V. G., & Callinan J. (1966) Experiments on two-dimensional base flow at subsonic and transonic speeds. ARC R & M 3427.

Poisson-Quinton, P. & Jousserandot, P. (1957) Influence du soufflage au voisinage du bord de fuite sur les caracteristiques aerodynamiques d'une aile aux grandes vitesses. La Recherche Aeronautique, No. 56, pp 21 - 32.

Reubush, D. E. & Putnam, L. E. (1976) An experimental and analytical investigation of the effect on isolated boat-tail drag of varying Reynolds number up to 130 x 10⁶. NASA TN D-8210.

Roshko, A. (1954) On the drag and shedding frequency of bluff cylinders, NACA TN 3169.

Sykes, D. M. (1969) The effect of low flow rate gas ejection and ground proximity on afterbody pressure distribution. Proc. 1st Symposium on Road Vehicle Aerodynamics, City University, London.

Tanner, M. (1965) Druckverteilungsmessungen an Kegeln, DLR FB 65 - 09.

Tanner, M. (1972) A method of reducing the base drag of wings with blunt trailing edges. Aero. Quart. Vol. 23, pp 15 - 23.

DISCUSSION

Prepared Discussion

T. Morel *(General Motors Research Laboratories)*

This discussion will concern itself with the effect of base cavities on the drag of an axisymmetric body. As Professor Mair mentioned in his paper, the subject was studied about 10 years ago by Goodyer who concluded that the effect of ventilated cavities is to increase drag. The purpose of this discussion is to point out that, while the conclusion arrived at by Goodyer is indeed correct for deep cavities, the opposite is true for cavities of smaller depth.

Our experimental setup consisted of a long circular cylinder, suspended on six wires and aligned with the flow, at the end of which were attached interchangeable afterbodies (Fig. 11). These afterbodies had cavities of three different types, and of six different depths. Altogether, 18 different configurations were tested. The three types of cavities were: solid-walled; slotted, with twelve longitudinal slots; and slitted, with 12 slots which ran out the end of the cavity. All the slots started right at the body base. The depths studied were 0.1, 0.2, 0.35, 0.5, 0.7 and 0.9 times the body diameter. I should point out that the single cavity tested by Goodyer, and which was mentioned by Professor Mair, had a depth of 0.96 times the cylinder diameter.

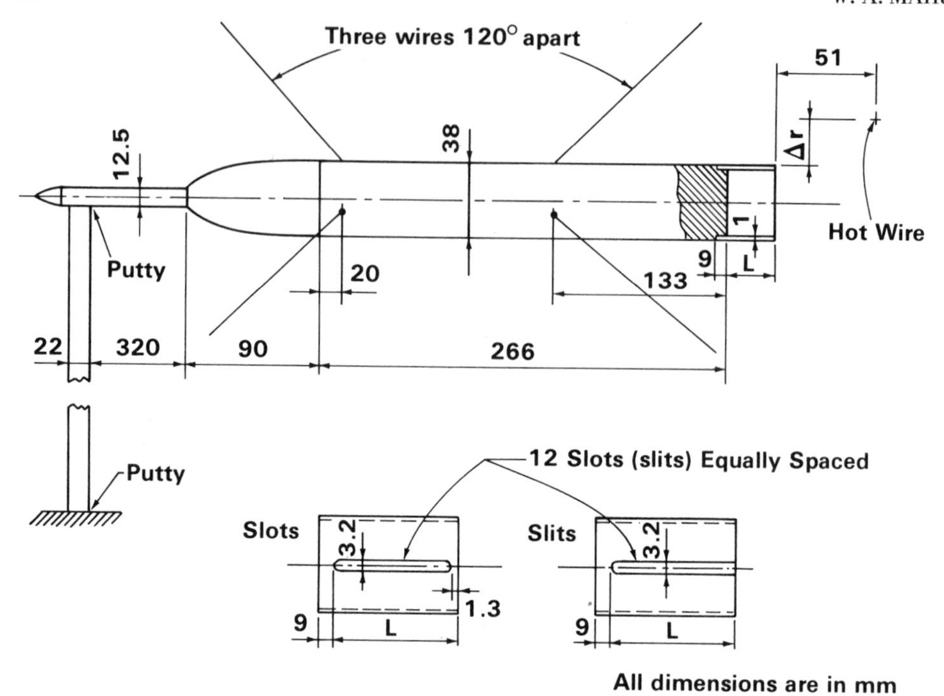

Fig. 11. Sketch of the experimental arrangement showing the test body and the dimensions of the interchangeable add-on cavities. All dimensions are in mm.

The variation of drag coefficient with cavity depth is shown in Fig. 12. At small cavity depths, there is a small but measurable drag decrease with all three types of cavities. At larger depths, the solid-walled cavity still shows some drag reduction even at the last point tested. However, the two ventilated cavities, the slotted and the slitted, both show a dramatic drag increase. The Goodyer data points for the zero depth case and for his deep ventilated (slitted) cavity are included for comparision. The base pressure, measured at a centerline tap inside the cavities (Fig. 13), had trends similar to those for the drag coefficient. However, the magnitude of the variation in C_{pb} was more pronounced than for the drag coefficient; we don't know quite why, but presumably the smaller effect on drag is accounted for by skin friction along the cavity walls.

One of the objectives of this study was to inquire into the mechanisms of drag generation. It has been proposed by Nash *et al* (1963)* that the reason why cavities reduce the drag of 2-dimensional bodies is the inhibition of vortex shedding. This is also what Peter Bearman, David Maull and Professor Mair mentioned briefly in their presentations. One may ask, is this the same mechanism which makes a cavity work in the axisymmetric case? To test this possibility we measured the periodic component in the near wake. A hot wire was placed at the shear layer edge about one diameter

**Nash, J. F., Quincey, V. G. & Callinan, J. (1963), Experiments on Two-Dimensional Base Flow at Subsonic and Transonic Speeds, Aeronautical Research Council, R&M 3427.*

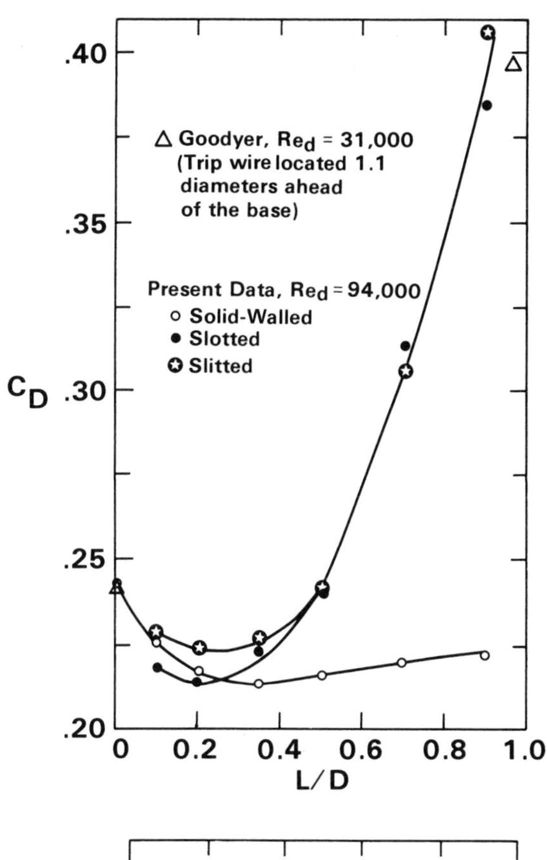

Fig. 12. Drag coefficient vs. cavity depth for the three cavity geometries.

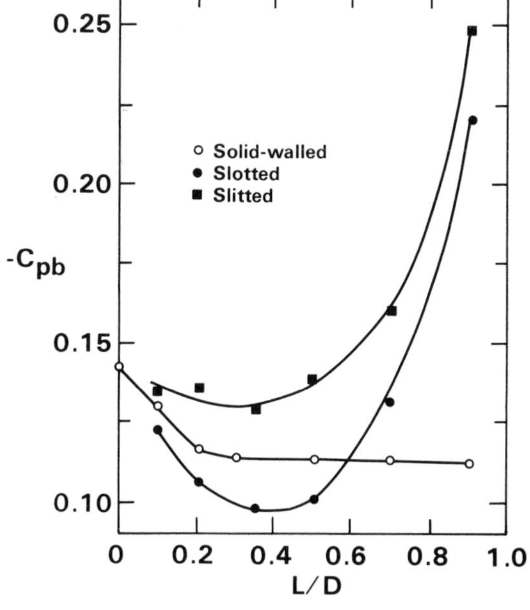

Fig. 13. Centerline base pressure coefficient vs. cavity depth.

downstream from the end of the cavity, and the velocity signal was processed to obtain its spectrum. The results for the solid-walled cavity are shown in Fig. 14, plotted against the Strouhal number S_D based on the cylinder diameter. The spectra displayed definite peaks, and so there is no question whether or not there is periodicity behind axisymmetric cylinders in high Reynolds number flow. I should point out that the separated boundary layer was fully turbulent and was relatively thick, with $\delta/D = 0.16$. For clarity, all spectra in Fig. 14 are displaced vertically from each other. If you would inspect them closely you would notice that the magnitude

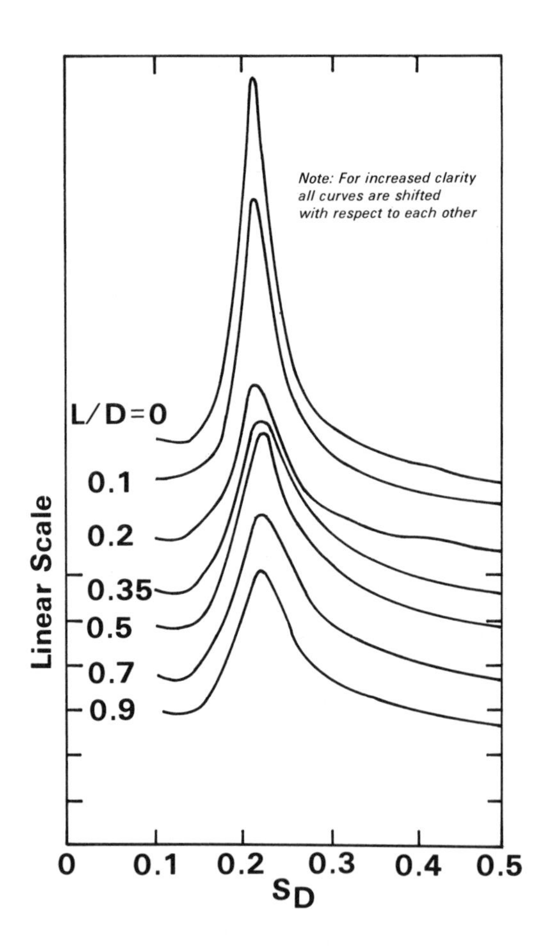

Fig. 14. Axial-velocity spectra for solid-walled cavities. Measured with a hot wire placed at the edge of the wake at $x/D = 1.33$ downstream from the cavity end.

of the spectrum peak follows roughly the same trend as the drag coefficient for the solid-walled cavity; that is, it decreases when the C_D decreases and vice versa. This seems to support Nash's idea about the mechanism involved in cavity drag reduction; however, Fig. 15 showing the spectra for the slotted cavity seems to dispute it. In this case the trend of the spectrum peak magnitude is monotonic, and the cavities succeed in wiping out the periodicity at L/D between 0.5 and 0.7. However, as we have seen in Fig. 12, the drag did not follow such a trend, but increased sharply beyond L/D = 0.35.

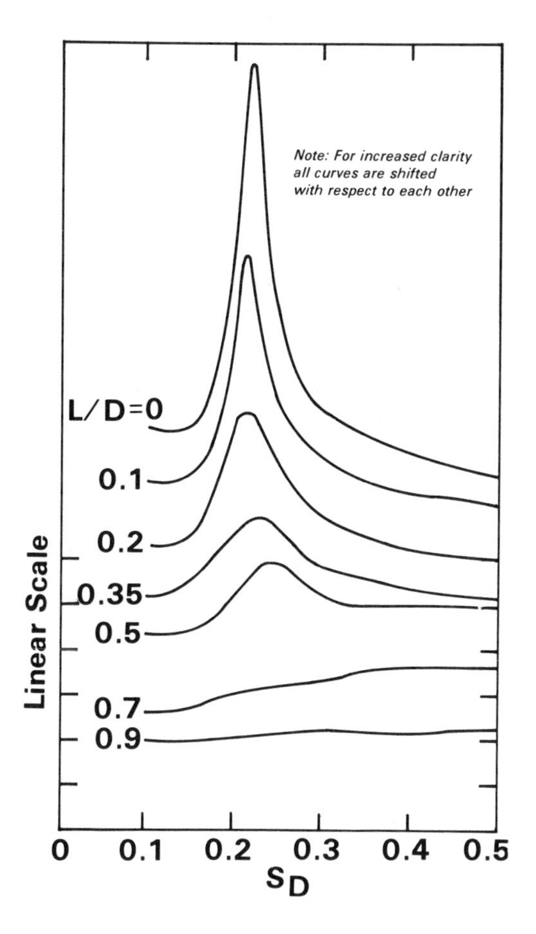

Fig. 15. Axial-velocity spectra for slotted cavities. Measured at the same location as for the solid cavities.

In summary, ventilated cavities can provide advantages over those with solid walls, but one has to pay attention to optimization of their geometry.

K. R. Cooper *(National Research Council, Canada)*

In some data I've seen, trapped vortices were used at the base of an axisymmetric body to reduce base drag. I wonder if you could comment on the effectiveness of this technique, using a disk displaced from the end of the body, or a circumferential groove machined in the boat-tailed surface near the base to lock the vortex in place, or some other similar technique.

W. A. Mair

I did some experiments with disks at the back of a body, either one disk or two disks; one could go on and have an entire grooved afterbody. I think the general conclusion was that this does work fairly well but not as well as a smooth boat-tail. That is why I didn't include it in my catalogue of drag reducing devices. It's interesting, perhaps, that it works nearly as well.

G. Heskestad *(Factory Mutual Research Corporation)*

Ten years ago I did some work on a drag reducing system for actual road vehicles that employed suction. Many people like to refer to it as boundary layer suction, but I don't think it should be called boundary layer suction. It is a principle which has been described in connection with other flow geometries, like the flow through a step expansion in a pipe going from a small diameter to a large one.* If you apply suction at the convex corner of the junction between the two diameters you get a great increase in the pressure recovery through the step. We wondered whether such a suction slot placed around the trailing perimeter of a constant area road vehicle would reduce its drag. We equipped a Volkswagen panel truck with a suction slot around its entire trailing perimeter and with some suction hardware. We ran deceleration experiments on the truck as a means of evaluating any drag reduction. Unfortunately, I never wrote up the work for publication, although an internal report can be made available. The drag coefficient, as measured through differences in deceleration, went from 0.54 down to 0.38. This is quite a large reduction even though there wasn't much reduction potential available, considering that the estimated base-pressure drag coefficient was on the order of 0.23. However, there is always the problem that power has to be provided for the suction flow. This made it less impressive, and instead of an actual reduction in drag coefficient of 0.16, we got 0.09. To get this

Heskestad, G. (1965), An Edge Suction Effect, AIAA Journal, October, 1965, pp. 1958-1961.
Heskestad, G. (1968), A Suction Scheme Applied to Flow Through Sudden Enlargement, Trans. ASME, Series D, Vol. 90, p. 541.
Heskestad, G. (1970), Further Experiments with Suction at a Sudden Enlargement in a Pipe, Trans. ASME, Series D, Vol. 92, p. 437.

number we had to make certain assumptions about the pumping power. The overall concept does seem worth keeping in mind as another item in the library of drag reducing techniques.

W. A. Mair

I think that is very interesting. I'm very glad to hear that you did get a net overall drag decrease, because that's always the worry – the power you save in relation to the power you use. I think it's very encouraging.

R. Sedney *(U.S. Army Ballistic Research Laboratories)*

It was very interesting to listen to your list of disappointments because it sounded like what I've been hearing in ballistics for 15 years or so. In fact the other thing that fascinated me was that just about every drag reducing technique that you mentioned has been considered in ballistics. Many of them are apt to reduce drag, but of course it is a matter of making them into a practical system. The best one in ballistics is the boat-tail. Base bleed has also been known for many years to reduce drag but it hasn't been made practical yet. In both cases, base bleed and the boat-tail, a great deal of cut-and-try is necessary. It is very easy to end up increasing the drag in some of the other cases you mentioned.

From what you said about base bleed, I wouldn't be convinced yet that it isn't a practical way to reduce drag. In the case of supersonic flow almost anything you do, as long as you don't inject too much momentum, will reduce drag. The largest gains, however, are obtained by injecting the mass in certain ways. For example, injection around the periphery of the base would be better than near the center. It isn't just the area over which the mass is introduced, but where it is introduced.

W. A. Mair

Yes, I'm sure that's true. I probably did dismiss base bleed a bit too hastily; there are various ways you can do it, certainly. The two investigations I referred to in my paper actually did leave out the case of injection around the edge of the base rather than further in. They both started with most of the base area being used for bleed, and then effectively blanked it in from the outside to change the bleed area. It is possible, of course, that some other arrangement might be more fruitful.

J. L. Stollery *(Cranfield Institute of Technology, England)*

Hearing Ray Sedney speak prompts me to ask him to confirm that it was tracer shells which showed base bleed to be so powerful in ballistics. Tracer shells are known to go far further than regular shells.

R. Sedney

Yes, at both subsonic and supersonic speeds the drag is lower with the tracer than without it.*

J. L. Stollery

To be more practical, we did some aerodynamic tests on a small delivery van; it was rather a blunt van, and we found a large difference in drag between road tests with the front windows open and shut. The window position changes the pressure distribution over the whole rear of the vehicle (Fig. 16). With the windows open a lot of air came into the van, and with leaky rear doors there was some base bleed at the rear, not intentional but unintentional. The figure shows the pressure distributions with windows open and closed. Notice that when the windows were closed the pressures were *negative,* but with open windows they were *positive.* Now, of course, there can be other explanations for this. For example, changes in the whole external flow field of the van may also have influenced the base pressure.

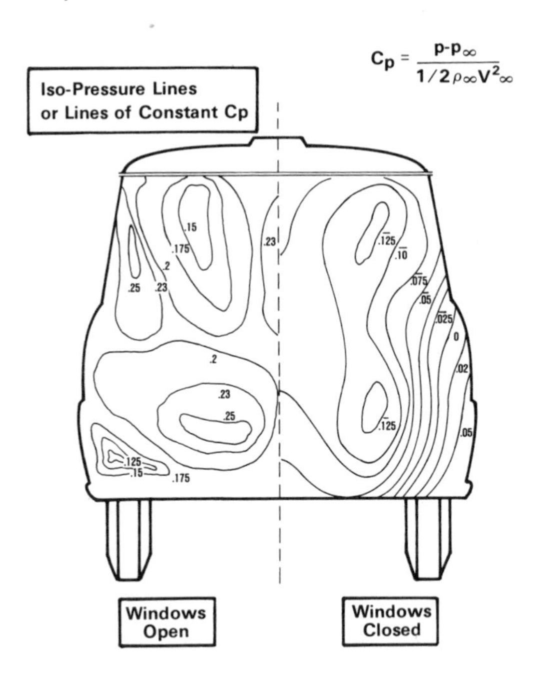

Fig. 16. Pressure distribution on the rear surface of a van, showing profound changes in pressure caused by opening of the van side windows.

*Sedney, R. (1966), *Review of Base Drag, Proceedings of AGARD Conference, AGARD-CP-10.*
Sedney, R. (1976), Aerodynamics of Base Combustion, Progress in Astronautics and Aeronautics, Vol. 40, AIAA and MIT Press.

H. H. Korst *(University of Illinois)*

You used C_q as a measure of "base bleed." One should interpret such a coefficient as representing the combined effects of both actual mass addition into the wake and that of the approaching boundary layer, the latter in the form of the equivalent bleed concept advanced by Sirieix, *et al.**

Some of the observed differences in the effectiveness of mass bleed are indeed due to variations in the relative importance of the boundary layer development upstream of the separation. In the case of high speed projectiles, the approaching boundary layer is not only thin, but it also expands sharply at the base so that its equivalent bleed contribution is small. Since such a boundary layer does not provide any effective "bleed" effect, mass bleed into the wake can provide some very significant benefits.

On the other hand, the momentum defect of the thick boundary layers found on road vehicles provides a strong initial bias level of the "bleed" effect. Now, in view of the diminishing returns of decreasing base drag by increasing bleed rate, one may expect that this bias level will reduce the effectiveness of any mass injection into the wake region.

W. A. Mair

I think this one needs some thought. It doesn't seem obvious that the boundary layer thickness would make a great deal of difference to the effectiveness, or otherwise, of base bleed. Perhaps it does, but I don't really see why it should. It does make some difference to the base pressure on a body of revolution, but not a very large one. You can double the boundary layer thickness and make only a comparatively small change in the base pressure coefficient. I don't really see why it should make much difference to the effect of base bleed, but perhaps we have to consider it. It's an involved subject.

W. A. Mair

Sirieix, M., Mirande, J. & Delery, J. (1966) "Experiences Fondamentales sur le Recollement Turbulent d'un Jet Supersonique," AGARD Conference, May 1966, Proceedings No. 4, Separated Flows, Part I, pp. 353-392.

SESSION III

Session Chairperson

A.M.O. SMITH

University of California at Los Angeles
Los Angeles, California

THE EFFECT OF BASE SLANT ON THE FLOW PATTERN AND DRAG OF THREE-DIMENSIONAL BODIES WITH BLUNT ENDS

T. MOREL

General Motors Research Laboratories, Warren, Michigan

ABSTRACT

The paper describes an experimental investigation concerning the effects of slanting the blunt base of three-dimensional bodies having either an axisymmetric or a rectangular cross section. It was found that base slant can have a very dramatic effect on body drag, particularly in a relatively narrow range of slant angles where the drag coefficient exhibits a large local maximum (overshoot).

Detailed study of the flow showed that the drag maximum is related to the existence of two very different separation patterns at the rear of either body. One pattern is similar to that found behind axisymmetric bodies with no base slant, and its main feature is the presence of a closed separation region adjacent to the base. The other pattern is highly three-dimensional with two streamwise vortices approximately parallel to the slanted surface, one at each side of the body. The drag coefficient maximum occurs in the slant-angle range where a changeover from one flow pattern to the other takes place. The observed phenomenon may be thought of as being associated with a broader category of "critical geometries," which is tentatively defined and discussed.

NOTATION

A	projected frontal area
AR	aspect ratio of the slanted surface; width/length
C_D	drag coefficient $\equiv$ drag force/$(\rho/2\ U^2A)$
C_L	lift coefficient $\equiv$ lift force/$(\rho/2\ U^2A)$

References pp. 216-217.

C_p	pressure coefficient $\equiv$ $(p - p_\infty) / \rho/2\ U^2)$
C_{pb}	base-pressure coefficient
D, d	body diameter
D_{eq}	equivalent diameter $\equiv$ $\sqrt{4\ \text{area}/\pi}$
H,h	height
ℓ	body dimension in the stream direction
L_S	length of a slanted surface
p	static pressure
p_∞	free-stream static pressure
Re_D	Reynolds number $\equiv$ UD/ν
r	radius
S_D	Strouhal number $\equiv$ fD/U
t	thickness
U	free-stream velocity
u'	rms turbulence intensity
W	width
x	streamwise coordinate
α	angle of inclination of a slanted base away from the normal to the stream direction
ν	kinematic viscosity
ρ	density
θ	momentum thickness

INTRODUCTION

One of the most important practical objectives of subsonic aerodynamics research is the determination of the overall pressure forces – in particular, of lift and drag. In the case of *streamlined bodies* the major interest is usually in lift, trends of which may be determined fairly well using inviscid flow models. Inviscid models can also be used to estimate the so-called "induced" drag of streamlined three-dimensional bodies with little or no separation. Other sources of pressure drag, e.g., separated regions or boundary layer displacement effects, are considerably more difficult to deal with but, for streamlined bodies, they usually are of only secondary importance. The *bluff-body* situation, where drag rather than lift is the major unknown, stands in sharp contrast to the successful prediction of streamlined-body flows. In this case there always are extensive regions of separation generating large drag, and there is no general theoretical model that can be used for its prediction.

There is a large amount of information in the literature about the pressure drag of various bluff bodies, and also about the mechanisms by which this drag is generated. Considering this wealth of information, one may inquire whether one can develop it into the form of building blocks, from which at least the *trends* of flow pattern and force variation with changes in body geometry may be predicted. The objective would be to progress to a point where educated guesses can be made to guide experimental programs. This, indeed, was one of the major ideas behind the program of this Symposium.

While an evolution towards a capability for better prediction of the flow behavior of bluff bodies is undoubtedly taking place, there will always be a need to check bluff-body flows experimentally (physically or numerically) through systematic perturbation of the geometrical parameters, in order to find the optimum configuration for any given purpose.

One of the reasons why a continued need for systematic experiments can be expected, is the existence of what can tentatively be called "critical geometries." The concept of critical geometries, as it stands now, is meant to deal with cases where drag exhibits a local maximum with respect to some geometrical parameter of a body shape. The fact that the variation of C_D with that parameter does not follow a monotonic trend creates a difficulty, since one is faced with two opposing trends and with some (unknown) critical value of the parameter at which the trends change.

A comprehensive survey of the bluff-body literature will reveal only relatively few carefully documented cases of such critical geometries. It is probable, that others may have been encountered but have not been reported for a variety of reasons, e.g., being unexpected and therefore suspect, or being of no interest to the purpose of the particular investigation and thus simply avoided. The few cases that we could identify in the bluff-body literature, and which are presented below, therefore show only some of the possible types of critical geometries that may exist.

Nash, Quincey & Callinan (1963) observed the existence of a local drag maximum on a two-dimensional body with a short splitter plate (Fig. 1). The change of trend at $\ell/h = 1$ was explained as being due to a change in flow regime, which was documented by shadow photographs showing differences in the separated shear layers in the two regimes.

References pp. 216-217.

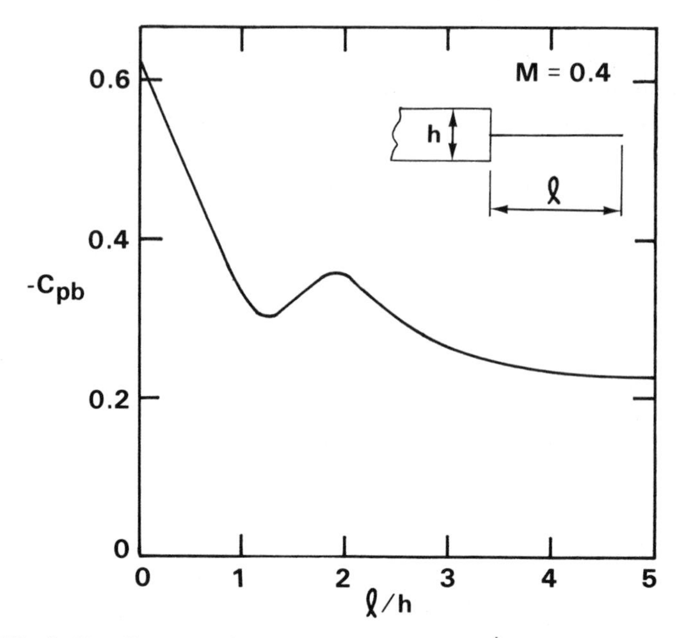

Fig. 1. Two-dimensional body with a splitter plate (Nash et al., 1963).

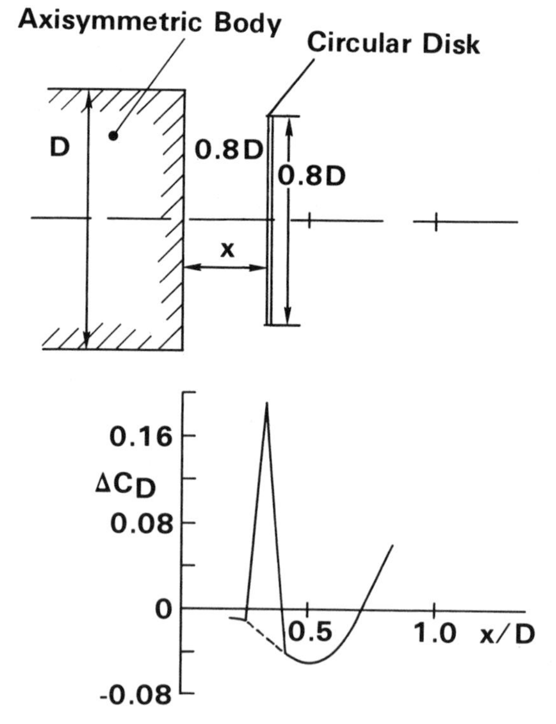

Fig. 2. Effect of a circular disk on afterbody drag, showing the occurrence of critical behavior at x/D = 0.3 (Mair, 1965).

Mair (1965) observed an unexpected drag increase while investigating the drag-reducing capability of circular disks placed concentrically in the near-wake of a blunt based body of revolution at $Re_D = 150{,}000$. He found the effect of the added disks beneficial in general, particularly so for a disk with d/D = 0.8 placed at about x/D = 0.5 (Fig. 2). However, when the disk that gave the largest drag reduction was moved from its optimum position towards the base, a new flow regime was formed which was highly unsteady and produced a large drag increase; it occurred in a relatively narrow range of x/D.

Critical geometries may also be found on smooth-shaped bodies. One such case was observed by Mair (1969) in his study of the effect of boat-tailing on the drag of axisymmetric bodies, at $Re_D = 460{,}000$. On three of eight boat-tails tested he noted critical boat-tail lengths at which the drag had a local maximum. Fig. 3 shows the drag curve of one of the three "critical" boat-tails, for which Mair noted: "For $L/D \approx 0.8$ the drag fluctuated considerably and separation was probably occurring close to the cut-off base. For $L/D < 0.65$ and again $L/D > 0.9$, there was no unsteadiness in the drag."

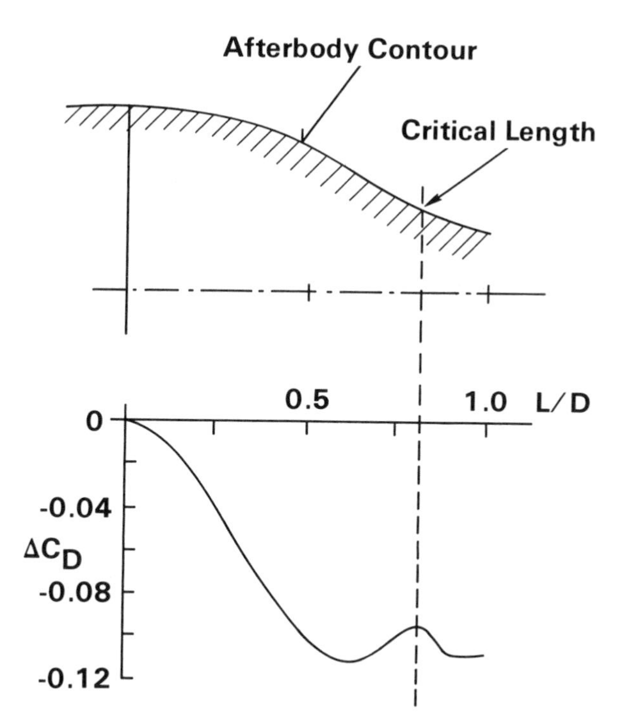

Fig. 3. Smooth-shaped afterbody exhibiting a critical behavior (Mair, 1969).

References pp. 216-217.

Bearman & Trueman (1972) studied one critical geometry uncovered earlier by Nakaguchi, Hashimoto & Muto (1968), with the objective of testing Bearman's vortex-street wake model. The tested bodies were rectangular 2-D bars with various thickness-to-height ratios (t/h = 0.2 - 1.2), at Re_h = 2-7 x 10^4 (Fig. 4). It was found that the drag had a local maximum at around t/h = 0.62, associated with a marked increase in the intensity of the regular vortex shedding. The magnitude of the increase was very large, with C_D going from about 2 at zero thickness to about 3 at the critical thickness. The whole drag increase was shown to be due to base-pressure decrease, the forebody drag being about constant for all t/h. Addition of a splitter plate behind the rectangular body entirely eliminated this behavior, showing that vortex shedding was playing a key role in the generation of the excess drag.

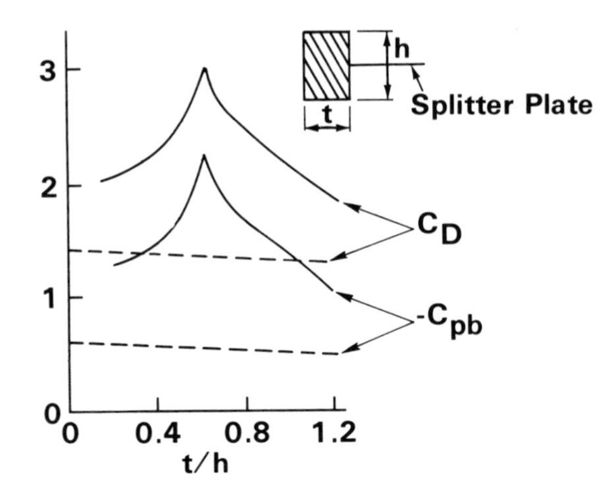

Fig. 4. Drag of 2-D rectangular bars of various thickness-to-height ratios (Bearman & Trueman, 1972).

One case of a critical geometry concerning road vehicles was documented by Janssen & Hucho (1974). Their experiment involved changes in the angle of the slanted portion of the roof of a car and their effect on drag. They observed that for a small range of roof angles (55-65°) the curve of overall drag exhibited a large overshoot (Fig. 5). They also observed a change in the extent of afterbody flow separation in the critical range, and documented it by the two sketches included in Fig. 5. They observed that the upper separation line was at the top of the slanted surface for $\alpha < 58°$, but that it moved to the bottom for $\alpha > 62°$; in between these two angles the point of separation was seen to pulse randomly from the top to the bottom and vice versa.

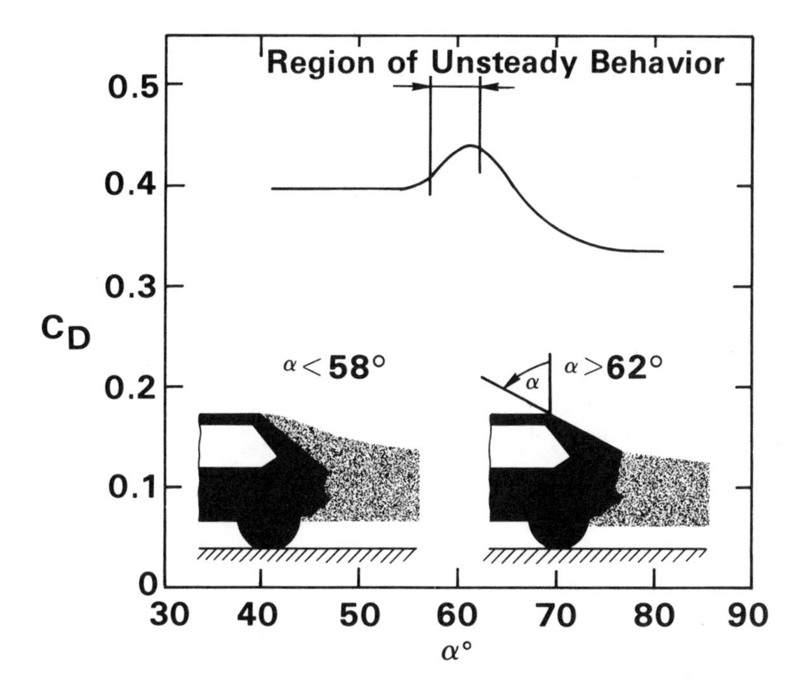

Fig. 5. Critical behavior of a hatchback automobile (Janssen & Hucho, 1974).

What do these five examples have in common? Can we consider them as one class? The definition of CG suggested earlier, which singles out the existence of a local drag maximum as the defining feature, is of a practical rather than fluid mechanical nature. It was intended to focus attention on geometries with local drag maxima mostly because they may be considered anomalous. They also have one thing in common: a small change of the critical geometrical parameter in either direction leads to a drag reduction. Since the necessary change can be in either direction, it should be feasible in many cases to avoid the high-drag region altogether. In fluid-mechanical terms the definition of CG is much more difficult. At this point the only common denominators seem to be the presence of flow separation, and the existence of two competing flow patterns. Excessive unsteadiness is often, but not always, present as well.

It is instructive to note that most of the mentioned examples of CG, if not all,

References pp. 216-217.

were encountered by chance with no prior expectation of the critical behavior. The fact that the drag overshoots are often unsuspected, coupled with the complex and difficult-to-analyze types of flow they tend to involve, means that one has to rely on systematic experimental studies for identification of critical geometries and for description of their behavior.

Finally, the above examples demonstrate that critical geometries can generate sizeable excess drag, and this should be of concern to designers of bluff-body hardware.

SLANTED BASE AS ONE PARTICULAR CRITICAL GEOMETRY

Of the critical geometries mentioned, the one that attracted our attention the most was the last one, concerning the effect of base slant on body drag. The observed drag variation with slant angle was of such a magnitude as to be of considerable practical interest. Consequently, it was decided to study this problem in detail, in order to search for the limits of its operation and for the maximum drag overshoot that may be expected. In addition, it was hoped that some light could be shed on the mechanisms involved in the generation of the excess drag, thus providing guidance for efforts to control or eliminate it.

The work of Janssen & Hucho (1974) provided two early clues to what was happening. First, they reported that near the critical angle the roof separation line moved from the top of the slanted surface to its bottom edge. Second, they stated that the increase in C_D was "due to strong edge vortices with a correspondingly large induced drag." Upon study of these observations we concluded that the drag overshoot found on the hatchback cars was most likely a consequence of a major change in the near-wake flow pattern. The conjecture was that below 58° the time-mean separation pattern was quasi-axisymmetric (or "closed"), with the external flow passing over a closed recirculating region adjacent to the base, in much the same manner as in the well-known cases of two-dimensional and axisymmetric base flows (Fig. 6a). Beyond 62° the picture was conjectured to have been very different; secondary flows, driven by pressure differentials from the sides of the body onto the slanted surface, rolled up upon separation from the side edges into two streamwise vortices extending into the wake. The resulting separation pattern was, thus, strongly three-dimensional and could be characterized as "open" (Fig. 6b). This type of separation is well known from delta-wing experiments. In terms of mean streamlines near the body surface, the quasi-axisymmetric pattern is one where the surface streamlines upstream of the base are approximately aligned with those expected in a potential flow, while in the 3-D case they may be substantially inclined away from the potential-flow streamlines.

The streamwise vortices appearing in the 3-D case may be expected to interact with the flow away from the edges and increase the tendency for reattachment in the

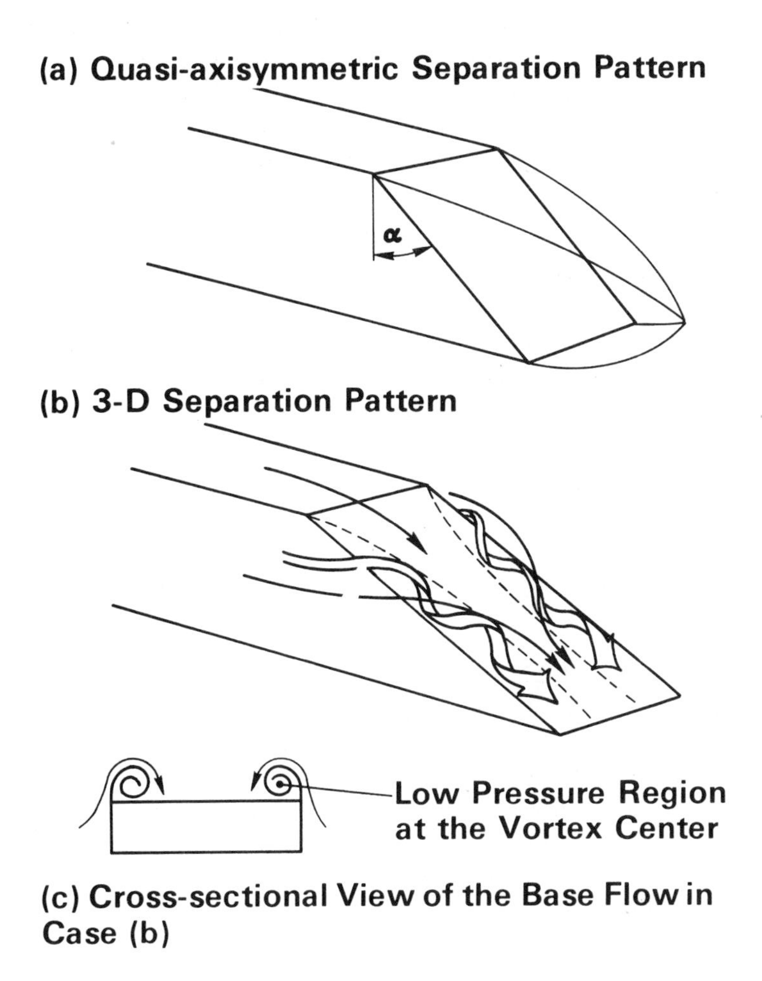

Fig. 6 Two types of separated-flow pattern on a slanted base.

central portion of the slanted surface. At first glance it may appear that such induced attachment should be beneficial from the drag point of view. However, it must be realized that the attachment need not mean that the wall flow is decelerating to the degree expected from 2-D thinking. Rather, the attached flow pattern is maintained by a supply of fluid from the sides of the body, which passes over the longitudinal vortices onto the slanted base, relieving the pressure rise along the central portion of the inclined surface. Near the side edges of this surface the wall pressure may be expected to be much lower than in the center, due to the proximity of the rolled-up streamwise vortices which may have very low pressure inside their cores. The contribution to drag which comes from these low-pressure regions is a manifestation

References pp. 216-217.

of the energy which is being continuously supplied into the edge-vortices. There is a contrast between the large separated region of almost uniform low pressure on the one hand, and the discrete vortices and localized very-low pressure regions on the other. Thus, from the drag point of view, the relative merits of the quasi-axisymmetric and 3-D separation are not immediately obvious and can be resolved only by experiments.

From a practical point of view, the critical-angle range should be avoided in situations where low drag is a high-priority design objective. However, since it may not always be possible to avoid this critical-angle range, one wonders whether small modifications to the body shape may be sufficient to suppress the drag overshoot. The answer to this question should be facilitated by a better understanding of the phenomena involved, and this was one of the main objectives of the investigation that is reported here.

SLENDER AXISYMMETRIC CYLINDER WITH SLANTED BASE

The details of the critical behavior of a slanted base, the amount of drag overshoot and the value of the critical angle may all be expected to depend on a number of parameters, for example: aspect ratio of the slanted surface, geometry of the afterbody, geometry of the forebody, ground proximity and Reynolds number. It is evident that no single experiment can answer all questions, and so one must seek a compromise. The most important question for this investigation was whether the conjecture about the flow patterns was correct, and so the first experiment was designed specifically so that all changes in the flow could be ascribed to the base slant angle alone.

The study concerned itself with the case where a base is preceded by a finite forebody. Because variation of the slant angle causes changes in the overall length/diameter ratio (ℓ/D), a rather long and slender body configuration with $\ell/D = 9$ was used to minimize this effect. The large ℓ/D also reduced the influence of the afterbody on the forebody flow. The body cross-section was chosen to be circular to facilitate the fabrication of exchangeable afterbodies, which were made from cylindrical pieces sliced off at the required angle; this resulted in the slanted surfaces being ellipses elongated in the streamwise direction.

Experimental Arrangement and Measured Quantities – The tests were conducted in a wind tunnel of the open-return, suck-down type, with test-section dimensions of 500 x 700 mm, maximum velocity of 55 m/sec, and free-stream turbulence intensity u'/U less than 0.1%. The main part of the model was a circular cylinder 254 mm long, with a diameter of 38 mm, suspended in the center of the test-section on six piano wires 0.35 mm in diameter (Fig. 7). The wires were inclined at 45° to the stream direction in order to suppress vortex shedding and to reduce flow disturbances; Reynolds number based on the wire diameter and on the nominal speed of the experiment was 875. The cylinder was preceded by a slender ogival nose-piece 90 mm long, and its downstream end was capped by interchangeable afterbodies with slanted bases having slant angles from 0° to 70°. Altogether, 17 different slant angles were tested; these were chosen to be at equal increments of 5° (excluding 65°), plus three others with angles of 46.5°, 47.25° and 48.5°.

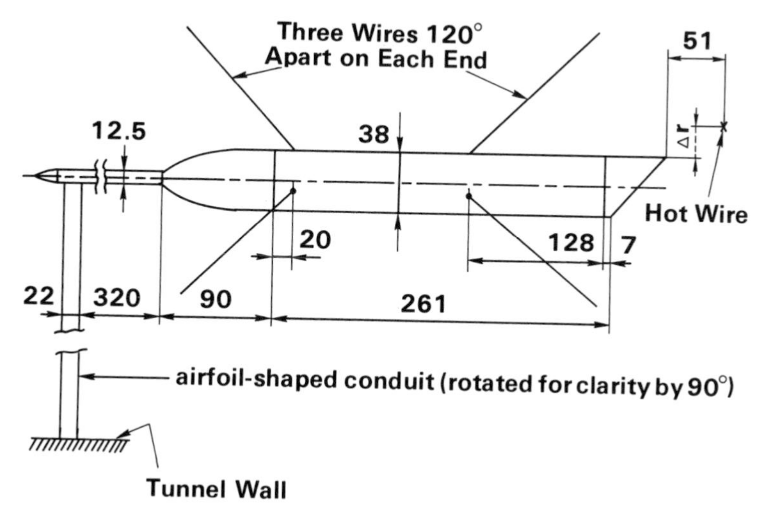

Fig. 7. Experimental arrangement for pressure and hot-wire measurements. Dimensions are in mm.

All tests were run at a nominal speed of 38 m/sec. The boundary layer was turbulent with a momentum thickness, at a point 0.5 diameter ahead of the base, of 0.70 mm ($\theta/D = 0.018$). Reynolds number based on the cylinder diameter was Re_D = 94,000. The suspension system was not entirely rigid on account of suspension-wire elasticity, but its natural frequency was more than an order of magnitude below the forcing (shedding) frequency. No oscillation of the cylinder was observed during runs. Blockage ratio of the cylinder in the wind-tunnel was about 0.3%.

A DISA 55M hot-wire system with one linearized single-wire miniature probe was used to investigate the unsteady properties of the near wake. The hot wire was positioned in the plane-of-symmetry of the base at a distance of 51 mm (1.33 diameters) downstream from the trailing edge. Its lateral position was always on the trailing-edge side of the wake (Fig. 7), at a point where u'/U, when high-pass filtered below a frequency corresponding to $S_D = f\,D/U = 1.0$, had a value of 0.5%. This was done to minimize the masking effect of the shear-layer turbulence on the periodic signal (S_D = 0.2 - 0.4). It was reasoned that if the rms of the filtered signal were maintained constant, the hot wire would always be located in about the same

References pp. 216-217.

position with respect to the shear-layer edge. As will be discussed later, increasing the base angle tended to turn the wake towards the trailing-edge side, and so this lateral location was not the same for all cases. The hot-wire signal was used to determine whether any regular motions were present in the wake, and to establish their frequency and intensity. To this end the signal was displayed on an oscilloscope for visual inspection, and also simultaneously processed by a narrow-band spectrum analyzer.

Surface static pressure was measured at three locations along a horizontal line placed midway down the slanted surface (Fig. 8). All three taps ended in one cavity, from which a single pressure line was led through the nose extension and a streamlined conduit to a Statham PM5TC pressure transducer. When making pressure measurements from any one tap, the other two were taped over, leaving only the one of interest open.

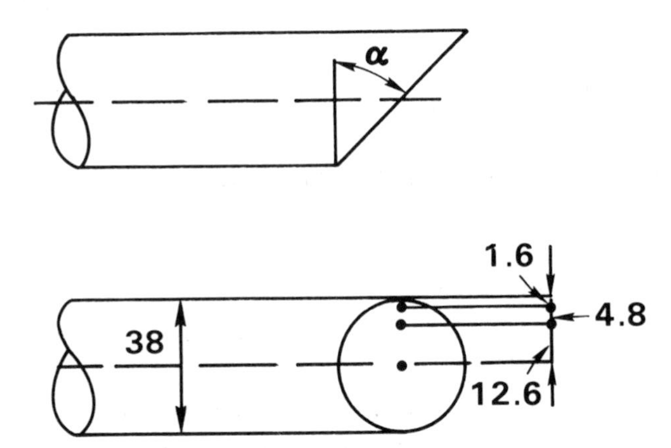

Fig. 8. Location of the three pressure taps on the slanted base. Dimensions are in mm.

Results – *Base Pressure* – The base-pressure development with base slant proved to be very dramatic, and confirmed the expected existence of two widely different base-flow patterns; the changeover occurred on this particular geometry at about 47.5° (Fig. 9). Below 47.5° (Regime I) the base-pressure coefficient was negative and fairly uniform across the base, with a minimum at the centerline tap (#1) and a maximum at the edge tap (#3). The difference between the coefficients at these two taps was about 0.02 at 0°, decreasing to 0.01 at 47.25°. The overall level of base pressure was seen to decrease slightly with increasing α up to 25°, and then was relatively constant. Above 47.5° (Regime II) the base-pressure distribution exhibited sharp pressure gradients across the base, with a maximum pressure at the centerline and a minimum somewhere in-board of the side edges.

The changeover from one flow pattern to another was very abrupt. The original set of end-pieces, manufactured in a sequence with 5° intervals, showed that the pattern

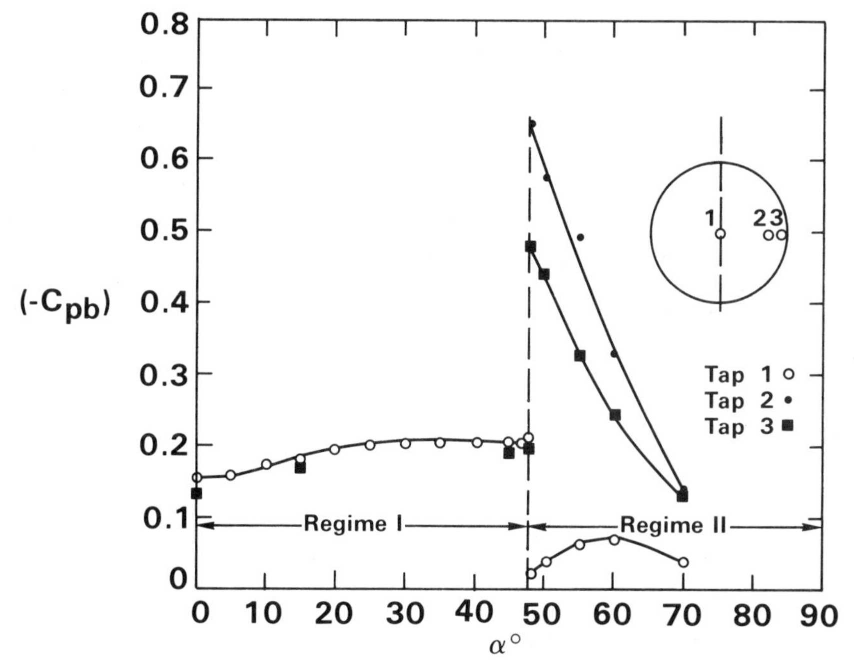

Fig. 9. Base-pressure variation with slant angle.

changeover occurred somewhere between 45° and 50°. To narrow down the interval where the change occurred, two other end-pieces were made with α = 46.5° and 48°. It was expected that one of these two end-pieces would be near enough to the changeover (within 1°) that an intermittent flow behavior would be observed, with the two flow patterns alternating more or less randomly. However, both pieces exhibited very stable flow patterns, 46.5° having the quasi-axisymmetric type (Regime I) and 48° having the 3-D type (Regime II). An even greater surprise was that the flow patterns were so stable that even large perturbations, (e.g., inclination of the whole model at $\pm$5° to the free-stream in either of two perpendicular planes) did not change their basic character. This indicated that each regime was inherently stable and that the changeover must be very abrupt indeed.

As a final step, an afterbody with α = 47.25° was manufactured, allowing the critical angle to be located within 0.375°. This configuration again showed a stable flow pattern, which was of the quasi-axisymmetric type; thus, the changeover was occurring somewhere between 47.25° and 48°. Inclination of this last model, in its plane of symmetry, with respect to the free stream did produce the expected changeover to the 3-D flow pattern. To achieve this it was necessary to incline the model tail downwards by about 5°, making the base angle about 52°. Once the changeover occurred the new pattern was completely stable. Returning the cylinder to the horizontal position did not bring back the original flow pattern; to achieve that, an inclination in the opposite sense by about 1° was required. This means that the flow had a hysteretical behavior over a range of some 6°.

References pp. 216-217.

One consequence of increasing the slant angle is an increasingly large projection of the base surface in the transverse direction. Since the base is subjected to a low pressure, there is a force on it acting in that direction. This also means that the flow experiences an equal but opposite force, making it deflect towards the trailing edge of the base; this deflection was actually observed using the hot wire, located as previously described 1.33 diameters downstream, at a lateral position Δr where the filtered $u'/U = 0.5\%$. The locus of Δr as a function of slant angle is shown in Fig. 10.

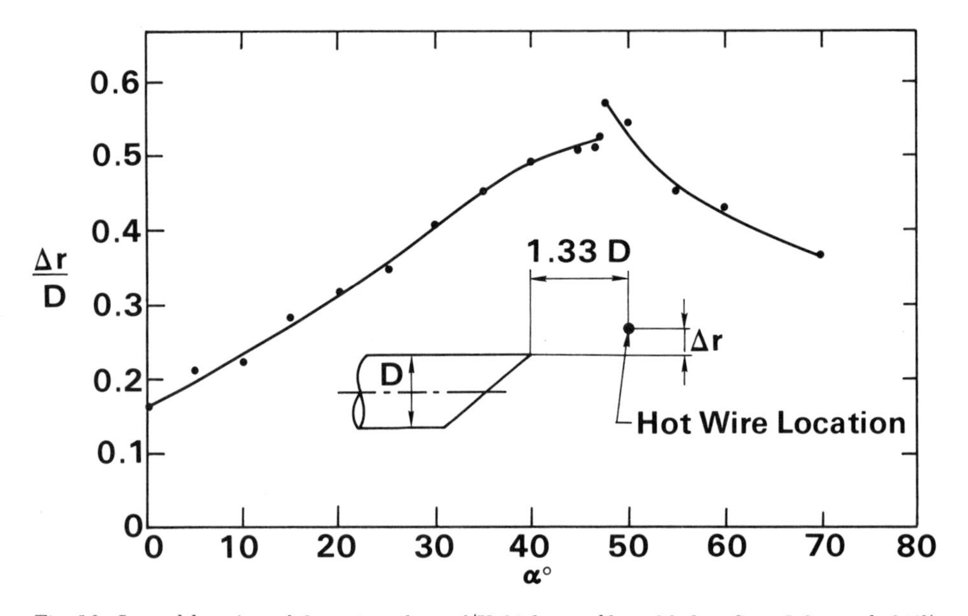

Fig. 10. Lateral location of the point where u'/U, high-pass filtered below $S_D = 1.0$, equals 0.5%.

Drag Force – The axial force on the body (drag) was measured using a different experimental arrangement. Following a suggestion of W. W. Willmarth of the University of Michigan, a pendulum method was used. The cylinder, minus the upstream pressure-line conduit, was suspended on four vertical 0.127 mm wires (two in front and two at the rear). The force on the body was calculated from the observed downstream displacement of the body caused by the air flow, as measured by a micropositioned transit. Good data repeatability ($\pm 1\%$ of C_D) was obtained using this simple arrangement.

The data obtained by the pendulum technique are presented in Fig. 11. In Regime I the C_D rose gradually with α, increasing by 0.06 from $\alpha = 0°$ to $\alpha = 45°$. The magnitude of this change is close to that observed in the base pressure (Fig. 9), as may have been expected since the base pressure, which is the only source of the drag increase, was fairly uniform across the base. Changeover to Regime II, at the critical angle, produced a dramatic increase in C_D. The total body drag increased by an impressive jump of $\Delta C_D = 0.325$ to more than double its value! Further increase in α gave a sharp drop in C_D, leading to values lower than the zero-slant drag beyond $\alpha \approx 69°$.

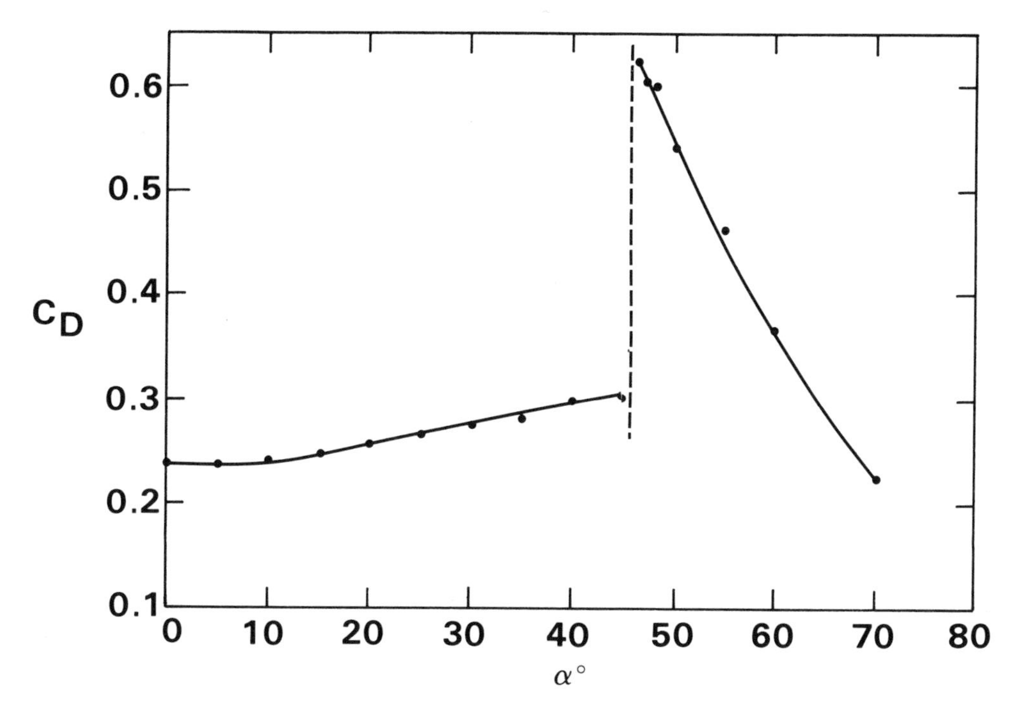

Fig. 11. Drag-coefficient variation with slant angle.

It is interesting to note that the jump from one regime to another occurred in this case between $\alpha = 45°$ and 46.5°, rather than between 47.25° and 48° as in the case of the pressure runs, a difference of some 2°. In both cases the body was placed in the test section in the same position and was aligned with respect to the same reference points. The only differences were in the suspending wires and in the pressure-line conduit, and this was apparently sufficient to produce the difference in the critical angle.

Wake Periodicity – Slanting the base from its original, vertical, position greatly increased the shedding intensity. In the vertical (0°) position, the near-wake hot-wire signal showed only weak signs of periodicity (Fig. 12), but the spectrum had a definite peak, and the hot-wire signal itself began to show a progressively better-defined periodicity. The spectral peak was highest and sharpest at $\alpha = 20°$; beyond this angle it decreased and broadened at the same time (Fig. 13b). The total intensity of the spectral "hump" increased up to about $\alpha = 30°$ (Fig. 14), where the hot-wire trace gave a very clear indication of a significant degree of regularity.

References pp. 216-217.

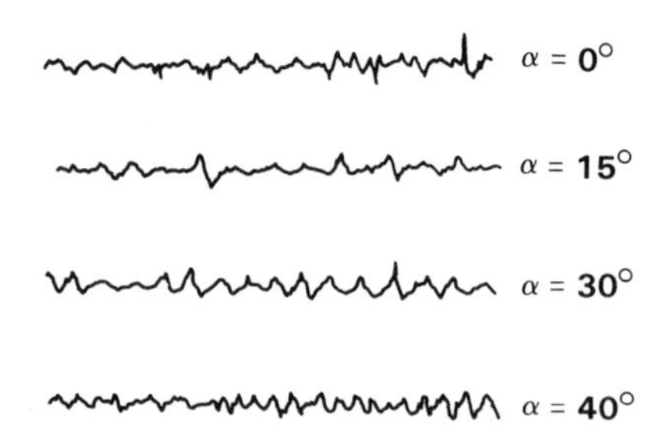

Fig. 12. Velocity traces for several different slant angles.

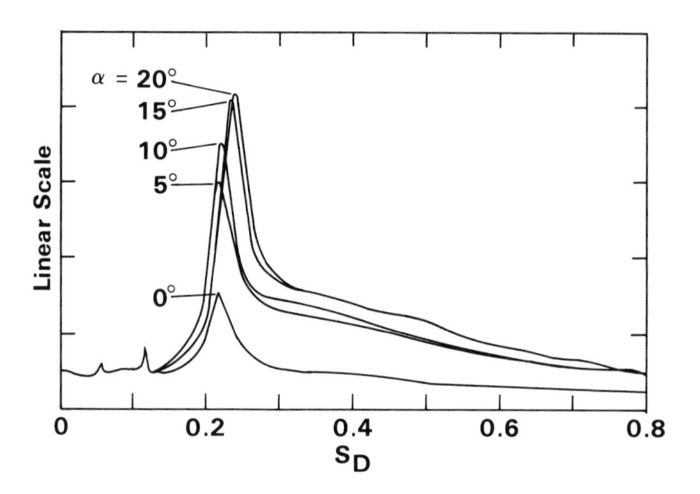

Fig. 13a. Velocity spectra at α = 0°-20°. Linear vertical scale.

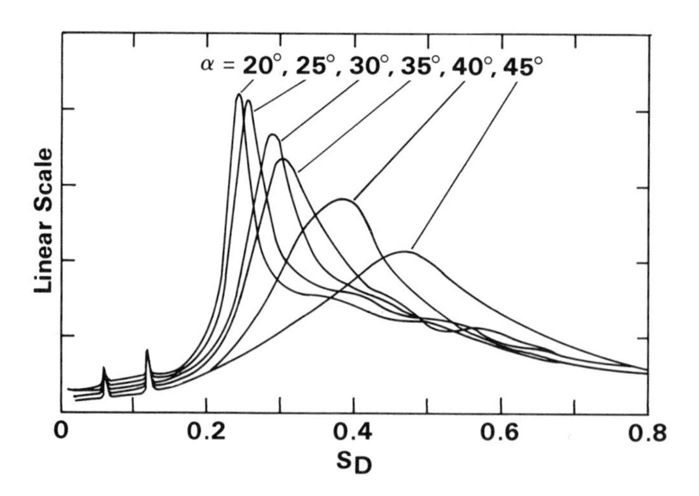

Fig. 13b. Velocity spectra at α = 20°-45°. Linear vertical scale.

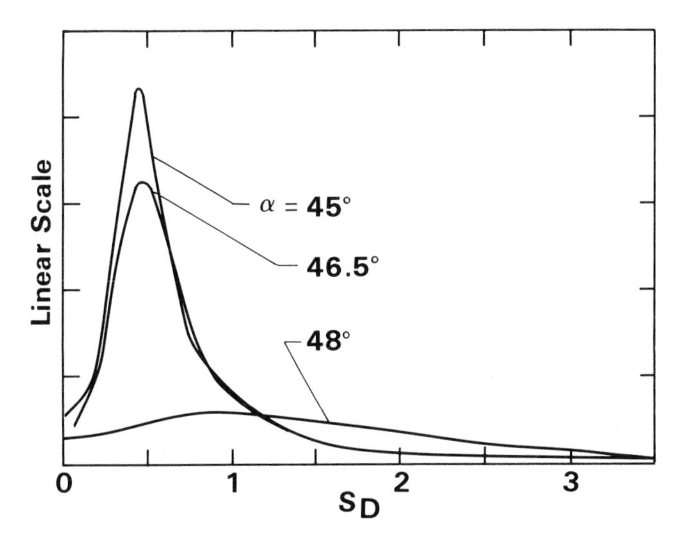

Fig. 13c Velocity spectra near the critical slant angle, α = 45°-48°.Linear vertical scale; note the change in the horizontal scale and in the slant angle increments.

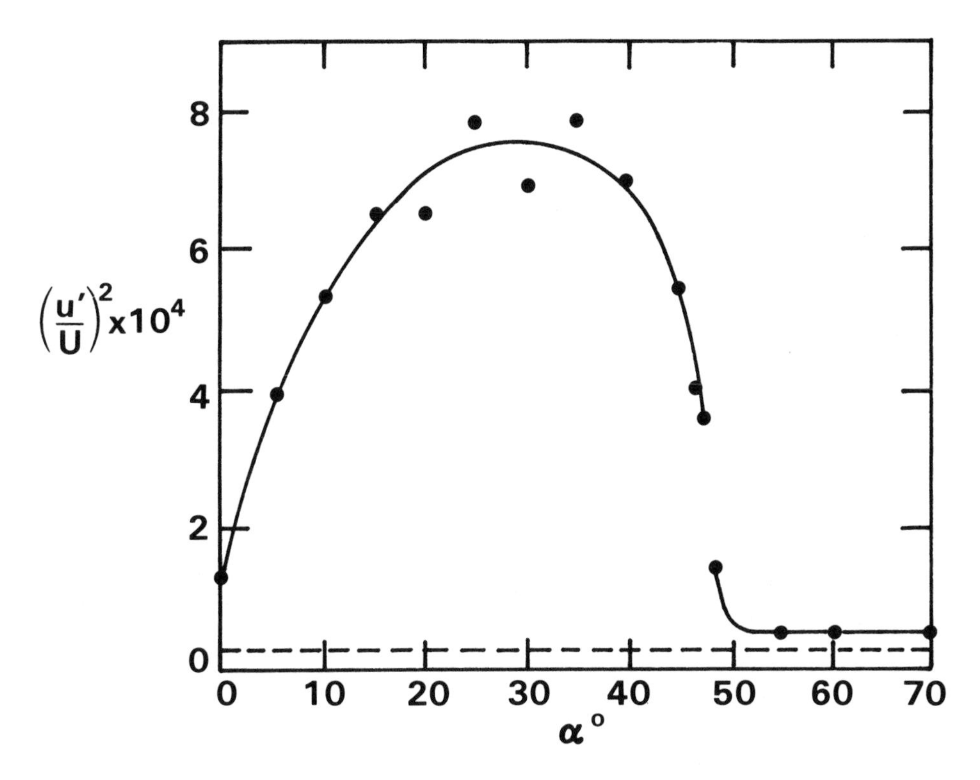

Fig. 14. Fluctuation intensity $(u'/U)^2$ vs. slant angle. –Total intensity, ----intensity of the filtered signal.

References pp. 216-217.

The Strouhal number associated with the mid-point of the spectral humps exhibited a monotonic increase with slant angle, and in the vicinity of the changeover point had a magnitude more than double that at 0° (Fig. 15). Above the critical angle, after the change in base-flow regime occurred, no regular shedding was visually detectable in the hot-wire signal, though the spectra for α = 48 - 55° still showed a spectral hump which was very broad and was centered around $S_D \approx 1.25$ (Fig. 13c). The magnitude of this hump decreased rapidly with α, and at 55° it was almost undetectable.

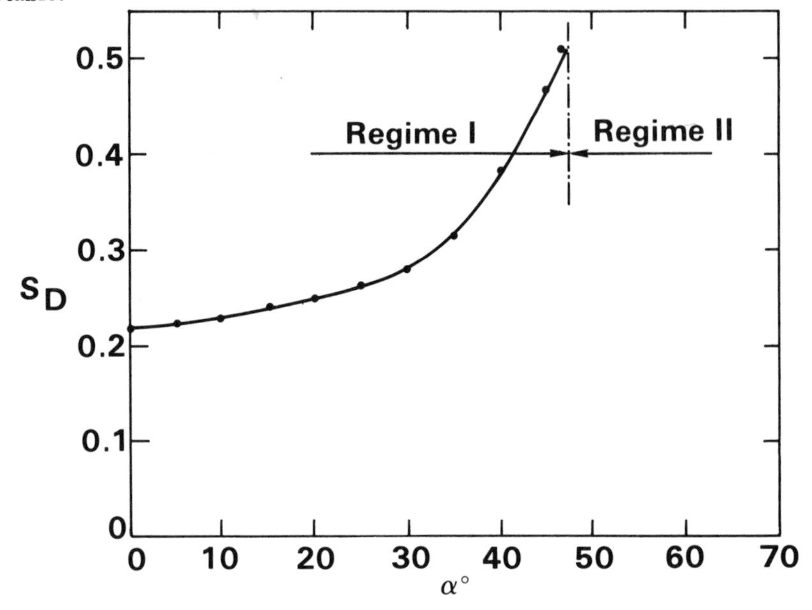

Fig. 15. Effect of slant angle on the shedding frequency (Strouhal number).

Discussion – The change in the slanted-surface pressure distribution at the critical angle is consistent with the proposed model of the flow behavior (Fig. 6). The pressure uniformity across the base in Regime I would be expected for the quasi-axisymmetric flow pattern. Similarly, the pressure distribution in Regime II agrees with what was expected; the large increase in centerline pressure is an indication of a tendency towards flow reattachment in the mid-portion of the base, while the very low pressure near the edge is due to the close proximity of a concentrated streamwise vortex rolling up off the side edge, in a fashion similar to that sketched in Fig. 6b.

Prior to the experiment it was expected that the flow pattern changeover would take place over a range of α, within which an intermittent switching back and forth between the two regimes would be taking place, as observed in the Janssen & Hucho experiment. The actual data proved to be a surprise as they showed the two patterns to be very stable. In fact, this observed stability was so strong that both patterns could exist and be completely stable on the same body (α =47.25°).

The basic experiment on the circular cylinder served well its purpose. It demonstrated the existence of two dramatically different separation patterns, one of which, the highly three-dimensional one, has the capability of large drag generation. At the critical slant angle the drag difference between the two patterns can be very large, giving this flow phenomenon and its understanding great practical importance. Another interesting effect of slant angle is its role in promoting near-wake periodicity. At slant angles around $\alpha = 30°$ the periodic motions in the near wake are much more pronounced in intensity than at 0°, and may cause significant periodic loading on the rear portion of a body. This effect is worth further study, as it may possibly couple with the natural frequency of the body's suspension system. (On an automobile traveling at typical highway speeds, the shedding frequency should be in the range of 3-6 Hz.)

VEHICLE-LIKE BODY NEAR AND AWAY FROM THE GROUND

The first experiment left unanswered two major questions: Are these results transferable to vehicle-like shapes, and what is the effect of ground proximity on the critical angle and drag overshoot? To provide the answers, a further experiment was conducted. The tested body had vehicle-like proportions (Fig. 16), with dimensions of 900 x 405 x 270 mm, i.e., length:width:height ratio of 3.33:1.50:1.00. The body was again equipped with interchangeable bases of various slant angles, but this time the area of the slanted surface (as well as its aspect ratio) was kept constant. This meant that the vortex-sheet-generating edge was of the same length ($L_S = H$, see Fig. 16) at all slant angles. The volume of the body was also kept constant, both to preserve the interior volume (a factor in road-vehicle applications) and to keep constant the effective length L_{eff} = volume/(W x H). This required lengthening the body's bottom surface at non-zero slant angles; the maximum elongation needed, at $\alpha = 45°$, was 67.5 mm or about seven percent.

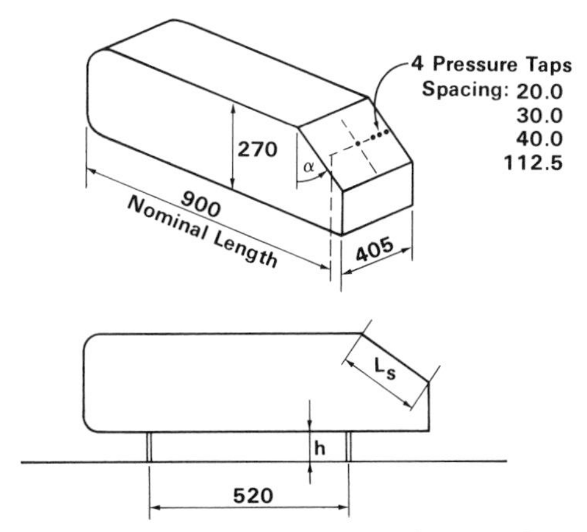

Fig. 16. Experimental arrangement for the vehicle-like body. Dimensions are in mm.

References pp. 216-217.

Two different heights above the ground were used, one to simulate a free-stream situation and one typical of a road vehicle. In terms of an equivalent diameter ($D_{eq} = \sqrt{4\,WH/\pi} = 373$ mm), the dimensionless ground clearances were $h/D_{eq} = 0.82$ and 0.12, respectively. The former value was chosen to be beyond the height where ground proximity starts to affect the force characteristics of this type of body.

Experimental Arrangement and Measured Quantities – The second experiment was conducted in a vehicle-aerodynamic wind tunnel used for partial-scale-model testing (Mason, Beebe & Schenkel, 1973). This facility is of the closed-path type, and is equipped with a free-standing test-section floor, which acts as a ground plane. Test-section dimensions are 1400 x 4570 mm; it has a maximum velocity of 256 km/h and a turbulence level $u'/U = 0.3\%$.

The model was suspended on two cylindrical stilts 15.9 mm in diameter. There were 19 different afterbodies, with slant angles between 0° and 85° (note that 90° is identical to 0°), which were spaced 10° apart in regions of lesser importance, but 5° and 2.5° apart in regions where rapid changes were occurring.

All tests were run at a nominal speed of 58 m/sec, giving a Reynolds number based on equivalent diameter of 1.4×10^6, or about 15 times that of the first experiment. Stilt Reynolds number was about 60,000, i.e., subcritical. Blockage ratio of the model at zero yaw was 1.7%.

Surface static pressure was measured at four locations across a line midway down the slanted surface (Fig. 16). In addition, all six forces and moments were measured using a strain-gage force balance to which the support stilts were anchored. Flow visualization using surface tufts, smoke and a vortex paddle-wheel were used to identify the near-wake flow patterns and to document their nature.

Results – *Drag and Lift* – The drag data was corrected for the tare drag of the stilts (both drag and lift coefficients are based on the body frontal area). In the data taken away from the ground (Fig. 17) one can observe trends very similar to those found in the first experiment; however, the critical angle shifted from 47° to 60°. In Regime I drag showed a slight increase ($\Delta C_D = 0.03$) with slant angle, followed by a sudden increase of $\Delta C_D = 0.16$ at the critical angle. Beyond this angle C_D dropped very rapidly, reaching a minimum at about $\alpha = 82°$, which was 0.21 below the maximum and about 0.03 below the zero-slant value. The drag decrease from $\alpha = 90°$ to 82° is apparently a type of boat-tail effect which, below 82°, is overpowered by the growing strength of the edge vortices. Lift data (Fig. 18) have an even more pronounced variation with α than the drag data. The fact that C_L was positive at all non-zero slant angles was expected, as the slant exposes an upward projection of the low-pressure afterbody surface. However, the magnitude of the lift coefficient variation was unexpected. Near the critical angle the maximum C_L exceeded greatly the maximum C_D, making the model a lifting body in that range (i.e., $C_L/C_D > 1$).

Drag-coefficient trends near the ground (Fig. 19) were quite similar to the free-stream trends, but the magnitude of C_D variations with the slant angle was smaller. The critical angle was the same as for the free-stream case. At $\alpha = 0°$ the drag coefficient was slightly higher than in the free stream, by 0.025. The drag rise in Regime I was less pronounced and the jump to Regime II at the critical angle was

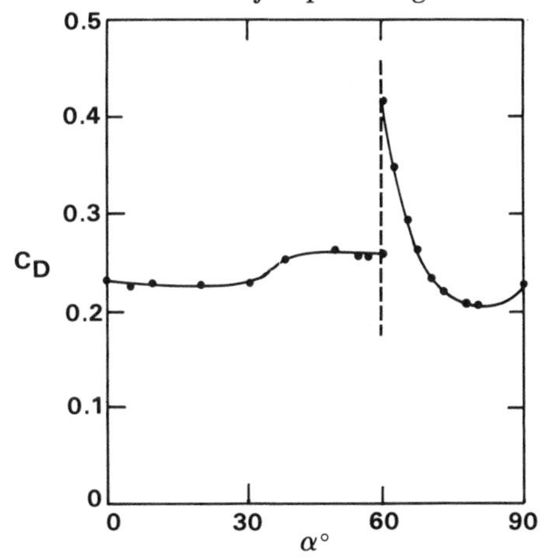

Fig. 17. Drag-coefficient variation with slant angle, $h/D_{eq} = 0.82$ (free stream). Data corrected for drag of stilts.

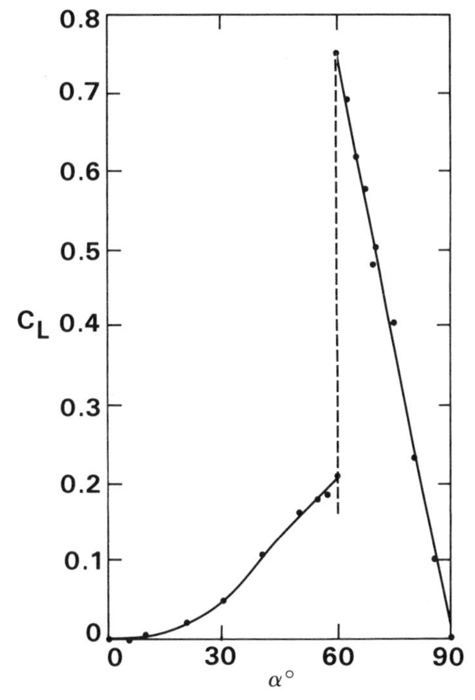

Fig. 18. Lift-coefficient variation with slant angle, $h/D_{eq} = 0.82$ (free stream).

References pp. 216-217.

only $\Delta C_D = 0.11$. The subsequent decrease to the minimum at $\alpha = 80°$ was $\Delta C_D = 0.155$. The trends of data of Janssen & Hucho (1974), drawn in a broken line for comparison, are seen to be in general agreement with the present results. The lift coefficient (Fig. 20) behaved very similarly to the free-stream case; in fact, Fig. 18 and Fig. 20 differ for the most part only by a constant vertical shift of $\Delta C_L = 0.2$.

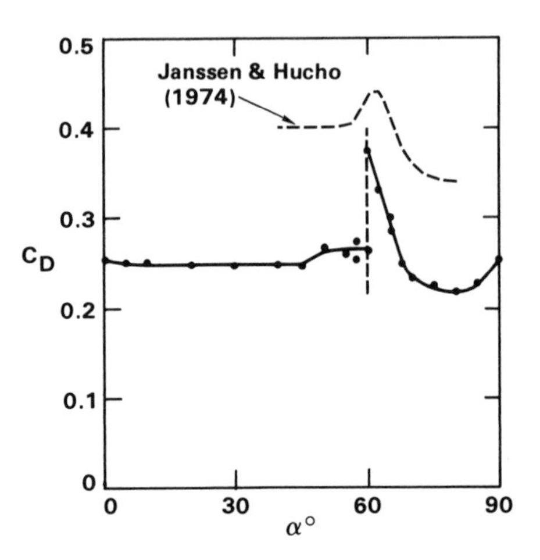

Fig. 19. Drag-coefficient variation with slant angle, $h/D_{eq} = 0.12$. Data corrected for drag of stilts.

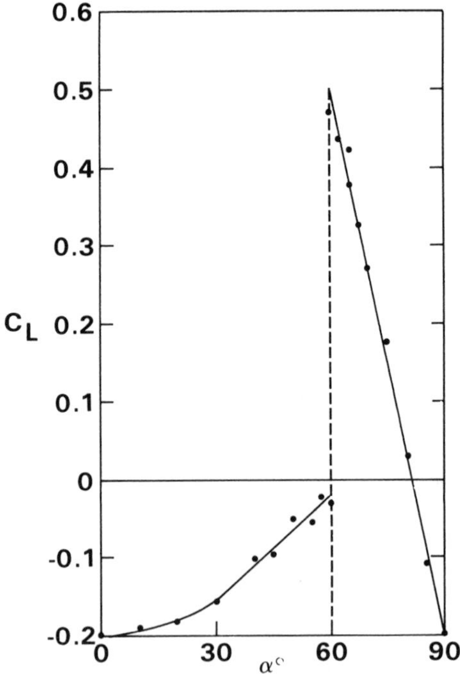

Fig. 20. Lift-coefficient variation with the slant angle, $h/D_{eq} = 0.12$.

Flow-Visualization – Flow visualization, through three different techniques, was used to study the nature of the two flow patterns near the critical angle. Flow at the body surface was explored by the use of tufts. Below the critical angle the slanted-surface flow was seen to be entirely separated, and the tufts on the sides of the body were quite well aligned with the free-stream direction. At $\alpha = 65°$, above α_{crit}, the upper third of the slanted surface showed signs of flow separation, but the lower two-thirds clearly had an attached flow. The tufts along the sides of the body were inclined upwards towards the slanted side edges due to cross flows from the sides onto the slanted base, indicating a presence of very low pressures near the side edges.

The wake was also probed with a vortex-paddle-wheel (25 mm in diameter) in an effort to compare the strength of streamwise vortex motions and to map their

(a)

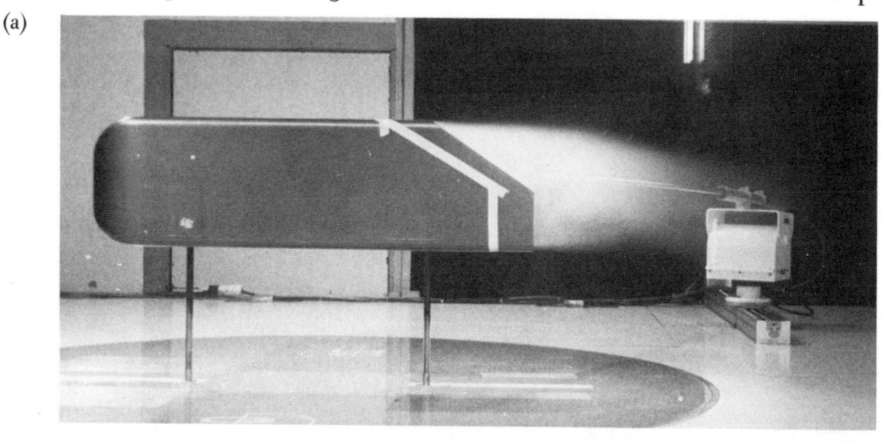

(b)

Fig. 21. Smoke visualization of the near wake, (a) $\alpha = 57.5°$, (b) $\alpha = 62.5°$. Point of smoke release is located on the body centerline, and in both photographs is in about the same position with respect to the base.

References pp. 216-217.

trajectory. The difference between the two patterns was very striking; while there was almost no vortex activity below α_{crit}, above it (α = 62.5°) two very strong vortices were observed. These vortices were approximately aligned with the two side edges, deviating from them by some 5-10° upward (away from the surface) and inward (towards the centerline).

Smoke visualization of the near wake displayed clearly the very different nature of the two flow regimes. In Regime I (α = 57.5°, Fig. 21a) the smoke showed a typical separated bubble, much like what one would expect behind a vertical base. In Regime II (α = 62.5°, Fig. 21b) one could see a strong downwash and no sign of separation. In both photographs the smoke probe was located on the body centerline in the same position with respect to the base. Moving the smoke probe sideways away from the plane of symmetry produced little change in the smoke behavior in Regime I, but in Regime II one could observe strong swirling motions near the body edges.

Slanted-Surface Pressure – Slanted-surface pressure data also showed the difference in the two regimes – a relatively uniform pressure in Regime I and a pronounced lateral variation in Regime II. This fact is well illustrated in Fig. 22, showing the two possible bi-stable pressure patterns recorded on the same (critical) afterbody placed near the ground. The two pressure curves complement the flow visualization pictures of Fig. 21, by showing the types of pressure distribution associated with the two smoke photographs. A comparison of the two figures also makes one realize that conventional visualization, using smoke released along the centerline, tends to encourage two-dimensional thinking, which may be quite misleading. In the case at hand, the smoke pictures suggest that the attached flow in Regime II means an increase in the base pressure as compared to Regime I. The pressure data show, however, that just the opposite is true.

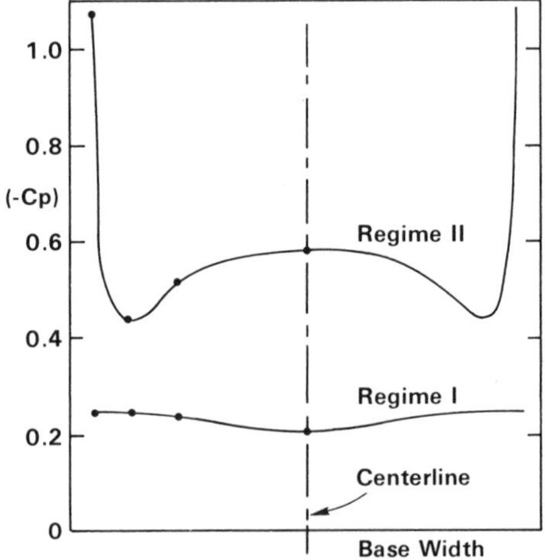

Fig. 22. Pressure distributions in Regimes I and II, measured on the same base with a 60° slant angle, at h/D_{eq} = 0.12. Pressure taps were placed on a horizontal line across the base, located midway down the base.

Discussion – The results obtained for the vehicle-like body were in general agreement with the cylinder experiment, but there were important differences, the most striking of which was in the value of α_{crit}, which was 60° instead of 47°. There were several important geometrical differences between the two models, e.g., a circular compared to a rectangular cross-section, but we believe that the main reason for the critical-angle shift was the change in the aspect ratio of the slanted surface. At 47° the base of the cylinder was an ellipse with AR = 0.68, while the rectangular body had AR = 1.5. The direction in which the critical angle changed with AR is consistent with the following reasoning. As AR increases, the effect of the side-edge vortices on the overall flow patterns gets progressively smaller, as only a relatively small portion of the base is exposed to them. Thus, the attached flow patterns, supported by inflow from the body sides which passes over the edge vortices, breaks down at larger and larger values of α (i.e., when the slanted surface is more closely aligned with the external flow). Also, as the relative importance of the edge vortices decreases with increasing AR, the value of the drag overshoot may be expected to decrease and this effect is probably at least partly responsible for the smaller jump in C_D at the critical angle, $\Delta C_D = 0.16$ for the rectangular body vs. 0.325 for the cylinder.

The effect of ground clearance (for the chosen clearance of $h/D_{eq} = 0.12$) was less important. The value of the critical angle appeared to be unchanged, but the variation in C_D with α was somewhat reduced. The shape of the lift-coefficient curve was affected only very little by ground proximity, which was a surprise, as lift would be expected to be very sensitive to ground clearance.

An important difference between the two bodies was in the stability of the two regimes. The vehicle-like body had less-stable flow in the vicinity of the critical angle; especially near the ground the force and pressure coefficients were not exactly repeatable. At $\alpha = 60°$ one could obtain either of the two flow patterns by yawing the model sufficiently far away from, and then returning to, the head-on position.

An interesting observation may be made concerning the slope of the almost-linear lift curve in Regime II. The slope of this curve $(-\partial C_L/\partial \alpha)$ is about 1.43 for the free-stream case and about 1.33 for the case near the ground. These numbers may be compared to the data for rectangular flat plates inclined with respect to a free stream. Interpolating the data presented by Lamar (1974), one obtains for AR = 1.5 a lift coefficient slope of about 2.35. Thus, the bluff and bulky block at zero angle of attack, but with a slanted rear surface, had fully 60% of the lift slope of a flat plate of the same size as the slanted surface. Furthermore, the lift-to-drag ratio for the bluff body at α_{crit} was only four percent less than for the flat plate.

CONCLUSIONS

1. Slanting the rear surface of a simple bluff-ended body away from the normal to the streamwise direction affects strongly its drag. The drag curve develops a

References pp. 216-217.

pronounced local overshoot (a bulge) over a relatively narrow range of slant angles, with a drag maximum at a "critical angle."

2. Base-pressure data and flow visualization show clearly that there are two very different separation patterns that can exist in the near wake. Below the critical slant angle a quasi-axisymmetric separation pattern is established (Regime I), while above it there is a 3-D separation pattern with two pronounced streamwise vortices forming on the side edges of the slanted surface (Regime II).

3. In Regime II, base slant tends to make a bluff-body a lifting body. For the vehicle-like body studied here, the slope $(-\partial C_L/\partial \alpha)$ of the lift coefficient curve (based on the area of the slanted surface) was 1.43 when the body was away from the ground, and 1.33 near the ground. The former value constitutes a full 60% of the lift slope measured for rectangular flat plates of the same aspect ratio as that of the slanted surface. The lift force has a maximum at the critical slant angle, where it may reach magnitudes larger than the drag force. In the case of road vehicles this means that a very sizeable rear-axle lift can be generated by improper roof (or window) slant.

4. The circular-cylinder data show that in Regime I base slant produces an increase in base-flow unsteadiness, dominated by quasi-periodic vortex shedding at a Strouhal number of 0.2 - 0.4.The increased intensity means increased transverse periodic loads, and they may couple with the natural frequencies of a body suspension system. The only data available are those obtained away from ground; therefore, a study of ground effect on shedding is needed in order to assess its importance to road vehicles.

5. The effect of the aspect ratio (AR) of a slanted surface on the critical phenomenon appears to be two-fold: First, increasing AR increases the value of the critical slant angle; second, AR has a strong effect on the magnitude of the drag overshoot.

6. The tests have indicated that ground proximity does not greatly affect the value of α_{crit}, but it tends to reduce the magnitude of the C_D variation with α. The trend of C_L with α (the lift slope)was unaffected by ground proximity, but the absolute value of C_L was lowered by a constant amount.

REFERENCES

Nash, J. F., Quincey, V. G. & Callinan, J. (1963), Experiments on Two-Dimensional Base Flow at Subsonic and Transonic Speeds, ARC R&M No. 3427, p. 17.

Bearman, P. W. & Trueman, D. M. (1972), An Investigation of the Flow Around Rectangular Cylinders, The Aeronautical Quarterly, Vol. XXIII, pp. 229-237.

Janssen, L. J. & Hucho, W.-H. (1974), Aerodynamische Formoptimierung der Typen VW-Golf und VW-Scirocco, **Kolloquium ueber Industrie-aerodynamik,** *Aachen, Part 3, pp. 46-69.*

Lamar, J. E. (1974), Extension of Leading Edge Suction Analogy to Wings with Separated Flow Around the Side Edges at Subsonic Speeds, NASA TR R-428, pp. 48 and 53-57.

Mair, W. A. (1965), The Effect of a Rear-Mounted Disk on the Drag of a Blunt-Based Body of

Revolution, The Aeronautical Quarterly, Vol. XVI, pp. 350-360.
Mair, W. A. (1969), Reduction of Base Drag by Boat-Tailed Afterbodies in Low-Speed Flow, The Aeronautical Quarterly, Vol. XX, pp. 307-320.
Mason, W. T., Jr., Beebe, P. S. & Schenkel, F. K. (1973), An Aerodynamic Test Facility for Scale-Model Automobiles, SAE Paper No. 730238.
Nakaguchi, H., Hashimoto, K. & Muto, S. (1968), An Experimental Study on Aerodynamic Drag of Rectangular Cylinders, Journal of the Japan Soc. of Aeronautical and Space Sciences, Vol. 16, pp. 1-5.

DISCUSSION

Prepared Discussion

J. E. Hackett *(Lockheed-Georgia Company)*

This is on behalf of Professor Rainbird. He would like to point out the similarity between the flows which have just been described and the flow on the upswept fuselage of an airplane. Both the Canadians and Lockheed use transport aircraft in which the aft end of the fuselage is upswept. This is pretty much the upside down of what we have just been talking about with respect to cars. Regime II is generally found in these cases because the upsweep is fairly modest, although, of course, there is a downwash field from the wings which tends to add to the cross-flow effects. Flows of this type have been studied quite widely by Peake.* The surface flow is very complex, and measurements of pressure distributions need to be very detailed indeed in order to make sense of what's going on. Another point that Professor Rainbird asked me to make was the fact that there has been some work done in Japan on semisubmerged ships** which showed very similar types of flows at the aft end of the ships. Studies have been done of the vorticity in the wake behind them, which is a valuable aid to interpretation.

My personal comments . . . I feel that it should be possible to get a much better understanding of how the drag occurs on a car by doing good wake traverses. I think it would be most useful to separate out the losses in total pressure; for example, to identify the viscous losses as opposed to the vortex losses - I don't like the term induced drag in this context. I think it would be possible to measure the cross flows (we've been doing some of this at Lockheed) and identify separately the vortex drag. Separate these two items and, hopefully, at some stage it ought to be possible to establish the drag floor for automobiles, because there does not have to be any vortex drag if cars are brought right down to the ground and the pressure distributions are suitably arranged. You don't have to have vortex drag! Practical considerations are

---

**Peake, D. J. (1976), Controlled and Uncontrolled Flow Separation in Three Dimensions, National Aeronautical Establishment, Ottawa, Aeronautical Report LR591.*

***Tanaka, H. (1974), A Study of Resistance of Shallow-Running Flat Submerged Bodies, Journal of the Society of Naval Architects of Japan, Vol. 136.*

what force you to have clearance under a car; it seems to me that it would be very helpful to car people to know how far they are from an absolute minimum as a function of that clearance.

Prepared Discussion

G. W. Carr *(Motor Industry Research Association, England)*

I'd like to make a prepared presentation which is related to what Tom Morel has just spoken about and also, to some extent, to what Eric Maskell was telling us yesterday about the types of flows we might expect to find. I shall be summarizing some work that has already been published by MIRA* and which was carried out in our 2 x 1 m wind tunnel with a simplified variable-geometry car model at a Reynolds number of about 3-1/2 million. This model had rounded edges to minimize flow separation on the forward end. The upper rear-end shape was varied, as shown in Fig. 23, so as to determine separately the effects of back-light rake angle and trunk length. Variations consisted first of taking a notchback shape and reducing the length of the trunk progressively; then for the notchback with the longest trunk, varying the back-light angle progressively until we had a complete fastback shape. We measured all six components of aerodynamic force over a yaw-angle range from 0° to 20°, and we also took photographs of surface oil-film patterns. From the surface patterns at zero yaw angle, three fundamentally different rear-end flow regimes were identified, with correspondingly different force characteristics.

G. W. Carr

First of all we have the *fastback* type of flow regime, which I think has already been covered pretty well by Tom Morel with his base slant (Fig. 24a). In this case, the

*Carr, G. W. (1974), *Influence of Rear Body Shape on the Aerodynamic Characteristics of Saloon Cars*, MIRA Report No. 1974/2.

flow sweeps up over the side edges, and large vortices disappear off downstream, with very strong downwash in the middle. The flow over the roof continues on downwards over the back-light, with no more than a short separation at the roof trailing edge, if this happens to be sharp. Sufficient downwash can be induced to maintain attached flow over a back-light inclined at up to 35 degrees to the horizontal. As one increases the slant of the back-light, making it more vertical, the flow ceases to follow this type of regime and separates straight off from the roof, again as has already been described. This type of flow regime is characteristic of the *squareback* car design.

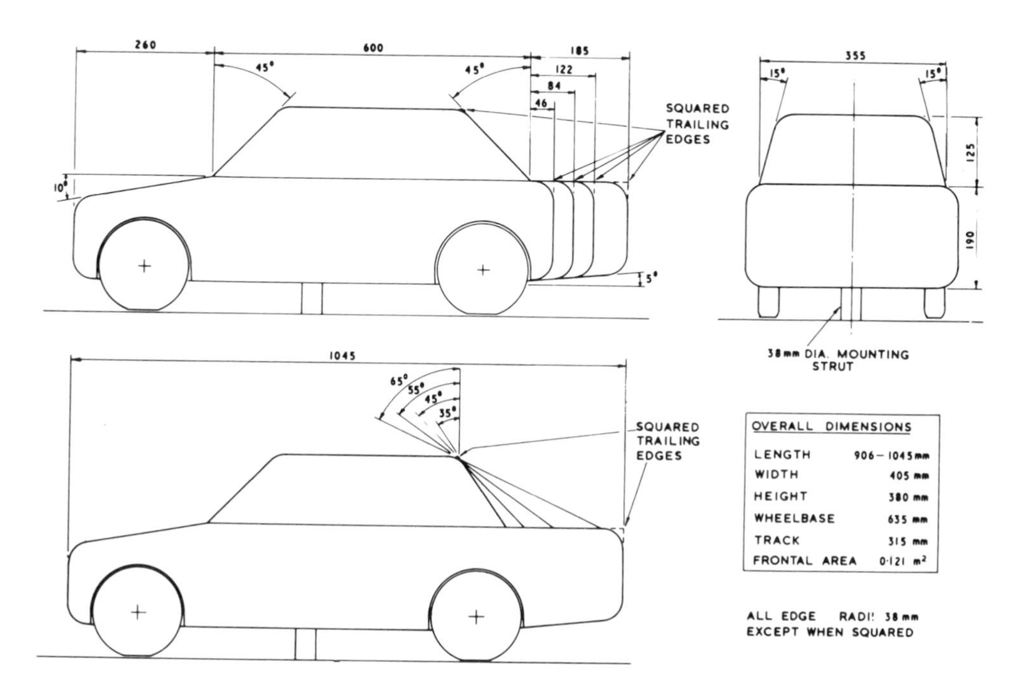

Fig. 23. Front and side elevations of model configurations tested, showing trunk length and back-light rake angle variations.

The third type of flow regime, which hasn't been touched upon during the course of this Symposium, occurs with the typical *notchback* rear end (Fig. 24b). In this case the flow leaving the roof separates and, if the length of the trunk is sufficient, vortex-induced downwash brings the flow leaving the rear edge of the roof down to the trailing edge of the trunk. In this case, however, there is a single vortex with a curved transverse axis, as sketched in Fig. 24(b), and the vortex is trapped in the step between the back-light and the trunk. The result is a type of flow similar to that found behind a rearward facing step.

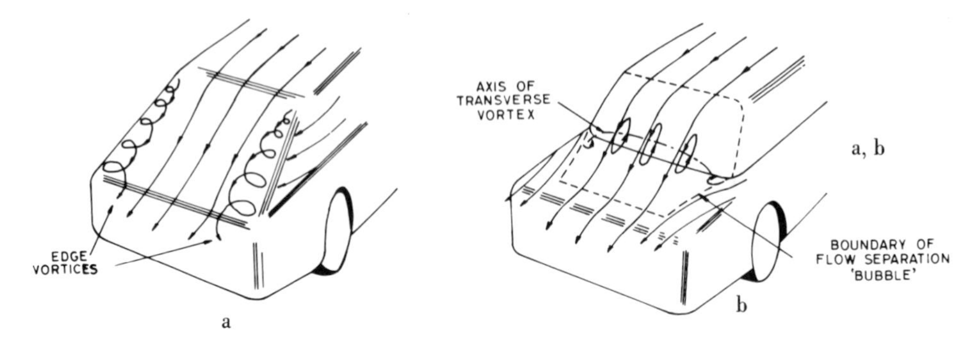

Fig. 24. Schematic representation of the possible flow patterns.

a) Sketch showing edge-vortex system on "fastback" rear end.

b) Sketch showing transverse vortex system within separation "bubble" on "notchback" rear end.

So we have three possible flow regimes on the back end. We have the notchback case with the trapped vortex, the fastback with the two edge vortices, or the squareback where the flow separates and just continues on, more or less parallel, without any reattachment. These three types of flow regime are naturally associated with quite distinctive force characteristics, which are summarized in Fig. 25. The most important differences are found in the values of the lift and side force acting at the rear-axle position, represented by the coefficients C_{LR} and C_{SR}, respectively. As far as drag is concerned, which after all is our main concern these days, the highest drag is in general given by the notchback shape. The lowest drag is given by the fastback in general, but it is only marginally better than the squareback with total separation.

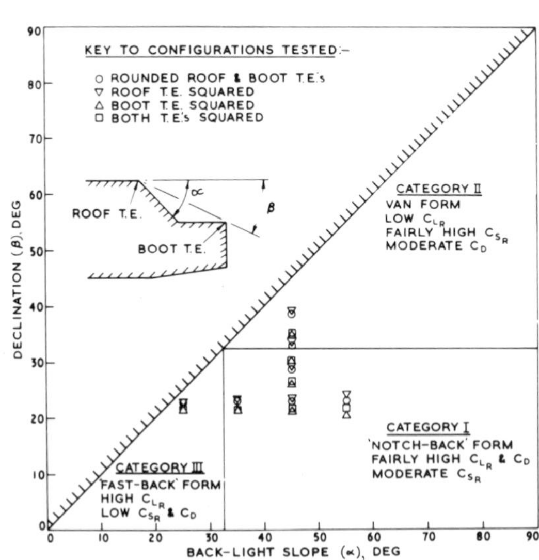

Fig. 25. Diagram showing categories of rear body shape in terms of declination (β) of trunk trailing edge and back-light slope (α).

D. J. Maull *(Cambridge University, England)*

Can I ask what the conjectured form of the instability is that you picked up with the hot wires? Is it a flapping of the longitudinal vortices in space, or is it really shedding from the body?

T. Morel

In Regime I, where shedding occurs, we have no evidence of longitudinal vortices – visualization indicates that the external flow passes over the afterbody parallel to the main stream. So we believe it really is some type of shedding. The peakiness of the spectra and the value of the Strouhal number certainly support this. What we are not quite sure about is what kind of shedding it is.

T. Morel

One possibility is that this shedding is similar to that observed by Calvert* behind inclined disks. Calvert studied a thin circular disk inclined to the flow, and he did some flow visualization behind it. He could see quite clearly elongated stretched vortex rings being shed by the inclined disk. I would expect similar things here though we did not try to visualize it, because the boundary layer on the body was turbulent and quite thick, about 0.16 times the base diameter. I don't really know what it looks like. We just assume that it's the same as Calvert's.

S. J. Kline *(Stanford University)*

Just a comment on the unsteadiness and criticality you observed. It is, of course, not new that unsteadiness appears when you get transition from one separated pattern to another, or from an unseparated to a separated pattern. We have had the suspicion that this is sometimes associated with an inability of the flow to sit steadily in either pattern, so it tends to go back and forth. What seems to be a little different

Calvert, J. R. (1967), Experiments on the Flow Past an Inclined Disk, Journal of Fluid Mechanics, vol. 29, pp. 691-703.

here, from some of the earlier cases we have seen, is the sharpness with which you go from one pattern to the other to get this awful jump in C_D and C_L. This is somewhat surprising.

Also, it seemed to me that there were several comments today and yesterday about not understanding how any of these base flow mechanisms work. We looked into it about 20 years ago and concluded that there *is* a way of seeing how the base pressure is set; you have to look at how much pumping out of the separated region you're doing, which must equal the backflow in. For example, vortices tend to scavenge the separated region more strongly than certain other flow mechanisms. In any event, some average continuity is set, even in the unsteady flow, and certainly in the steady flow, and the mean base pressure has to be at such a value that continuity is satisfied – that is, the backflow equals the outflow. That somehow sets the pressure, and that in turn reflects as base drag. Many of the things we're hearing at this Symposium, although they may be varied, and in some cases the details are very hard to understand, are fairly simply explained just by that continuity argument.

T. Morel

In the first part of your comment you talked about the critical behavior. There clearly are at least two different types. One is represented by a disk behind an axisymmetric body, where the drag comes from the unsteadiness; the pattern cannot decide which way to go, there probably is a lot of pumping out of that flow region, and drag is increased. The other one, the one I talked about, is quite different. There is no unsteadiness involved at all. There are two completely stable patterns, as you would appreciate if you would try to measure drag by suspending the body from four thin wires as we did. In the critical region the body was very steady; there didn't seem to be any significant variation in drag. Also, the base pressure measurements showed that the pressure was quite steady. The pressure can obviously never be entirely steady in a separated flow, but it didn't indicate changes jumping from one regime to another.

As for your second comment, I agree with you. What sets the base pressure is the equilibrium which has to be set up in the near wake along the lines that you mentioned. It is unfortunate that while we can perhaps understand *why* the base pressure is set to the observed values, there seems to be very little we can *do* about it in the case of three-dimensional bodies, as Professor Mair and others clearly indicated yesterday.

R. Sedney *(U.S. Army Ballistic Research Laboratories)*

You didn't show any smoke pictures for the case where the vehicle-like model was close to the ground. I'm curious about what they looked like, and specifically I wonder whether any horseshoe vortices were set up in the boundary layer on the ground plane. If there were, I would then expect them to interact with the vortices that you were describing. What did the smoke pictures show in that case?

T. Morel

We did not take any smoke pictures near the ground, but judging from the force data we expect them to be similar to those taken away from ground. In any case, the boundary layer on the plate was relatively thin and I would not expect it to form a large horseshoe vortex around the whole body, only two small ones around the stilts.

S. Saunders *(Rudkin-Wiley Corporation)*

I'd like to ask about the spectra of the oncoming stream. In particular, I suspect that turbulence intensity and scale can broaden the range where you see critical geometries.

T. Morel

Yes, I would expect that the range may be broadened, and we want to look into it. Most of the energy of that 0.1 percent turbulence level in the smaller wind tunnel is below 50 Hz.

R. T. Jones *(NASA-Ames Research Center)*

One thing that is unsettling about this to airplane aerodynamicists is that bodies that have complete separation can have less drag than bodies in which the flow is partially attached. I don't know quite what to do about it. Evidently the cure we've always used with the airplane, namely just to avoid separation as much as possible, does not work here.

T. Morel

In the majority of cases separation is a drag producing thing. You want to avoid it. But when you see a slanted edge, you have to be cautious.

R. T. Jones

You've got the highest drag with a partially attached flow.

T. Morel

Yes. I believe the high drag is caused by the concentrated edge vortices, which you don't see in the smoke picture (this is how deceptive 2-dimensional visualization techniques can be), impressing very low pressure on a small part of the base. The surface pressure coefficient was on the order of minus 2 near the edge. Even though it acts on only a small piece of the base, it contributes very significantly to the drag.

Thus, one could perhaps say that partially attached flow is not good enough, but complete attachment should tend to eliminate the pressure drag entirely.

K. R. Cooper *(National Research Council, Canada)*

With reference to some of the work of Wickens that was mentioned earlier, he found that on upswept airplane fuselages which generate the same kind of vortices, Reynolds number had a large effect on how close the vortices sat relative to the surface. This had a huge effect on the surface pressures and, of course, this is the particular point you're talking about. Also the vortices were unsteady in that they seemed to move around. Putting strakes on a fuselage to fix the location of the separation point, so to speak, got rid of the unsteady forces. So you still had, perhaps, the high drag condition but you did not have unsteady forces.

W. A. Mair *(Cambridge University, England)*

I would like to draw attention to one sub-class of a critical geometry involving cavities, as for example a disk mounted in the wake of a round cylinder. As Tom Morel said, in that case there is a pulsation, or gulping, of air in and out of the cavity. It occurs to me that this is a dangerous situation which should be avoided in the tractor-trailer situation, where you have a cavity between the tractor and the trailer, and you could have gulping in and out, leading to a very high drag and high dynamic loads. I wonder if anybody has encountered it?

T. Morel

Yes, this is possible, although I recollect the comments of last night's banquet speaker, Bobby Allison. He said that when side windows were removed from racing cars, tremendous pressure fluctuations were experienced inside the cars but the drag, as judged by the top speed, was apparently unchanged.

W. W. Willmarth *(University of Michigan)*

With reference to the aerodynamicist, I would like to comment on what happens when you have a slanted back and the angle is increased till the flow attaches to the slanted surface. At this point the lift jumps upwards and the drag jumps up as well, so it's induced drag.

W. W. Willmarth

T. Morel

T. Morel

Let's define induced drag. If we define it as in finite airfoil theory with C_{Di} proportional to C_L^2, I would say it was not an induced drag. If you define induced drag as a drag which is *somehow* related to lift, then I agree.

W. W. Willmarth

I always thought induced drag was rotational kinetic energy in the Trefftz plane due to the trailing vortices.

T. Morel

It is a matter of the exponent. If what you mean by induced drag is Prandtl's theory of a drag proportional to the square of C_L, I would say no. There has been analytical work done at NASA, Langley by Lamar*, and elsewhere on sharp-edged bodies, say rectangular plates, inclined to the flow like finite span airfoils. For them, the component of drag associated with the side-edge vortex is proportional to lift with an exponent of about 1.5, because the lift coefficient goes like the square of the angle of incidence while the drag coefficient goes like the cube. So it's a matter of definition. This is something we haven't studied sufficiently yet. We haven't defined what we mean by induced drag**; thus, it is difficult to decide what is and what is not an induced drag.

R. T. Jones

One of the most difficult things to overcome is a theory. The induced drag theory establishes a necessary, essential, relationship between lift and drag, involving of course the dimensional proportions of the shape of the wing. The examples that I showed yesterday indicate that there is not an essential or necessary relation in those situations between the lift and the drag. Experimentally we observe that very often when you produce lift you get drag that goes along with it. But inasmuch as there is not really an essential relationship, then I think we want to try very hard to draw away from the old induced drag theory.

T. Morel

What would you pick? Would you call it a lift-related drag? Or a vortex drag?

R. T. Jones

In your situation it certainly was a lift-related drag. I don't know that it helps to call it vortex drag, but perhaps it does.

**Lamar, J. E., (1974) Extension of Leading-Edge Suction Analogy to Wings with Separated Flow Around the Side Edges at Subsonic Speeds, NASA TR R-428.*

***Whether we simply mean by it what is predicted by the Prandtl-Lanchester finite airfoil theory, or whether it is broader than that.*

W.-H. Hucho *(Volkswagenwerk AG, Germany)*

I would like to point out that one of the figures in my presentation (Fig. 19) showed that you can have high drag with low lift, and vice versa, i.e. there is not an essential relationship between lift and drag.

I would like to express my views on where induced drag fits into the overall drag picture. If you integrate the pressure distribution on a car, and if you also integrate the surface shear stresses, and then you add both these pieces together, you will have

W. H. Hucho

all the drag that is exerted on the car. The rotational energy which is within the vortices and which can be found in the Trefftz plane, is a part of this overall drag. (In the Trefftz plane the rotational part of the drag can be calculated either by momentum or energy considerations; you can call it whatever you like – I would prefer vortex drag). You should *not* add it to the drag obtained from the pressure and shear stress distributions. I think that is very important to realize. By contrast, as I tried to outline in my paper, in airfoil theory things are kept separate. Profile drag, which is the integral of pressure and shear stresses, is evaluated in a strictly 2-dimensional sense; then you calculate the induced drag in a inviscid flow; then you add them.

Finally, I would like to congratulate the author on the work he has carried out. It's exactly the kind of work I am looking for. It's what we in industry expected from this Symposium. We at VW observed the phenomenon of the hatchback car but we did not have the time to explore it in depth. Furthermore, when you work with a real car you can't study all the geometrical variations you would like. But if one uses an idealized model as you did, you can go much, much further.

I would recommend that more of this kind of work be done. Another geometry is the one Geoffrey Carr drew on the blackboard. We also have observed critical drag behavior on notchback cars. It is another subject which could be studied in the same depth as you have done.

RECENT JAPANESE RESEARCH ON THREE-DIMENSIONAL BLUFF-BODY FLOWS RELEVANT TO ROAD–VEHICLE AERODYNAMICS

H. NAKAGUCHI

University of Tokyo, Japan

ABSTRACT

The introductory part of the paper contains a very brief discussion of recent research on automobile aerodynamics in Japan, touching on the progress and the prospects of aerodynamics in automobile design.

The major part is devoted to bluff-body research relevant to truck and bus aerodynamics. It mainly focuses on three-dimensional single-body configurations – bars of square cross-section having length-to-width ratios, L/W, from virtually 0 to 5. All edges of the bars were sharp, the longitudinal axis of each was aligned in the flow direction, and testing was performed in the absence of a groundplane. The drag coefficient changed considerably as L/W was varied from 0 to 5. The character of the variation of drag coefficient with angle of incidence, as well as that of lift coefficient with incidence, changed drastically when L/W reached 1.6. Flow visualization showed that this was accompanied by a change in the flow pattern, just as in the two-dimensional case of rectangular bars aligned perpendicular to the flow direction that was previously studied by the author.

The results of a preliminary experiment on two-body configurations representative of tractor-trailer systems are also presented. In this case the bodies were in tandem, each having its axis aligned in the streamwise direction, and the configuration was in close proximity to a simulated groundplane. The shape of the forebody and the rear of the afterbody were changed, while the gap and the afterbody length were varied. Only drag data at zero yaw angle is presented for these configurations.

References pp. 245-246.

INTRODUCTION

Before discussing the main subject the author has been asked to comment very briefly on the progress and the future prospects of aerodynamics in automobile design in Japan. Except for a few scattered experimental studies, research in automobile aerodynamics was started in the early Sixties. Since then, the greatest attention has been directed to the safety of driving at high speed, especially with respect to the response to side gusts. This was accentuated by the fact that the majority of cars produced in Japan were compact and of light weight, which made them naturally more sensitive to aerodynamic forces than larger and heavier cars. Of all the aerodynamic characteristics, therefore, lift, pitching moment, side force and yawing moment were the main concerns. Full-scale wind tunnels for studying automobile aerodynamics were built in the late Sixties and early Seventies, and wind-tunnel testing has since been an indispensable part of new vehicle development. Details of these facilities can be found in Kimura (1971), Ohtani, Takei, & Sakamoto (1972), and Muto & Ueno (1976).

Since the oil crisis, however, drag reduction has become another concern in view of fuel conservation. To reduce road-load power is now one of the most acute demands, in order to compensate for fuel economy penalties that have resulted from treatments to meet antipollution requirements.

The relationship between wind tunnel test results of scale-models (usually 1/4-scale clay models) and those of full-scale vehicles has been reexamined, and empirical correction procedures have been determined temporarily. Studies to get more realistic drag data from scale-model tests, and to reduce the drag of external protrusions and that due to internal flows, leakage, etc. have continued.

An understanding of the structure of the airflow around vehicles is, of course, fundamental to generating aerodynamically favourable vehicle shapes, but it is also very important from other practical points of view, e.g. to avoid getting the body surfaces dirty from mud splash, etc. For flow-field studies, relatively large (1 m^2 test section) two and three-dimensional smoke wind tunnels have been built in the laboratories of automobile manufacturers. Examples have been reported by Oda & Hoshino (1974) and Takagi & Hayakawa (1975).

The field of Japanese automobile aerodynamics seems to be developing in the following directions:

1. Research to determine vehicle shapes of low drag without impairing driving qualities, cost and consumer appeal,

2. Research to reduce aerodynamic noise,

3. Research on local flows, e.g. engine-cooling and passenger compartment ventilation, etc.

Since the research that has been done in Japan on the aerodynamic drag of automobiles has been generally restricted to particular cars, presented here are two more-or-less fundamental studies, one of which treats the three-dimensional single-body case and the other the two-body case.

NOTATION

C_D wind-axis drag coefficient, drag $/\frac{1}{2}\rho U^2 S$

C_{D_b} base-drag coefficient = mean base-pressure coefficient

C_L wind-axis lift coefficient, lift $/\frac{1}{2}\rho U^2 S$

C_m pitching moment coefficient about the center of a bar, moment $/\frac{1}{2}\rho U^2 SW$; positive for nose up

C_x axial-force (body-axis drag) coefficient, axial force $/\frac{1}{2}\rho U^2 S$

C_z normal-force (body-axis side force) coefficient, normal force $/\frac{1}{2}\rho U^2 S$

g gap between tractor and trailer

L length of bar

L_T length of trailer

S cross-sectional area, W^2

U air velocity

W width of bar, tractor, and trailer

α angle of incidence; positive for nose up

ρ air density

ν kinematic viscosity

STUDY OF THE AERODYNAMIC CHARACTERISTICS OF BARS OF SQUARE CROSS-SECTION ALIGNED WITH THE FLOW

Scope – Very few studies have been reported on three-dimensional bluff-body flows, except for those of spheres, see Goldstein (1938), Rosenhead (1953), Taneda

References pp. 245-246.

(1956), Maxworthy (1969), Calvert (1972), and Achenbach (1972, 1974). The aerodynamic force and moment characteristics and the flow field of another simple three-dimensional bluff-body, a single bar of square cross-section aligned with the airstream, was studied by Nakaguchi & Hirose, University of Tokyo, and will be reported in this paper. Aerodynamic force, moment, and base pressure were measured in a low-speed wind tunnel, and the structure of the flow in the wake was studied by means of flow visualization techniques.

Test Equipment – The force and moment tests were carried out in a closed-circuit wind tunnel with a test-section diameter of 1.5m. Fifteen models, all of 115 x 115mm square cross-section, W x W, but of varying length, L, were used. Each model was mounted on a T-bar which, in turn, was attached to the wind-tunnel balance with 0.5mm diameter piano wire, as shown in Fig. 1. Tests of a bar of L/W = 4.2 at several wind speeds between 10 and 30 m/s showed no appreciable change in the force and moment coefficients. The subsequent tests were carried out at a wind speed of 22 m/s, and the corresponding Reynolds number, UW/ν, was 1.7 x 10^5. Angle of incidence was varied from -3° to 15°.

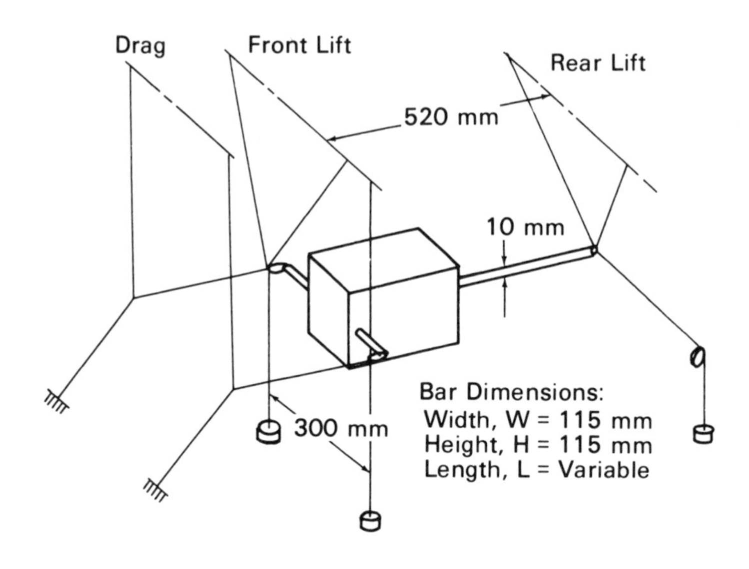

Fig. 1. Experimental set-up for measuring aerodynamic forces.

The study of flow in the wake was carried out in a blow-down-type smoke tunnel for the convenience of flow visualization. The test section of this tunnel was 0.6 x 0.6m, and similar models of smaller cross-section, 50 x 50mm, were used. Flow visualization techniques consisted of either vaporized-oil smoke, or titanium dioxide-oleic acid applied to the model surfaces.

Results – Fig. 2a shows the relationship between drag coefficient and angle of incidence for bars of various L/W ratio. For bars of L/W ratio less than 1.2, the drag coefficient remains fairly constant throughout the angle of incidence range tested. On bars of L/W ratio more than 1.6, however, the drag coefficient increases parabolically as angle of incidence increases. The same data are presented in Fig. 2b in a body-axis coordinate system. The variation of the axial force (body-axis drag) with angle of incidence (yaw angle) is more appropriate for road vehicles.

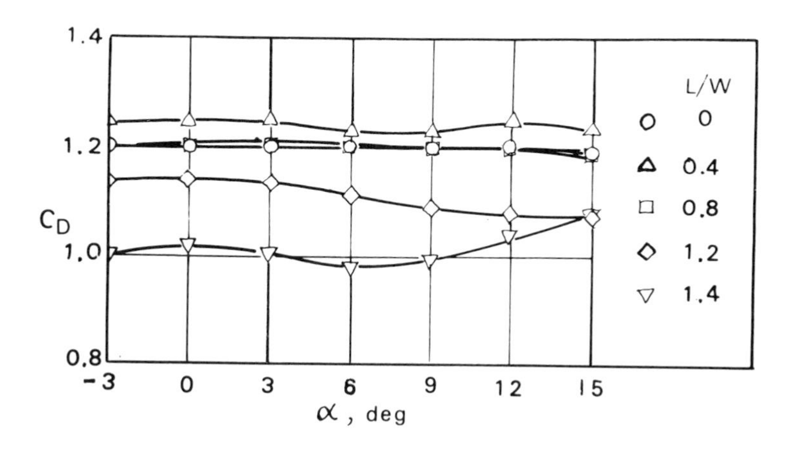

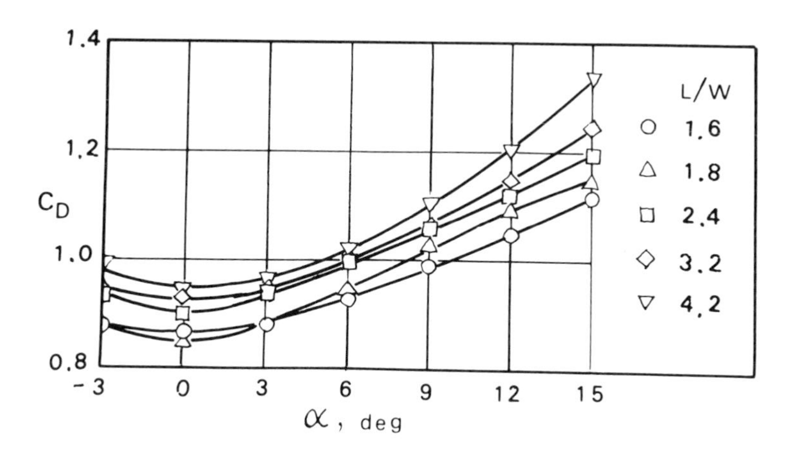

Fig. 2a. Wind-axis drag coefficient vs. angle of incidence.

References pp. 245-246.

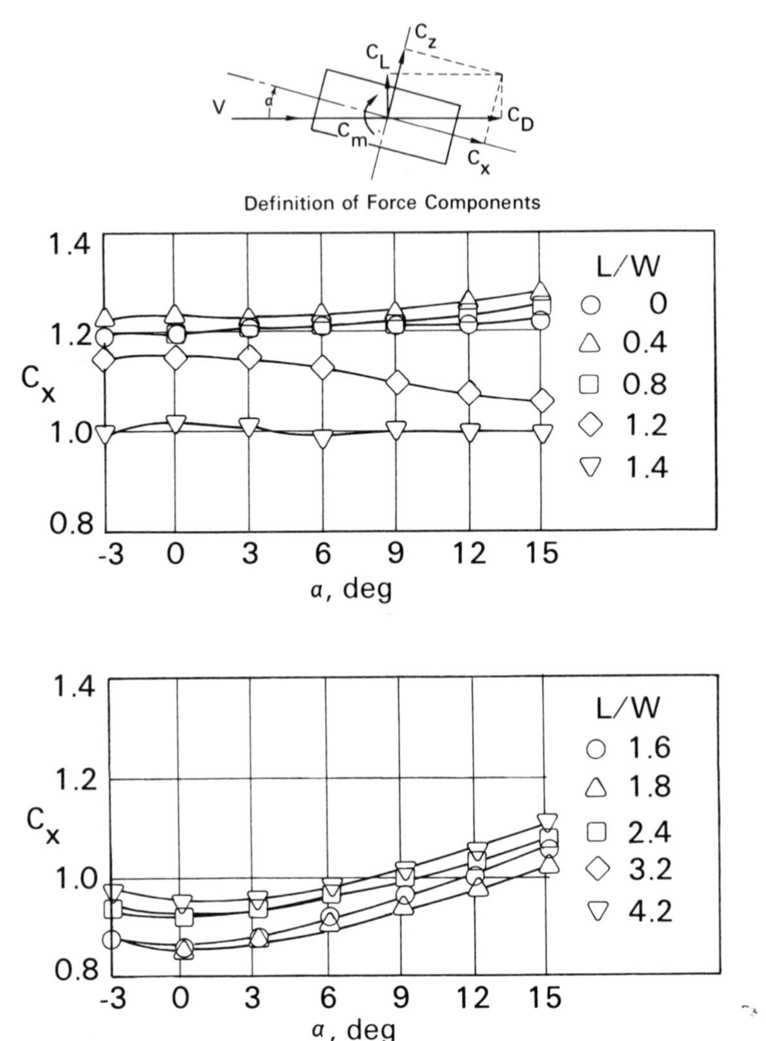

Fig. 2b. Axial-force (body-axis drag) coefficient vs. angle of incidence.

The drag coefficient at zero angle of incidence, and the base-drag coefficient (estimated from the mean value of base pressure) are shown as a function of L/W in Fig. 3. As the L/W ratio increases from zero, the drag coefficient increases from 1.20 (about 3 percent greater than the known value of 1.17 for circular and square plates; see Simmons, 1930) to a maximum value of 1.25 at L/W = 0.5. Then the drag coefficient decreases to a minimum of 0.84 at L/W = 1.8; from that point on, the drag coefficient increases gradually. The base-drag coefficient varies in parallel with the total drag coefficient; therefore, the forebody drag remains roughly constant over the range of L/W ratio tested. Known results of previous studies are also shown in Fig. 3, and it should be pointed out that some discrepancy exists for L/W ratios less than 1.8.

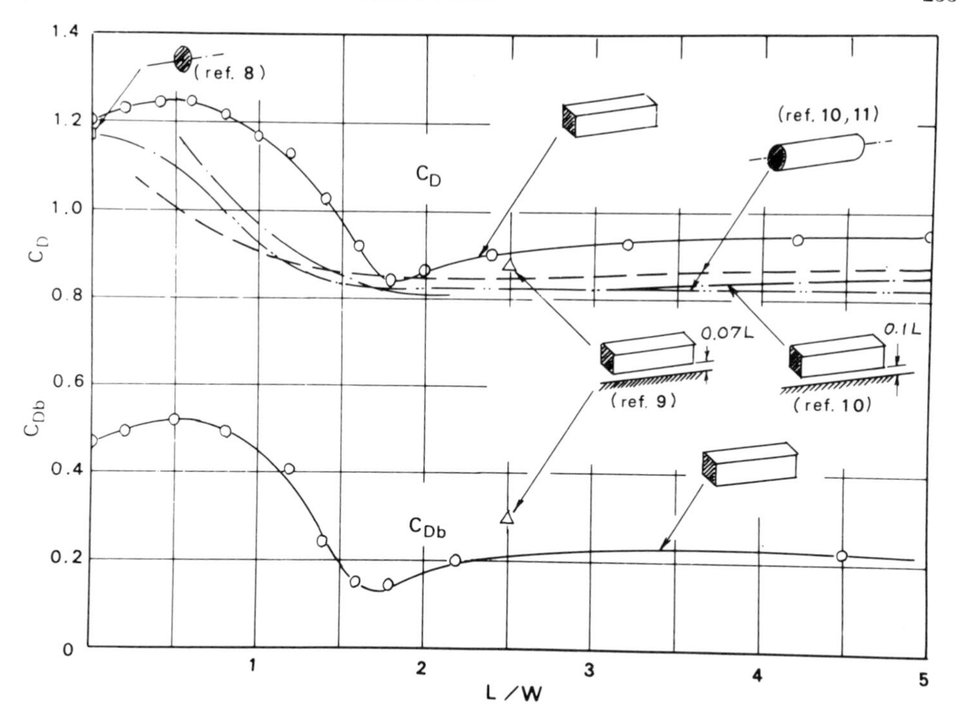

Fig. 3. Variation of total and base drag coefficients at zero angle of incidence with L/W ratio.

A study of the associated flow patterns has led to the conclusion that the above-mentioned variation of drag coefficient with L/W ratio can be attributed to interference between the flow separation from the front edges and the downstream side surfaces. On bars of small L/W ratio (e.g. less than 1.2), the separated flow at the rear forms a bulb-like wake, and a ring vortex is formed close to the back surface of the bar; a heavily turbulent flow follows downstream. Tests to detect any periodic shedding of the ring vortices were unsuccessful. In this range of L/W the separated flow at the front never reattaches to the adjoining surfaces (upper, lower, and side) and the drag coefficient remains constant throughout the angle of incidence range tested. On bars of L/W ratio more than 1.8, the separated flow at the edges of the front surface reattaches to the adjoining surfaces, and then separates again at the downstream end. The size of the wake becomes considerably smaller than in the former case, and the drag coefficient becomes correspondingly smaller.

Fig. 4a shows the relation between lift coefficient and angle of incidence. The relation between normal force (body-axis side force) and angle of incidence (yaw angle) is shown in Fig. 4b. On the bars of L/W ratio less than 1.4 the slope of the lift curve is negative, but for L/W ratios more than 1.8, it turns positive. This inversion may be explained by the same reasoning used in conjunction with the drag coefficient. That is, below 1.4 the separated flow at the front never reattaches to the

References pp. 245-246.

side surfaces; the drag doesn't vary with α and the corresponding $dC_L/d\alpha$ is negative (the bar behaves like a flat plate). Above 1.8 the flow reattaches to the side surfaces and the bar behaves like a lifting body, developing suction on the upper (leeward) surface. On bars of L/W ratio between 1.2 and 1.8, however, the flow on the upper surface is unstable and the lift force shows a strong tendency to oscillate about its mean value.

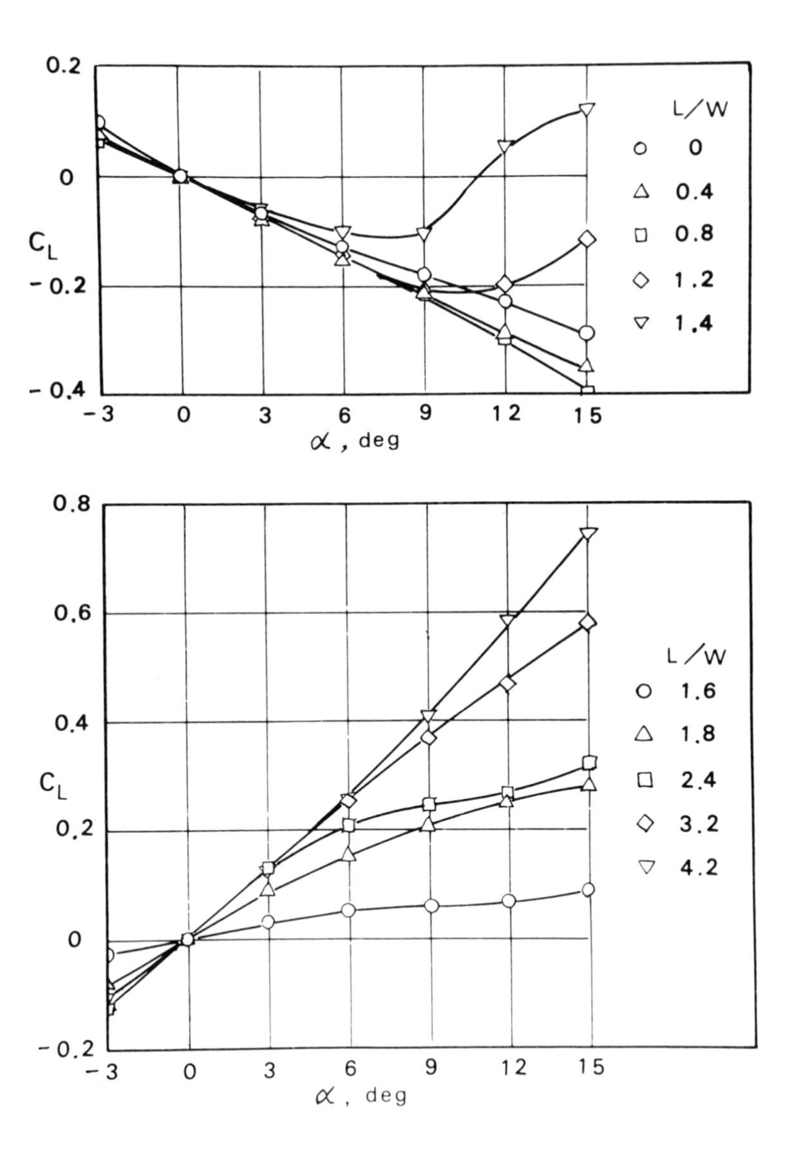

Fig. 4a. Wind-axis lift coefficient vs. angle of incidence.

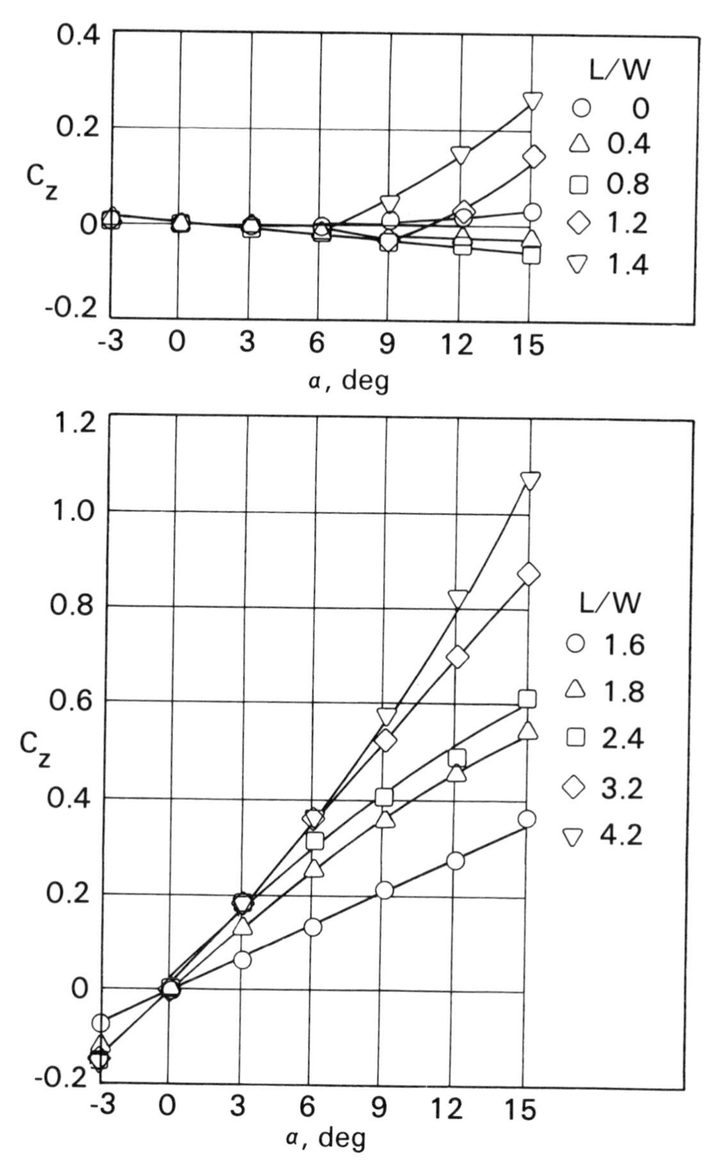

Fig. 4b. Normal-force (body-axis side force) coefficient vs. angle of incidence.

Fig. 5 shows the relation between the pitching moment about the center of the bar and the angle of incidence. The pitching-moment coefficient for bars of L/W ratio between 1.2 and 2.4 is erratic because of the unstable nature of the flow on the upper surface mentioned above. In this range the reading from the rear lift balance (see Fig. 1) varied irregularly.

References pp. 245-246.

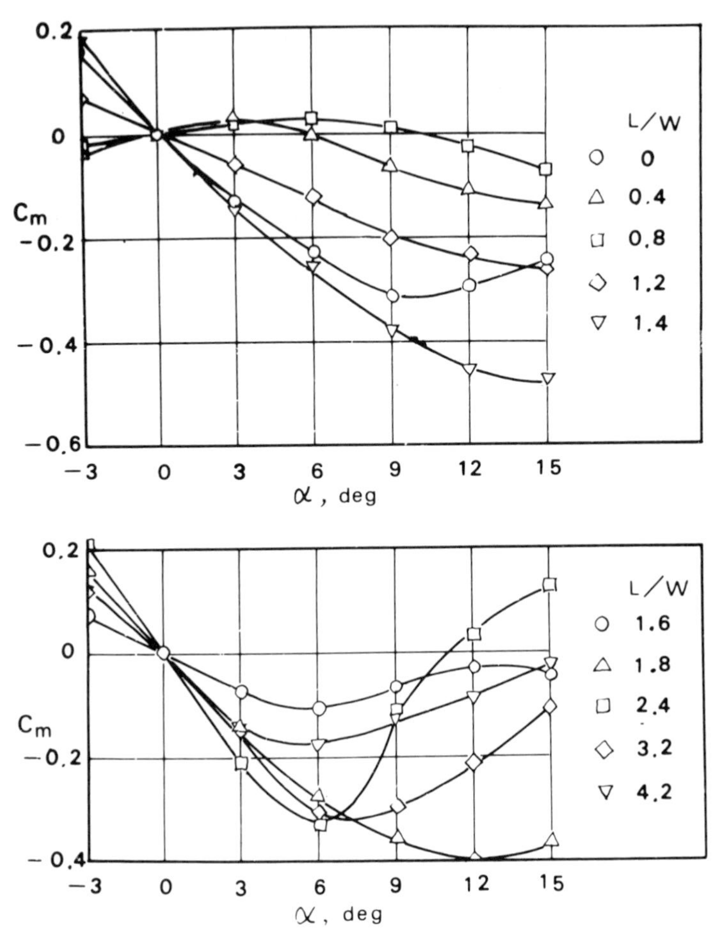

Fig. 5. Pitching-moment coefficient about the center of bar vs. angle of incidence.

Structure of Flow in the Wake – Because of the complexity of the structure of the flow in the wake, no single flow visualization technique resulted in successful recognition of the flow pattern. Some examples of visual observation are shown in Figs. 6 and 7. In Fig. 6 the region of separated flow from the leading edge is shown by means of smoke injection. Smoke was fed through a pipe either in front of the bar (a), or behind it (b and c). Illumination was from two slide-projectors with narrow slits adjusted to illuminate the vertical plane of symmetry. The height of the separated region was maximum in this plane and minimum at the side edges (not shown in the figure). On the bars of L/W ratio less than 1.4, smoke fed from behind goes upstream to the forward edges, but on bars of L/W ratio more than 1.8 smoke does not flow upstream over the rear edge. This means that the separated flow from the front edges reattaches on the sides in the latter case. On the bar of L/W ratio 1.6 smoke flows upstream occasionally.

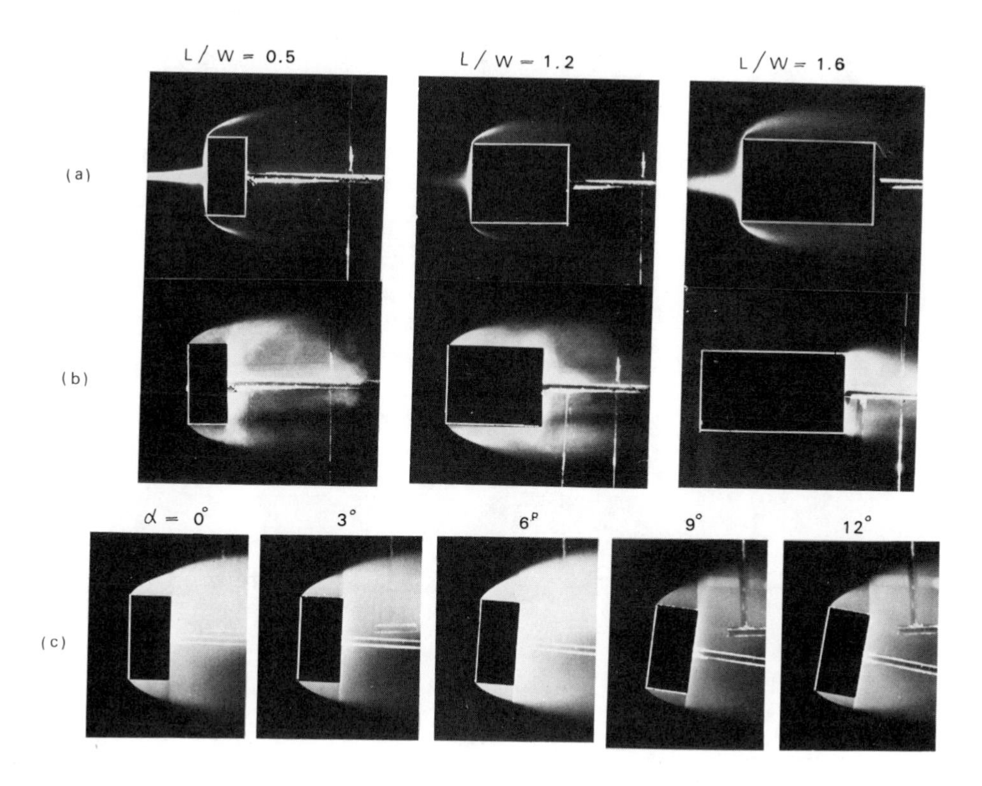

Fig. 6. Regions of separated flow and the wake. (a) $\alpha = 0$, smoke fed in front of the bar, (b) $\alpha = 0$, smoke fed behind the bar, (c) L/W = 0.5, smoke fed behind the bar at different angles of incidence.

In Fig. 7a the flow on the surfaces of the bar is shown by means of the titanium dioxide-oleic acid technique. These photographs show the flow direction only on the surface; the flow direction at points a short distance above it can be greatly different, so that care must be taken in utilizing such a technique for inferring the structure of the *whole* flow field. Nevertheless, the complexity of the flow field can be clearly recognized from these figures. At zero angle of incidence, no appreciable surface flow occurs on bars of L/W ratio less than 0.6; backflow from the rear edge is visible on bars of L/W ratio between 0.6 and 1.6; and reattachment of the separated flow from the forward edge occurs on bars of L/W ratio more than 2. The line of reattachment (shown by the chain line in Fig. 7b) is convex-downstream, corresponding to differences in the thickness of the separated region. On the bar of L/W ratio 1.8, reattachment occurs only in the vicinity of the side edges.

References pp. 245-246.

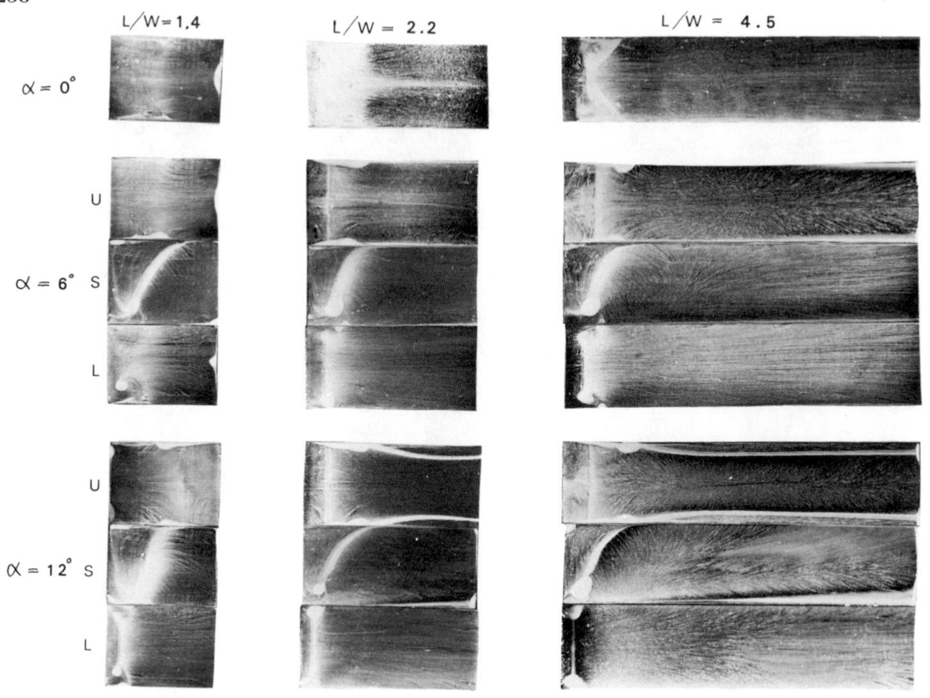

Fig. 7a. Patterns of surface flow. Titanium dioxide-oleic acid photographs.

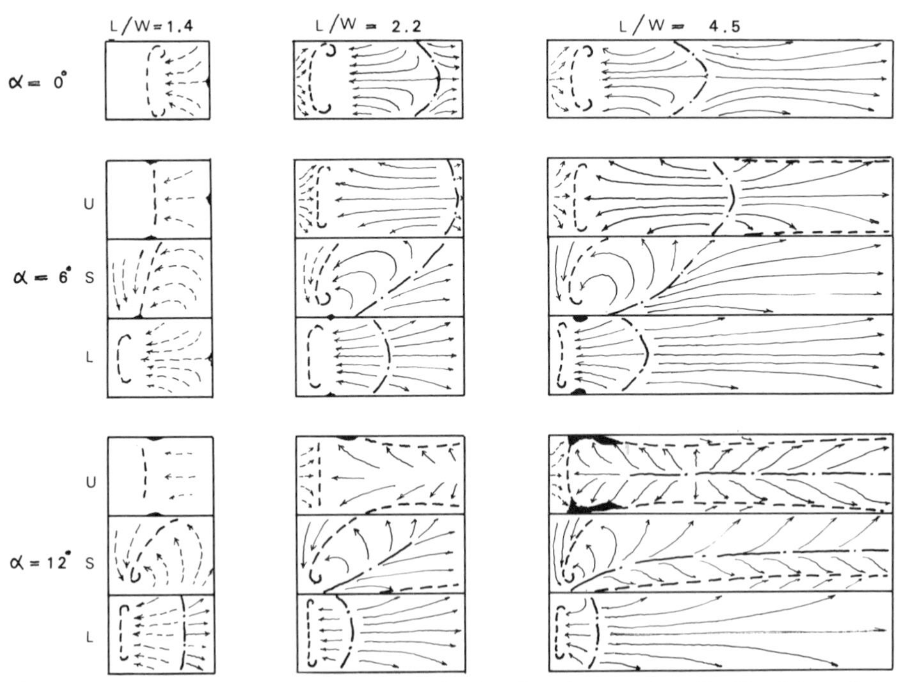

Fig. 7b. Patterns of surface flow. Schematic sketches based on titanium dioxide-oleic acid photographs. Chain line: line of reattachment; thick dashed line: line of separation.

For non-zero angles of incidence the flow patterns are far more complicated. For instance, in the case of the bar of L/W ratio 4.5 at an angle of incidence of 12° (bottom right of Figs. 7a and 7b), reattached flow on the lower (windward) surface expands laterally, rolls up, strikes against the side surfaces, and is then divided in two directions (upward and downward). The flow directed upward on both side surfaces makes a pair of vortices in the separated region above the upper (leeward) surface. A schematic view of the flow in a plane perpendicular to the axis of the bar at its middle point is shown in Fig. 8. The flow pattern noted above suggests the existence of a strongly skewed velocity field around the bar.

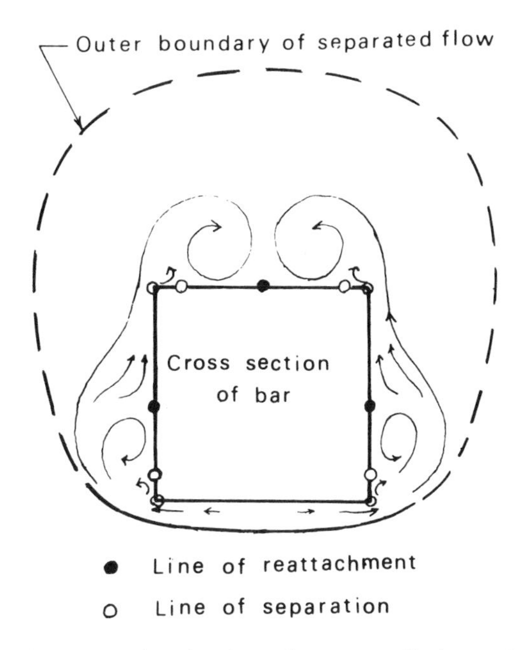

Fig. 8. Flow inside the separated region in a plane perpendicular to the axis of the bar at its middle point, looking upstream; L/W = 4.5, α = 12°.

PRELIMINARY STUDY OF AERODYNAMIC CHARACTERISTICS OF TRACTOR–TRAILER COMBINATION

Scope – A preliminary study of the aerodynamic characteristics of tractor-trailer combinations was recently carried out by Muto, Yoshida & Imaizumi at the Japan Automobile Research Institute (JARI) and is reported here. The object of this study was to investigate the effects of various parameters of such combinations over wide ranges, and qualitatively rather than quantitatively. Although three-component forces were measured in all cases, the material presented here is limited to the drag, and to zero angle of yaw.

References pp. 245-246.

Test Equipment – The study was carried out in a 0.6 x 0.6m closed-circuit open-throat wind tunnel. The models consisted of two separate bodies, namely a forebody simulating the tractor and an afterbody simulating the trailer. As shown in Fig. 9a, the tractor models were of a box type (A), a simulated cab-over-engine (COE) type (B), and a streamline type (C). They all had a 110 x 110mm square section at their aft ends. The trailer models also had a 110 x 110mm square section, and had either a box-type or a streamlined tail, as shown in Fig. 9b; there were two or three trailer lengths. The tractor and trailer models were connected by a round tube of 20mm (0.18W) diameter, and the gap between them could be changed. Vertical stagger was also variable.

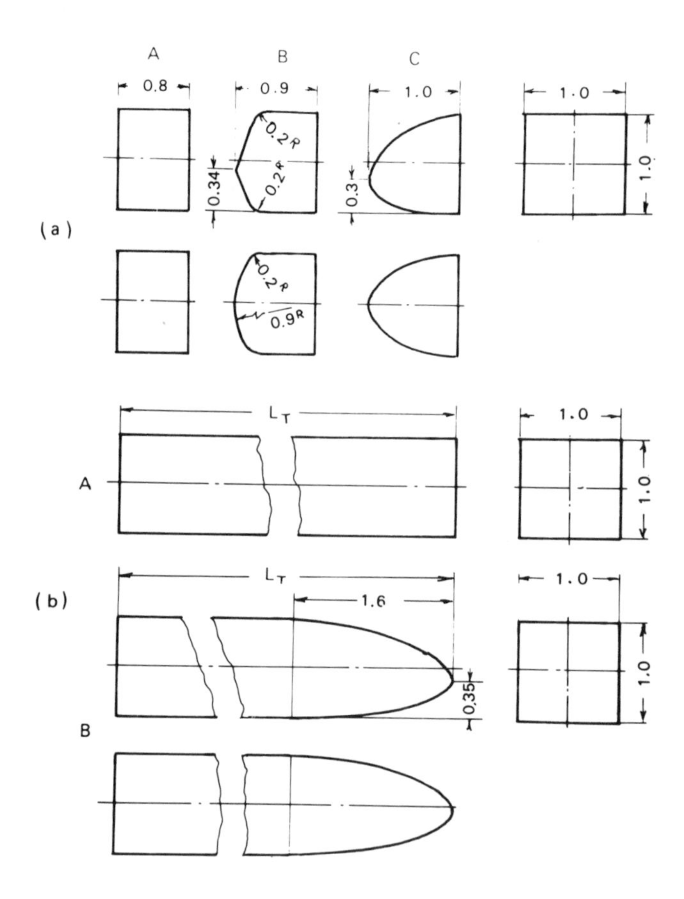

Fig. 9. Models of simulated tractor-trailer combinations, dimensions are relative to W, where W = 110mm. (a) Tractors – A: box-type, B: simulated COE type, C: streamline type. (b) Trailers – A: box-type, B: streamline type.

The trailer model was mounted on the strut of the wind-tunnel balance and the strut, in turn, extended through a clearance hole in the simulated groundplane, as shown in Fig. 10. The width and length of the groundplane were both 1200 mm. The area blockage of the airstream above the groundplane was 4.03 percent. The ground clearance of the trailer was kept at 50 mm (0.45W) throughout the test. The thickness of the groundplane boundary layer over the center of the balance at an airspeed of 20 m/s was 3.0 mm without the model installed. The tests were carried out at a wind speed of 21 m/s; the corresponding Reynolds number, UW/ν, was 1.6×10^5.

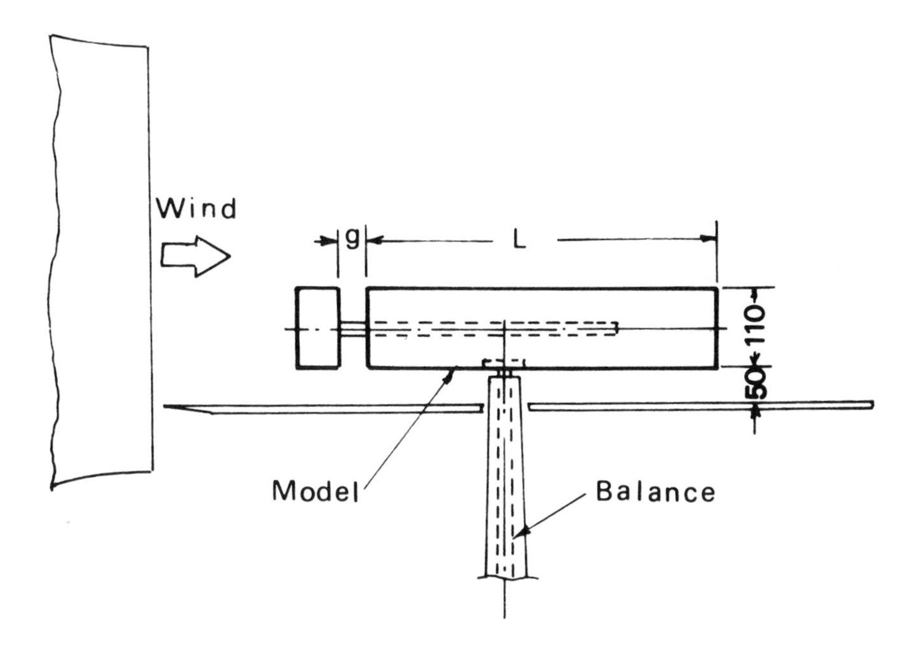

Fig. 10. Experimental set-up, dimensions in mm.

Results – *Effect of gap* – Fig. 11 shows the effect of gap on the drag coefficient at zero angle of yaw. Except for the box (tractor) - box (trailer) combination, gap length plays an important role in the determination of total drag. This suggests that

References pp. 245-246.

the gap should be minimized; much study is needed on the flow through and around the gap.

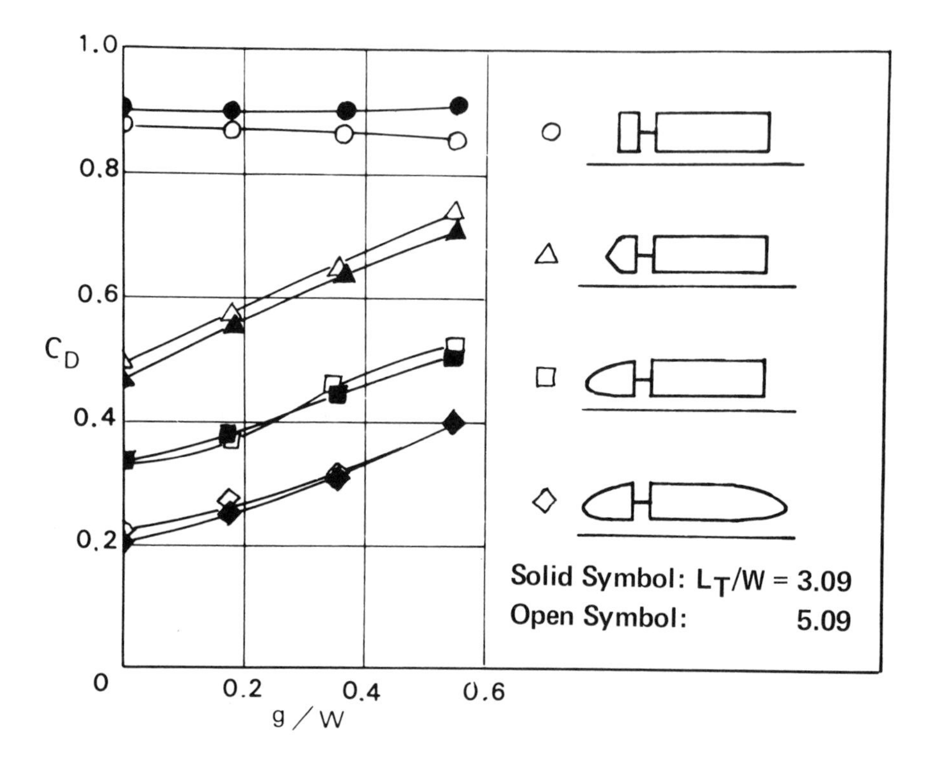

Fig. 11. Drag coefficient at zero yaw angle. Effect of gap and tractor-nose shape.

Fig. 12 shows the drag coefficient of the trailer alone in the presence of two different kinds of tractor. In these cases only the trailer model was set on the balance, and the dummy tractor model was held independently at its proper position. In the case of the box-box combination, there is a thrust (negative drag) acting on the trailer and it decreases as the gap ratio increases; however, this is compensated by a decrease in tractor drag and the total drag remains constant. In the case of the COE-box combination, the drag of the trailer increases with gap ratio faster than the corresponding decrease of tractor drag, and the total drag increases steadily with the gap ratio.

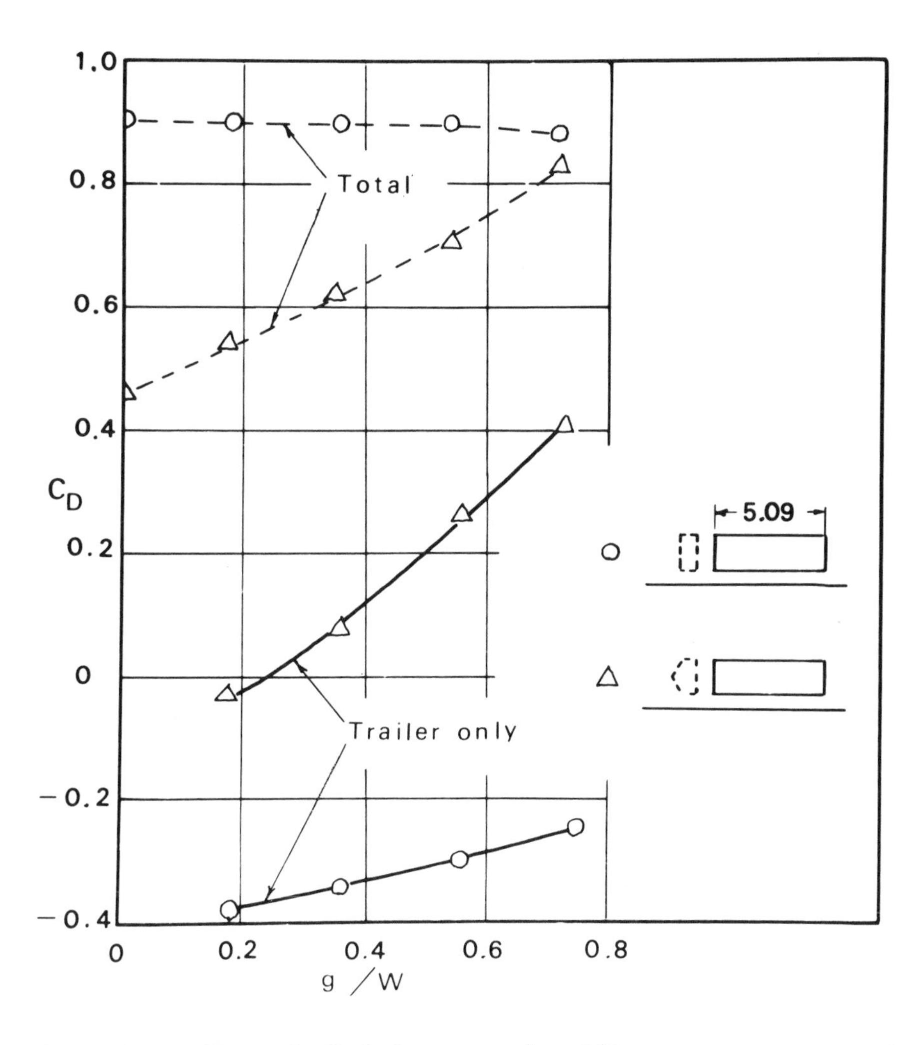

Fig. 12. Drag coefficient of trailer in the presence of two different tractors at zero yaw angle.

Effects of tractor-nose and trailer-tail shape – In Fig. 11 the effect of tractor shape is shown, as well as that of gap ratio. When the gap is zero (the single-body case), the drag of the COE-box combination is only about 50 percent of that of the box-box combination; and the drag of the streamline-box combination is 37 percent of that of the box-box combination. For non-zero gap ratios the drag reductions are smaller, and decrease with increasing gap.

References pp. 245-246.

Fig. 13 shows the effect of trailer tail-shape on the drag coefficient of combinations with three types of tractor. Drag reduction by streamlining the trailer tail is of about the same absolute value for all types of tractor, but the percentage reduction is largest for the streamline-streamline combination.

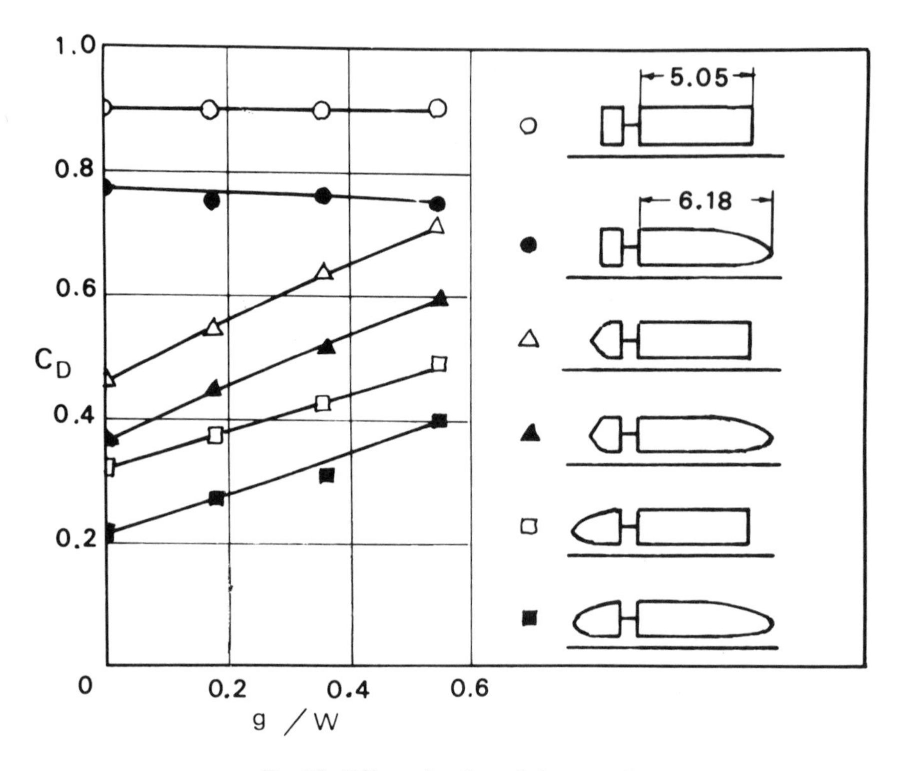

Fig. 13. Effect of trailer-tail shape on drag.

Effect of vertical stagger – Fig. 14 shows the effect of vertical stagger between tractor and trailer. In this representation the reference area for defining the drag coefficient is still taken as the cross-sectional area of the trailer as in the case of zero stagger. When the gap is zero, the drag coefficient remains unchanged with increasing stagger ratio, in spite of the fact that the projected area of the tractor-trailer combination increases.

For larger gap ratios the drag coefficient even decreases as the stagger increases. Although a definite explanation has not been obtained, complicated interference between the ground clearance under the tractor and trailer, the gap between them, and the relative position of the top of the tractor and the forward edge of the trailer seem to play an important role in the flow pattern.

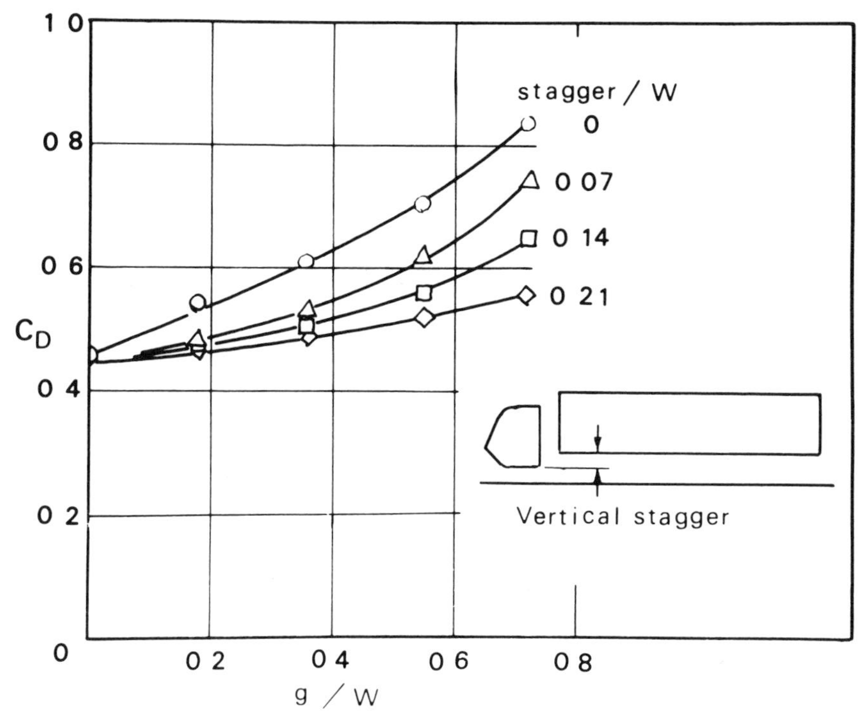

Fig. 14. Effect of vertical stagger on drag, C_D, based on trailer area, W^2.

CONCLUDING REMARKS AND ACKNOWLEDGEMENTS

Experimental studies of two examples of flow around simple three-dimensional bluff bodies have been presented. Much effort will be needed to find a way to attack these particular problems, and to unify the knowledge that has been obtained from them.

The author wishes to express his sincere thanks to Professor I. Tani and Professor T. Ishihara for their continual encouragement. He also thanks Mr. T. Hirose, University of Tokyo, Mr. S. Muto, Mr. Y. Yoshida and T. Imaizumi, JARI, for their eager assistance in contributing material to this paper.

REFERENCES

Achenbach, E. (1972), Experiments on the Flows Past Spheres at Very High Reynolds Numbers, J. Fluid Mech., Vol. 54, pp. 565-575.

Achenbach, E. (1974), Vortex Shedding from Spheres, J. Fluid Mech., Vol. 62, pp. 209-221.

Barth, R. (1956), Einfluss der Form und der Umstroemung von Kraftfahrzeugungen auf Widerstand, Bodenhaltung und Fahrtrichtungshaltung, VDI Zeits., Bd. 98, Nr. 22, pp. 1265-1312.

Calvert, J. R. (1972), Some Experiments on the Flow Past a Sphere, Journal of the Royal Aero. Soc., April, pp. 248-250.

Carr, G. W. (1967), The Aerodynamics of Basic Shapes for Road Vehicles, MIRA Rep. No. 1968/2.

Goldstein, S., ed. (1938), Modern Developments in Fluid Dynamics, Vol. II, Oxford Press.

Hoerner, S. (1958), Fluid-Dynamic Drag, published by the author, p. 3-12.

Kimura, Y. (1971), Toyota's All Weather Wind Tunnel, the Toyota Engineering Vol. 22, No. 2, (in Japanese).

Maxworthy, T. (1969), Experiments on the Flow Around a Sphere at High Reynolds Number, J. Appl. Mech., Trans. A.S.M.E. E36, pp. 598-607.

Muto, S. & Ueno, H. (1976), Report on JARI's Full-Scale Automobile Wind Tunnel, JARI Technical Memo. No. 56, (in Japanese).

Oda, N., & Hoshino, T. (1974), Three-Dimensional Airflow Visualization by Smoke Tunnel, SAE 741029.

Ohtani, K., Takei, M., & Sakamoto, H. (1972), Nissan Full-Scale Wind Tunnel – Its Application to Passenger Car Design, SAE 720100.

Rosenhead, L. (1953), Vortex Systems in Wakes, Advances in Applied Mechanics, Vol. 3, Academic Press, pp. 185-195.

Simmons, L. F. G. & Dewey, N. S. (1930), Wind-Tunnel Experiments with Circular Discs, ARC R & M No. 1334.

Takagi, M. & Hayakawa, Y. (1975), The Development of a Smoke Tunnel and Its Application to Aerodynamic Studies, Nissan Technical Review, No. 10, (in Japanese).

Taneda, S. (1956), Studies on the Wake Vortices, III. Experimental Investigation of the Wake Behind a Sphere at Low Reynolds Numbers, Res. Inst. Appl. Mech., Kyushu Univ. Rep. 4, pp. 99-105.

DISCUSSION

A. E. Perry *(University of Melbourne, Australia)*

I have a question about Fig. 14. At zero gap ratio the drag coefficient did not vary with stagger ratio. Since the frontal area of the combination was changing, how could the drag coefficient remain constant?

H. Nakaguchi

The drag coefficient based on a constant reference area, that of the trailer, did not vary. This means that the drag *force* was constant.

A. Roshko *(California Institute of Technology)*

I suggest that the reason the drag *force* stays constant is that you're getting beneficial interference topside between the tractor and the trailer as the stagger ratio increases. I think this beneficial interference contributes to cancelling the drag associated with the rearward facing step on the underside.

F. T. Buckley, Jr. *(University of Maryland)*

To add to Professor Roshko's comment, I think that a general interpretation of the limited results presented here would be erroneous because the mutual interference effect is going to be strongly dependent on the shape of the tractor. There can be situations where opposite effects will occur compared to the null effects you report.

H. Nakaguchi

You are probably correct.

P. W. Bearman *(Imperial College of Science and Technology, England)*

I don't think the constant drag with increasing stagger ratio at zero gap has been adequately explained. Is it possibly due to increases in the boundary layer thickness along the top and bottom of the model? Thicker boundary layers *can* change the base pressure. Were there any changes in base pressure when the stagger ratio was varied?

H. Nakaguchi

Unfortunately, the base pressure was not measured during those tests.

K. R. Cooper and H. Nakaguchi

K. R. Cooper *(National Research Council, Canada)*

Compared to the *zero* gap results of Fig. 14 I think it's perhaps even more interesting that at the *non-zero* gap ratios the drag coefficient based on a constant reference area actually *decreased* with *increasing* stagger ratio. I'm going to present some related results* for more realistic tractor-trailer models which were 1/10-scale,

Cooper, K. R. (1976), Wind Tunnel Investigations of Eight Commercially Available Devices for the Reduction of Aerodynamic Drag on Trucks, Roads and Transportation Association of Canada National Conference, Quebec City, Sept., p. 15.

and tested at a Reynolds number of 1.2×10^6 based on an average of the trailer width and height. The full-scale trailer box was 102 in. wide and 114 in. high. The drag coefficient is shown as a function of the tractor-to-trailer separation distance (see Fig. 15). Two pairs of curves are plotted. The top pair are for a cab-over-engine tractor, and the bottom pair are for a conventional tractor. The reference area in each case (107 ft^2 for the C.O.E. and 104 ft^2 for the conventional) was held constant, and the height differential was increased by lowering the tractors on their chassis. For each tractor the drag coefficient *decreased* when the stagger ratio *increased.* So, in fact, for realistic tractor-trailer models you *do* find an effect of stagger ratio similar to that reported by Professor Nakaguchi for simplified models.

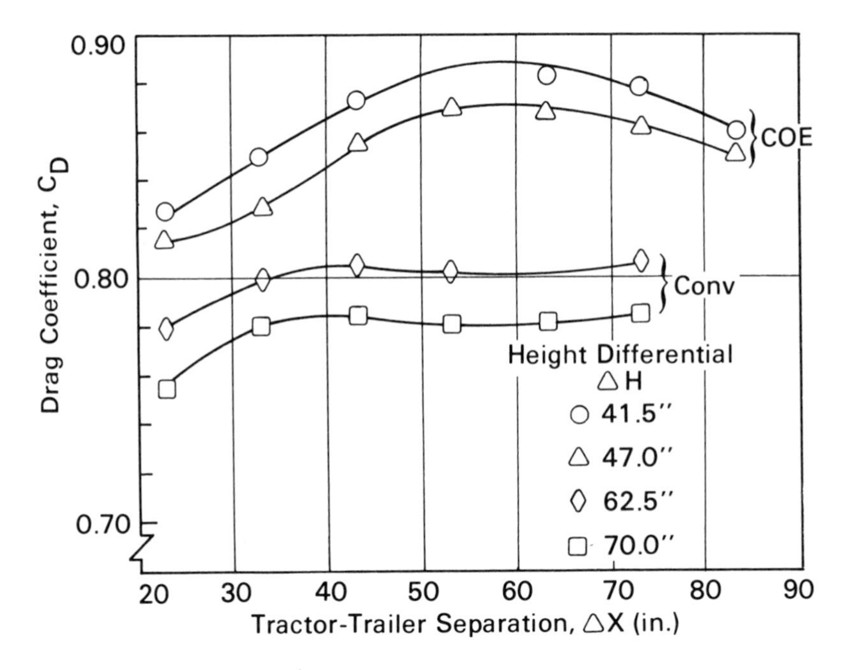

Fig. 15. Effect of height differential (vertical tractor location varied) on drag of realistic tractor-trailer models (1/10-scale).

Editors' Comment

The effect of stagger ratio, as defined and varied by Nakaguchi and Cooper, is certainly of interest, but it should be pointed out that its primary relevance has to do with tractor *design*, i.e. for a given height trailer, *where* should the tractor be placed vertically with respect to its chassis for minimum drag of the combination? When the stagger ratio is varied, the tractor-to-trailer roof *and* underbody height differentials *both* change by the same amount. From an *operational* standpoint *changes* in roof height differential can occur with *fixed* underbody differentials because of differences in trailer height. Some data on the effect of changing only the roof height differential can be found in Mason (1975)*. The drag coefficient based on overall trailer height times trailer width was found to *increase* with *increasing* height differential (see Fig. 16). This is *opposite* to the effect which occurs when the height differential is increased by increasing the stagger ratio.

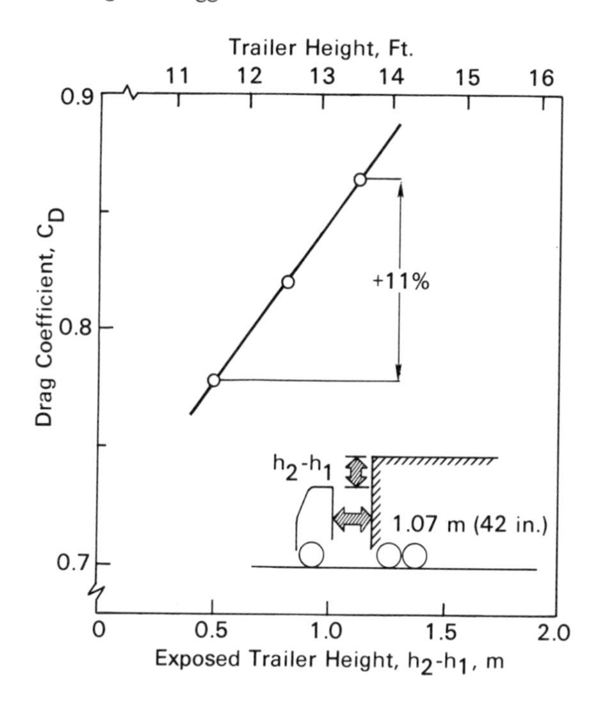

Fig. 16. Effect of height differential (fixed vertical tractor location) on drag of 1/7-scale tractor-trailer.

I. S. Gartshore *(University of British Columbia, Canada)*

I presume that the critical value of L/W, i.e. 1.8, that you showed in Fig. 3 for the single bar of square cross-section, would be very much influenced by turbulence level. Could you tell us what the turbulence intensity of the wind tunnel was, and whether or not you investigated the effects of turbulence?

H. Nakaguchi

I don't know the turbulence level of the tunnel, but the contraction ratio is about 4, so it is probably not too high. The effects of turbulence were not investigated.

---

*Mason, Jr., W. T. (1975), *Wind Tunnel Development of the Dragfoiler – A System for Reducing Tractor-Trailer Aerodynamic Drag*, SAE 750705, Seattle.

A.M.O. Smith *(University of California at Los Angeles)*

This is just an educated guess based on personal experience, but it seems to me that the ordinary effect of turbulence level won't affect these large separations very much. It takes a tremendous turbulence level, such as that from the wake of an upstream rod, to produce a significant effect on such flows.

H. Nakaguchi and A. M. O. Smith

R. T. Jones *(NASA-Ames Research Center)*

Isn't it true that for bodies with all sharp edges, the influence of turbulence level as well as Reynolds Number is minimized?

I. S. Gartshore

For a 2-dimensional body with sharp edges, normal turbulence levels of small scale can quite strongly affect drag, particularly as the length-to-height ratio changes. I think the same will be true for 3-dimensional bodies, certainly for short ones.

A. T. McDonald *(Purdue University)*

There's a paper* by Roberson that focuses on some work related to building aerodynamics in which pronounced effects of freestream turbulence level were observed on both the drag level and the fluctuating characteristics of the forces. The bodies had effectively sharp corners.

**Roberson, J. A., Lin, C. Y., Rutherford, G. S., & Stine, M.P. (1972), Turbulence Effects on the Drag of Sharp-edged Bodies, Journal of the Hyrdraulics Division, American Society of Civil Engineers, Volume 98, No. HY7, July, pp. 1187-1203.*

K. W. Wolffelt *(SAAB-SCANIA, Sweden)*

We tested models of three tractor-trailer combinations in 1/10-scale at the Stockholm Institute of Technology's Wind Tunnel, and in 1/2-scale at the MIRA full-scale wind tunnel. For one configuration the tractor and trailer were of equal height with a small gap between them. There was a tremendous effect of Reynolds number (see Fig. 17). The drag coefficient of the 1/10-scale model was much greater than that of the 1/2-scale model. The other two configurations had a tractor-to-trailer roof height differential. The height differentials were the same, but the gaps were different. The influence of the air gap depended on the Reynolds number. At the low Reynolds number (1/10-scale models) it was negligible; at the high Reynolds number (1/2-scale models) there was considerably more drag with the larger gap. All three configurations had very sharp leading edges. The small wind tunnel was an aeronautical type with a low turbulence level and the MIRA tunnel's turbulence level was probably greater.

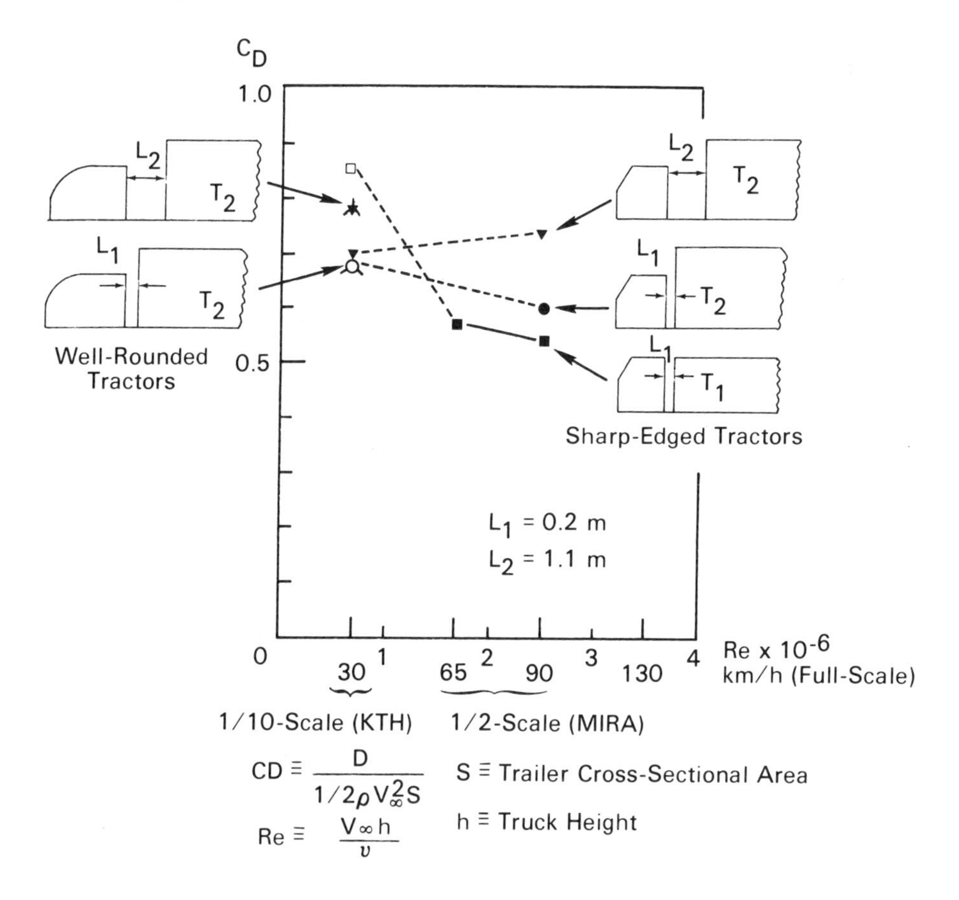

Fig. 17. Influence of Reynolds number on tractor-trailer drag.

Interestingly, when the tractors of the 1/10-scale models with the roof differential were provided with well-rounded noses, the effect of gap previously observed only at high Reynolds number, with the sharp-edged tractor, was now observed at low Reynolds number. It seems to me that this can be explained by the fact that the sharp-edged tractor gave such a tremendous flow separation at low Reynolds numbers that the difference in air gap had no effect. The point of flow reattachment was downstream of the trailer face.

In later low-Reynolds-number tests with sharp-edged tractor configurations we found that drag reducing devices mounted to the tractor roof, such as the Rudkin-Wiley Airshield™ and the General Motors Dragfoiler ™ , had *no* influence. However, with the well-rounded nose on the tractor, the drag reductions *were* similar to those reported at higher Reynolds numbers during the Symposium. Subsequently, with a 1/2-scale model test at MIRA we experienced almost exactly the same performance as other investigators.

A. E. Perry

I have a comment which relates to the surface flow patterns you showed on Figs. 7a and 7b. I think that the patterns are very revealing. More work like that should be done on bluff bodies. If you look at the patterns carefully, you can see that they're full of singularities such as saddles and nodes. These are carefully classified in Lighthill's section of Rosenhead's book* on Laminar Boundary Layers.

**Rosenhead, L., Editor (1963), Laminar Boundary Layers, Oxford University Press, pp. 72-82.*

INTERACTION EFFECTS ON THE DRAG OF BLUFF BODIES IN TANDEM

A. ROSHKO and K. KOENIG

California Institute of Technology, Pasadena, California

ABSTRACT

The objective of this study is to obtain better understanding of the flow over two tandemly positioned bluff bodies in close enough proximity to strongly interact with each other. This interaction is often beneficial in that the drag of the overall system is reduced. Prototypes for this problem come from tractor-trailer and cab-van combinations, and from various add-on devices designed to reduce their drag.

The primary object of the present investigation is an axisymmetric configuration which seems to have first been studied by Saunders (1966). A disc of diameter d_1 is coaxially placed in front of a flat-faced cylinder of diameter d_2. For a given ratio d_1/d_2, there is a value of gap ratio, g^*/d_2, for which the drag of the forebody system is a minimum. In the most optimum configuration, $d_1/d_2 = 0.75$, $g^*/d_2 = 0.375$, and the corresponding forebody drag coefficient is 0.02, a remarkable reduction from the value of 0.75 for the cylinder alone. For each value of d_1/d_2, the minimum drag configuration, g^*/d_2, appears to correspond to a minimum dissipation condition in which the separation stream surface just matches (joins tangentially onto) the rearbody. Support for this idea is furnished by comparison with some results derived from free-streamline theory and from flow visualization experiments. However, when g^*/d_2 exceeds a critical value of about 0.5, the value of $C_{D_{min}}$ is almost an order of magnitude higher than for subcritical optimum gap ratios. The increase seems to be connected with the onset of cavity oscillations.

For non-axisymmetric geometry (square cross-sections) the separation surface cannot exactly match the rearbody and the subcritical minimum values of drag are higher than for circular cross-sections.

References p. 273.

NOTATION

A_1, A_2	frontal area of frontbody and rear body, respectively
C_D	drag coefficient of forebody system based on A_2 and freestream dynamic pressure
C_{Dmin}	minimum drag coefficient for fixed A_1/A_2
C_{D_1}	drag coefficient of frontbody based on A_1
$C_{D_{1_f}}$	drag coefficient of frontbody face based on A_1
C_{D_2}	drag coefficient of rearbody face based on A_2
C_p	local rearbody-face pressure coefficient
C_{p_s}	constant-pressure surface or free-streamline pressure coefficient
$C_p{}^*$	average cavity-pressure coefficient at optimum gap
d_1, d_2	diameter of frontbody and rearbody, respectively
$(d_1/d_2)_{cr}$	frontbody to rearbody diameter ratio at critical g^*/D_2
g	gap between frontbody and face of rearbody
g^*	optimum gap for a given d_1/d_2
$(g^*/d_2)_{cr}$	optimum gap ratio of critical geometry
q_∞	freestream dynamic pressure
r	radius of corner on rearbody face
$r_s(x)$	radial position of the separation surface
Re	Reynolds number based on q_∞ and d_2
τ_s	shear stress on separation surface
U_s	flow velocity outside separation surface
U_∞	freestream velocity

x coordinate parallel to freestream velocity

y radial location on rearbody face

INTRODUCTION

In analyses of the problem of reducing the drag of trucks and, especially, of tractor-trailers, the importance of the front end of the vehicle is always apparent. How to shape the cab and trailer to reduce overall drag, taking into account their mutual interaction, is an important question. The invention of devices to deflect and guide the flow from the tractor over the trailer has demonstrated that beneficial interference between otherwise high-drag bodies can be obtained. The technique of drafting used in automobile racing (described by Mr. Bobby Allison at this Symposium) provides another example of beneficial interference.

We thought it would be interesting and useful to investigate interference between bluff bodies of much simpler geometry than those of real road vehicles, the objective being to identify drag mechanisms and to better understand the conditions that lead to drag reduction of the system. We concluded that an *axisymmetric* configuration would have the needed simplicity and would also better represent the vehicle problem than would a two-dimensional one, and we were thus led to study the configuration shown in Fig. 1.

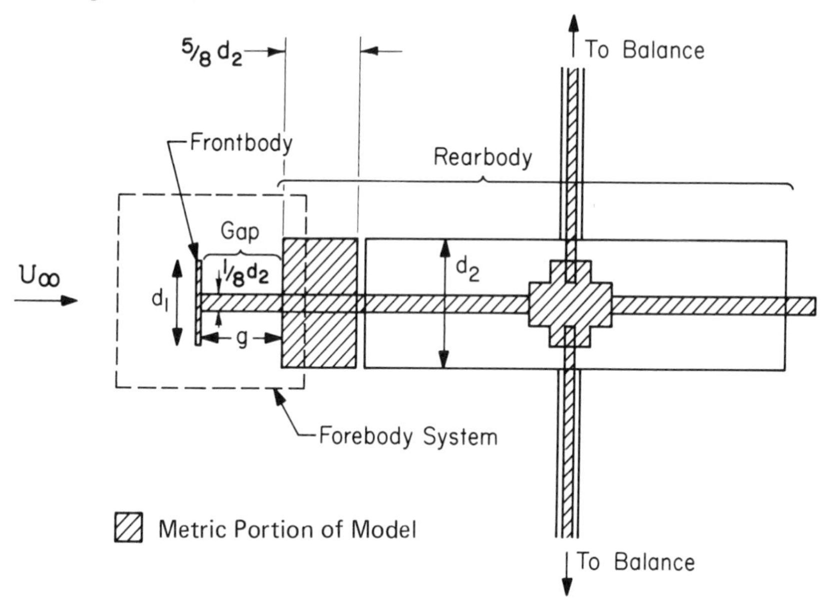

Fig. 1. Experimental model

References p. 273.

The object here is to investigate the drag of a semi-infinite half body, in this case a circular cylinder with a flat face. In *potential* flow the integral of the pressure over the area of this face is zero; in the *real* flow, separation from the edge leads to loss of the high suction near the edge and results in a face pressure drag whose coefficient $C_D \cong 0.75$, based on the cylinder's cross-sectional area. This is comparable to the drag coefficient of the front of a trailer without a tractor. If now a disc is placed coaxially in front of the cylinder, the pressure and thus the drag of the cylinder will be modified. In turn, the drag of the disc will be altered from the free body value of $C_D = 1.2$, Hoerner (1965), based on its own area. Denoting the disc and the face of the cylinder by subscripts 1 and 2, respectively, the total drag of the forebody *system* (frontbody and face of the rearbody) will be

$$D = D_1 + D_2 = (C_{D_1} A_1 + C_{D_2} A_2)\, q_\infty \tag{1}$$

where A_1 and A_2 are the respective cross-sectional areas, and q_∞ is the freestream dynamic pressure. In what follows, the drag coefficient for the system will be based on A_2, i.e.

$$C_D = \frac{D}{q_\infty A_2} \tag{2}$$

A configuration similar to this was investigated by Saunders (1966). His results, described briefly in his patent application, showed how the drag of the system can be minimized by suitable choice of diameter ratio d_1/d_2 and gap ratio g/d_2. The system in Saunder's experiments included all of the rearbody, i.e. its base pressure drag as well as skin friction on the sides, in addition to the frontbody. In our experiment we arranged to exclude the sides and base of the rearbody from the drag system by suspending only a short section of the front portion of the rearbody from the drag balance, as shown in Fig. 1. The pressure distribution in the slot between this active front section and the dummy part of the cylinder could be measured in order to correct for contributions to drag from the model internal pressure. These pressures were observed to depend in a consistent manner on the drag of the forebody system, i.e. C_p became more negative as C_D increased. Skin friction on the sides of the small active segment of the rearbody could contribute no more than 0.01 to C_D. (Measurements on a similar segment, but with a well-rounded leading edge to restore the edge suction, gave values of C_D less than 0.01 when tested without a frontbody.) With the forebody system isolated this way, the interaction effects are revealed much more clearly than when measurements are made on the complete system.

EXPERIMENTAL DETAILS

The configuration shown in Fig. 1 was built with d_2 = 8 in. (20.3 cm) for installation in the GALCIT* Merrill Wind Tunnel which has a test section 32 in. x 46 in. (81 cm x 117 cm) in cross-section and 104 in. (264 cm) long. The central sting supporting the forebody system was mounted on a unique force balance which had been designed and built by Professor F. Clauser, providing direct analog readout of the three aerodynamic forces and the three moments acting on the metric part of the model. This balance has an effective time constant of approximately 1 second. C_D repeatability for these experiments was approximately $\pm$ 0.006. Pressure distributions on the face of the rearbody and in the slot between the metric and dummy portions of the model were measured using 1/32 in. (0.08 cm) pressure orifices and a 0-100 mm Hg Barocel pressure transducer. Tygon tubing used to transmit the pressures and the Barocel gave a frequency response of approximately 17 Hz. The pressures were measured on a Hewlett-Packard Timer-Counter DVM with a variable integration time up to 10 seconds. Measurements were made at speeds from 25 to 190 ft/sec (7.6 to 57.9 m/sec); corresponding values of the Reynolds number, Re, based on d_2 were from 1×10^5 to 8×10^5.

The model could also be arranged so that only the frontbody was connected to the balance, all of the rearbody then becoming nonmetric. This permitted decomposition of the forebody system drag into contributions from the frontbody and the face of the rearbody.

A second model, with d_2 = 4 in. (10.2 cm), was built for installation in the GALCIT Free Surface Water Tunnel, Ward (1976), mainly for flow visualization with dye. The dye, diluted food coloring, was injected into the flow from the face of the frontbody, the face of the rearbody, or from an upstream probe, allowing the flow field to then be observed and photographed. These tests were made at a water speed of 3.5 ft/sec (1.1m/sec) giving a Reynolds number of 1×10^5.

In addition to the models with circular cross-section, cylinders of square cross-section were also constructed and tested in both facilities, the discs in front now being replaced by square plates.

RESULTS FOR CIRCULAR CROSS-SECTION

Measurements of the drag of the forebody system are shown in Fig. 2 for several frontbody-to-rearbody diameter ratios, with the gap ratio g/d_2 as the independent variable. Most remarkable is the great reduction in drag that can be achieved for certain combinations of d_1/d_2 and g/d_2. In particular, for $d_1/d_2 = 0.75$ the value of C_D is only 0.02 for $g/d_2 = 0.38$, compared to a value of about 0.75 for $g = 0$. This value is hardly higher than what could be achieved for a well shaped solid forebody system without separation. It is so low that its accuracy of measurement is perhaps no

* *Graduate Aeronautical Laboratories, California Institute of Technology.*

References p. 273.

better than ±30 percent. Minima are also reached for other values of d_1/d_2, even for $d_1/d_2 = 1.0$ whose minimum occurs at $g/d_2 = 1.5$. For $g/d_2 \rightarrow 0$, C_D tends to the value for the front face drag of the rearbody, namely 0.75; for $g/d_2 \rightarrow \infty$, the drag ought to approach the sum of the free field values of each body, i.e. using (1) and (2)

$$C_D \rightarrow C_{D_2} + C_{D_1} \left(\frac{d_1}{d_2}\right)^2 \cong 0.75 + 1.2\left(\frac{d_1}{d_2}\right)^2 \qquad (3)$$

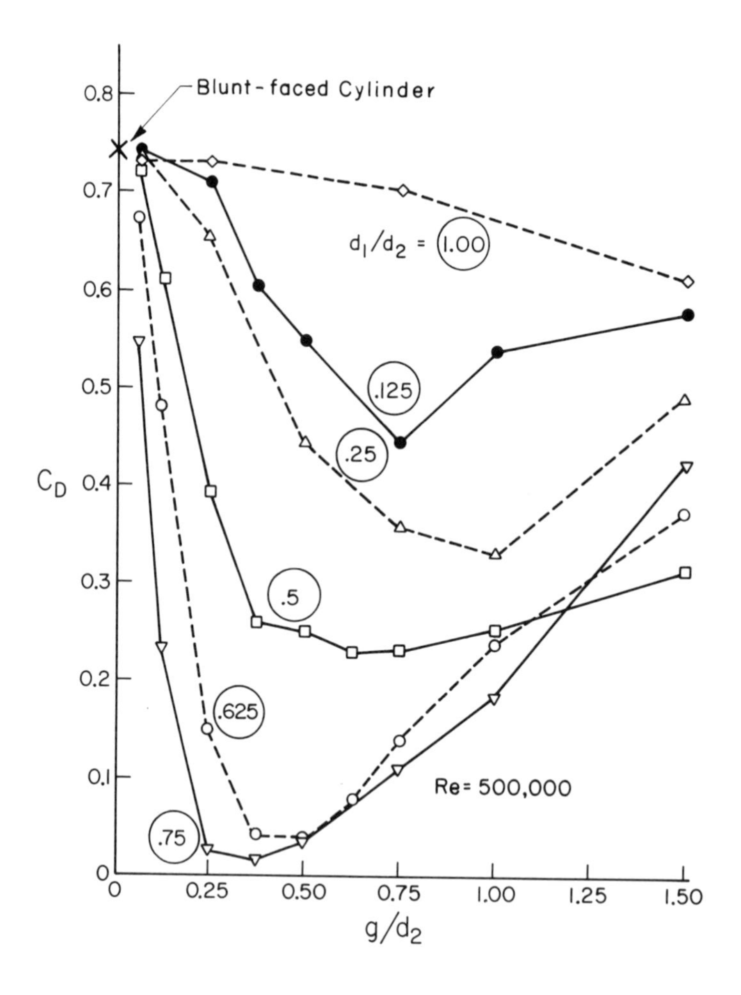

Fig. 2. Effect of gap on system drag (axisymmetric).

Fig. 3 shows some pictures obtained with dye visualization in the water tunnel. The exposure time for these photographs was 1/1000 sec. First, as a baseline for comparison, the rear cylinder without the front disc or its sting support is shown in Fig. 3a. The large separation from the edges accounts for the high value of $C_D \cong 0.75$. By contrast, in Fig. 3b is shown the configuration corresponding to $C_D = 0.03$, not quite optimum, while Fig. 3c shows the flow corresponding to the minimum value, $C_D \cong 0.02$, for this d_1/d_2. In both (b) and (c) it may be seen that the separation surface from the disc joins smoothly onto the edge of the rearbody, becoming a boundary layer on that body. Curiously, for the optimum case (c), the separation surface does not appear to join on quite as smoothly as in (b); there is a small overshoot before the point of reattachment. These minute details of the flow may be due to Reynolds number effects, however.

Fig. 3a. Blunt-faced cylinder. The perfect symmetry is the result of this picture being a composite. $C_D = 0.75$.

Fig. 3b. $d_1/d_2 = 0.75$, $g/d_2 = 0.25$, $C_D = 0.03$.

References p. 273.

Fig. 3c. $d_1/d_2 = 0.75$, $g/d_2 = 0.375$, $C_D = 0.02$.

Figs. 3d and 3e show flows for which the gap is too small and too large, respectively, for optimum drag reduction; the values of C_D are 0.24 and 0.18.

Fig. 3d. $d_1/d_2 = 0.75$, $g/d_2 = 0.125$, $C_D = 0.24$.

Fig. 3e. $d_1/d_2 = 0.75$, $g/d_2 = 1.0$, $C_D = 0.18$.

SOME THEORETICAL CONSIDERATIONS

As has been noted by Saunders (1966) and is evident from the flow pictures, favorable interference and low total drag occur when the separated flow from the frontbody, i.e. the separation stream surface, joins smoothly (in some sense) onto the front edge of the rearbody, so that the total extent of the separation region in the gap and beyond is minimized. Assuming that the separation surface is nearly a constant-pressure surface, a free-streamline analytical model can be used to calculate the gap corresponding to the condition of smooth flow onto the rearbody. In the free-streamline model, this is simply the downstream distance at which the separated stream surface becomes parallel to the freestream, i.e. tangent to the side of the rearbody, as shown in the inset sketch in Fig. 4. The free-streamline model can be analytically formulated in two-dimensional flow with the help of conformal mapping (Roshko, 1954), but for the axisymmetric case the computations have to be carried out numerically (Strück, 1970).

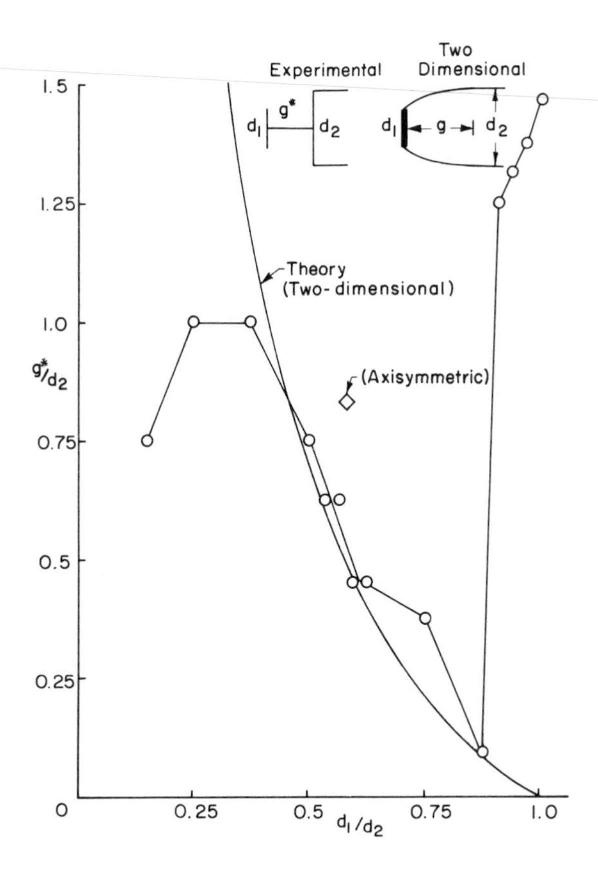

Fig. 4. Optimum gap.

References p. 273.

Shown in Fig. 4 are the results of the two-dimensional theory in the form of consistent combinations of gap and height ratio for a constant-pressure streamline joining tangentially onto the rearbody. The parameter varied to trace out the curve is the pressure coefficient on the constant-pressure separation stream surface (which is the same as the base pressure coefficient of the disc because the separated region is assumed to be a quiescent, constant-pressure region). The one computed case from Strück's work is also shown for comparison with the two-dimensional case. Also shown are our experimental values of g^*/d_2 for each value of d_1/d_2. These are taken from Fig. 2 where it may be seen that there is some uncertainty in determining precisely the value of g/d_2 for which minimum C_D occurs. The favorable comparison between the measured and calculated gap ratios over the range $0.38 < d_1/d_2 < 0.88$ suggests that the separated flow pattern for minimum drag corresponds fairly well with a free-streamline model. For values of d_1/d_2 closer to unity, the good correspondence does not hold up because the pressure coefficient on the separation streamline would be tending to negatively infinite values. For values of d_1/d_2 less than 0.38 the minimum values of C_D occur at values of gap width considerably smaller than for the free-streamline model. This is because, for larger gaps, the separation surface begins to close itself ahead of the rearbody and the flow field is quite different from that which is modeled by the free-streamline theory.

Some other results from the free-streamline analytical model may also be used for comparison with our experimental results. In particular, the free-streamline constant pressure associated with each d_1/d_2 can be compared to the measured pressure in the gap between the two bodies of our test configuration. To use the two-dimensional analytical result would be greatly in error; the axisymmetric result is the appropriate one and it can be estimated semi-empirically, as follows, without resort to the complete numerical calculation.

Temporarily regarding the free stream surface as a solid surface, the semi-infinite half body so defined has zero drag in potential flow, i.e.

$$D = D_{1_f} + D_s = 0 \tag{4}$$

where D_{1f} and D_s are the contributions from the front of the disc and from the constant-pressure surface, respectively, and $D_{1f} = C_{D1f} q_\infty A_1$ while $D_s = C_{ps} q_\infty (A_2 - A_1)$. This relationship ($D_s = -D_{1_f}$, or $C_{p_s}(A_2 - A_1) = -C_{D1f} A_1$) can be applied to the free-streamline model. If the relationship between C_{D1_f} and C_{p_s} can be established, (4) will yield an expression for C_{p_s} in the model. A suitable relationship can be determined from experimental results as follows.

For isolated discs,

$$C_{D1f} = C_{D1} + C_{p_s} \tag{5}$$

where C_{D_1} is the drag coefficient of the disc. An empirical expression for the dependence of C_{D_1} on C_{p_s} may be obtained from measurements of the drag of bodies with cavity wakes collected by Perry & Plesset (1953). The data for discs, for values of C_{p_s} down to -0.25, are well fitted by the expression

$$C_{D_1} = 0.80(1 - C_{p_s}) \tag{6}$$

Thus from (5), $C_{D_{1f}} = 0.80 + 0.20\, C_{p_s}$ which, when substituted in (4), gives for the stream-surface pressure coefficient the result

$$-C_{p_s} = \frac{0.80}{\left(\frac{d_2}{d_1}\right)^2 - 0.80} \tag{7}$$

This semi-empirical result is shown in Fig. 5 (down to values of C_{p_s} as low as -3!) together with the single value determined by Strück from his numerical calculations.

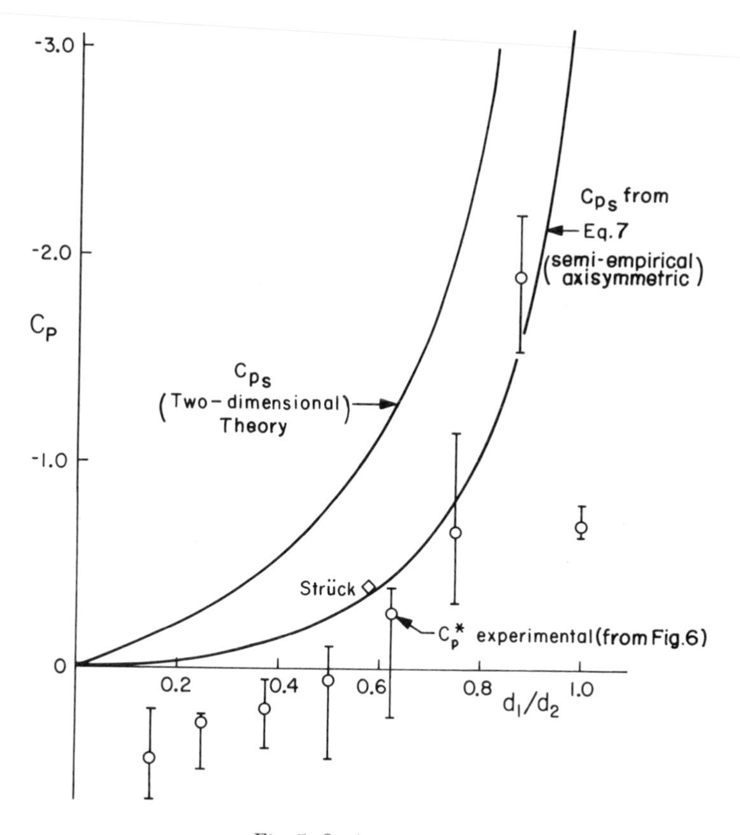

Fig. 5. Optimum-gap pressure.

References p. 273.

In our experiments, the base pressure on the disc was not measured, but pressures in the gap were estimated from pressure distributions measured on the face of the rearbody for optimum configurations, and shown in Fig. 6. Average values of the gap pressure coefficient, C_p*, that were estimated from these distributions are plotted in Fig. 5; they are weighted toward values deeper in the cavity, the excursions near the top being attributed to dynamic effects associated with the reattachment region and with a trapped vortex (these effects are discussed later). The maximum excursions about the selected average values are indicated by the bars in Fig. 5. Again, as in Fig. 4, a rough correspondence between measurements and the free-streamline model is observed. For the optimum axisymmetric test configurations, C_{p_s} of the free-streamline model compares fairly well with the measured average gap pressure. The most notable difference is that in the theoretical model C_{p_s} is always negative, but in the experiment C_p* becomes positive for $d_1/d_2 < 0.5$.

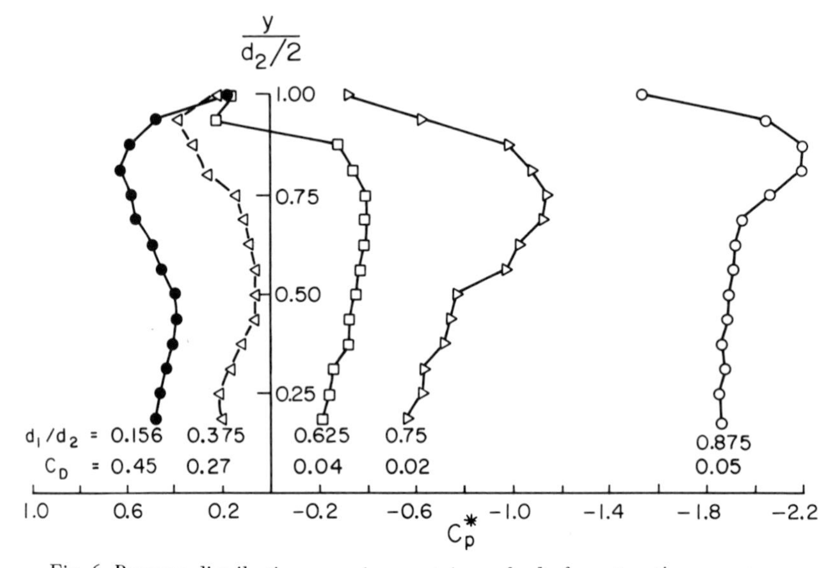

Fig. 6. Pressure distribution on axisymmetric rearbody face at optimum gap.

As a final comparison between theory and experiment, we have attempted to obtain an estimate of the forebody system drag for the optimum geometries. We note that the flow at optimum geometry appears to be a more general example of the flow over a simple gap or cutout in a flat surface with uniform flow over it, and it is constructive to compare the two cases as follows. For the cutout in uniform flow, the free-streamline solution is the simple one, $C_p = 0$ (i.e. $p = p_\infty$) on the streamline spanning the gap and in the gap itself, so that $C_D = 0$. Now the effect of viscous and diffusive effects on the free streamline (which is now the dividing streamline in the free shear layer) is to produce a shear stress along the dividing streamline and a departure from $p = p_\infty$ on the cutout walls. The integral of the perturbed pressure over the walls, together with a small negative contribution from the shear stress on the bottom, gives the drag of the cutout. A simple momentum balance on the fluid enclosed by the cutout and the dividing streamline shows that the drag

$$D = \int_s \tau_s \, d\,A_s$$

where the integral is taken over the free surface and τ_s is the shear stress on that surface. Extending this idea to our case of the axisymmetric forebody system, we assume that the shear along the free surface is similarly related to a pressure perturbation $p - p_s$ on the cavity walls (i.e. the front of the rearbody and the back of the frontbody or disc), and we neglect any pressure perturbation on the free surface. Again, a momentum balance gives

$$D = 2\pi \int_s \tau_s r_s \underset{\sim}{n}_x \bullet \underset{\sim}{ds} = 2\pi \int_s \tau_s r_s dx \tag{8}$$

where $\underset{\sim}{n}_x$ is the direction cosine for the free surface and $\underset{\sim}{ds}$ is a vector element of length along it. The momentum balance properly takes into account the fact that, in the inviscid, free streamline, zero-drag case ($C_p = C_{p_s}$ throughout the gap), the drag on the frontbody, including pressures on its front and back, is just balanced by the negative drag ($C_{p_s} < 0$) on the face of the rearbody.

To evaluate the drag from the integral in (8), we assume that the free shear layer across the gap is a completely developed (self-similar) turbulent flow at constant pressure. For such a flow the value of τ_s is constant on the dividing streamline. The assumption that the path of integration, s, in (8) coincides with the dividing streamline introduces some approximation into the equation, as does the assumption that the value of τ_s is constant all the way to the reattachment point. With these assumptions we use the value which can be determined from the measurements of Liepmann & Laufer (1947), namely, $\tau_s = 0.0115 \rho U_s^2$, where U_s corresponds to the flow velocity outside the free shear layer and hence is related to the corresponding pressure coefficient C_{p_s} (which is also $C_p{}^*$ for the gap) by $(U_s/U_\infty)^2 = 1 - C_{p_s}$. Making these substitutions, (8) becomes

$$D = 2\pi (0.0115) \rho U_\infty^2 \, (1 - C_{p_s}) \int_{gap} r_s \, dx$$

The integral could be evaluated numerically if the shape of the free stream surface, $r_s(x)$, were available. Instead, we simply use an average value $r_s = ½(r_1 + r_2)$, which is

References p. 273.

equivalent to assuming that the free stream surface is a truncated cone. With this approximation the result is

$$D = \tfrac{1}{2}\pi(0.0115)\rho U_\infty^{2}(1 - C_{p_s})(d_2 + d_1)g \tag{9a}$$

and

$$C_D = 0.046(1 + \frac{d_1}{d_2})(1 - C_{p_s})\frac{g}{d_2} \tag{9b}$$

To obtain C_D as a function of d_1/d_2, we use (7) for C_{p_s} and the two-dimensional free-streamline calculation for g^*/d_2 shown in Fig. 4 since it appears to fit the measurements in the low-drag range. The results are plotted in Fig. 7. Also shown there is one point for which the data from Strück's single numerical axisymmetric calculation were used.

Fig. 7 suggests that our approximations are not unreasonable. For $d_1/d_2 \to 0$, it is clear that $C_D \to \infty$ because, in (9b), $1 - C_{p_s} \to 1$ while $g/d_2 \to \infty$ since the radius of curvature of the separation stream surface varies inversely with C_{p_s}. However, for $d_1/d_2 \to 1$, $1 - C_{p_s} \to \infty$ while $g/d_2 \to 0$ and it is not clear *a priori* what their product will be. For the two-dimensional case, where an exact theory is available, we find $C_D \to \infty$, but for the present calculation $C_D \to 0$.

Whatever the limiting behavior for $d_1/d_2 \to 1$, the agreement of C_D with the magnitude of the experimental values of C_D in the range $0.625 \leq d_1/d_2 \leq 0.825$ is convincing confirmation that in this range the flow over the gap is a simple cavity flow in which the shear layer is the classic turbulent mixing layer. On the other hand for $d_1/d_2 < 0.6$, (i.e. $g^*/d_2 > 0.5$, from Fig. 4) there is a large discrepancy, the measured C_D being much higher than the theoretical value, suggesting a departure from the flow conditions assumed in the model. This is discussed in the following section.

CRITICAL GAP RATIO

For values of g^*/d_2 which are less than about 0.5, i.e. $d_1/d_2 > 0.6$, the flow, as experienced in the wind tunnel and as observed on the flow pictures from the water tunnel, is well behaved. The separation surface is, of course, a turbulent free shear layer, and it appears to develop normally. For larger gaps, however, it becomes unsteady on a larger scale, suggesting some kind of gap-coupled oscillation. For optimum geometries having $g^*/d_2 > 0.5$, the conditions in the turbulent mixing layer are no longer those of a normal layer such as that studied in the Liepmann-Laufer experiment. As may be seen from Figs. 2 and 7, for $d_1/d_2 \geq 0.625$, values of $C_{D_{min}}$ are very low, less than 0.05, while for $d_1/d_2 \leq 0.6$ values almost an order of

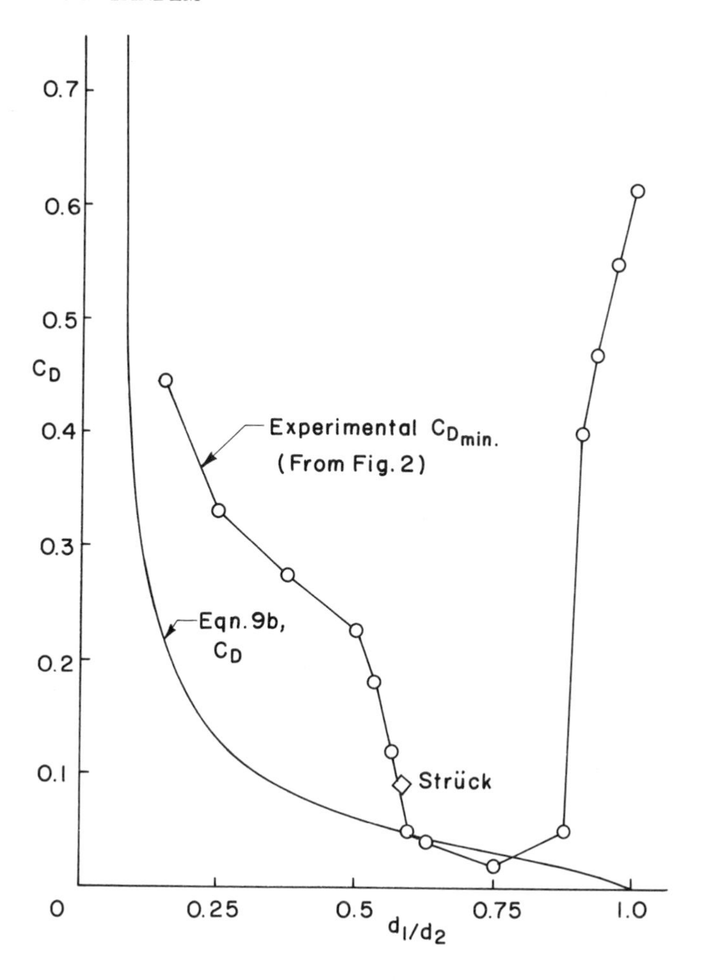

Fig. 7. Comparision of $C_{D\,min}$ and axisymmetric cavity momentum balance.

magnitude higher are attained for optimum configurations. We shall call these the critical values and, temporarily, assign them the values $(d_1/d_2)_{cr} = 0.6$ and $(g^*/d_2)_{cr} = 0.5$. Smaller values of gap ratio will be called subcritical and larger ones supercritical.

Evidence of a critical gap ratio can also be seen in some of the figures previously discussed. For example, it is possible that the critical condition is reached when the gap pressure coefficient passes through zero (Fig. 5); this occurs at $d_1/d_2 = 0.5$ rather than 0.6, but as has been discussed, there is some uncertainty in the correct representative values for gap pressures.

References p. 273.

Also, the pressure distributions in optimum gaps (Fig. 6) show some interesting effects. In particular, the positions of the minima are believed to correspond to the positions of a vortex, which is especially stable and visible in the flow visualizations for subcritical gaps. From Fig. 6 it may be seen that the radial location of this minimum changes rather abruptly at the critical gap (and at $C_p = 0$); the vortex becomes less well-defined, perhaps because of unsteadiness. An interesting observation is that the existence of a steady vortex in the gap is not at all incompatible with low drag and, indeed, may be necessary in some cases.

Also of interest in Fig. 6 is the locus of the pressure maximum, which should correspond to the mean stagnation point of the separation streamline. For subcritical gaps, i.e. the very low values of C_{Dmin}, this position is at or close to the edge of the rearbody face while for supercritical gaps (higher values of C_{Dmin}) it moves some distance inside the edge toward the centerline.

For d_1/d_2 greater than 0.88 the measured values of optimum drag also depart strongly from the values given by (9b). The reason is that the free shear layer does not reattach smoothly because the adverse pressure gradient on the rearbody is now too high.

To conclude this section, there is strong indication of a critical gap ratio below which unusually low values of optimum drag can be achieved. It may be quite useful to consider this behavior when designing for practical applications. The change in flow that occurs for optimum gap ratios larger than critical may correspond to the appearance of a cavity oscillation, but this requires further investigation.

RESULTS FOR SQUARE CROSS–SECTION

Many of the measurements made on the axisymmetric system were repeated for a system with square cross-section, i.e. a box with a square plate in front of it. The question naturally arises whether the same large reductions of drag can be realized as in the axisymmetric case. One might expect the situation to be less favorable since the separation surface leaving the front plate will not retain a square cross-section and so will not reattach smoothly everywhere onto the leading edges of the rearbody. A similar situation would exist in most practical applications. The following figures show the main results.

In Fig. 8 the variation of C_D with g/d_2 is shown for only two values of d_1/d_2, namely 0.25 and 0.75, although measurements were also made for other values. The trends are roughly similar to those for the circular cross-section, with which they are compared in Fig. 8. However, there are some important differences. For the subcritical gap ($d_1/d_2 = 0.75$) the minimum value of C_D is considerably higher than for the circular section, about 0.07 compared to 0.02, although it still represents a full order of magnitude decrease in the drag. That is, in the region where smooth flow onto the rearbody is crucial, the mismatch between the separated surface and rearbody has a large effect. For the supercritical gap ($d_1/d_2 = 0.25$), there is rather little (percentage) difference between the two cases, suggesting that the mismatch is not so important as for the subcritical gap. This would be consistent with the existence of a large scale oscillation at the supercritical values; that is, one can suppose that a mismatch of amplitude comparable to the amplitude of oscillation of the reattachment point would make little difference.

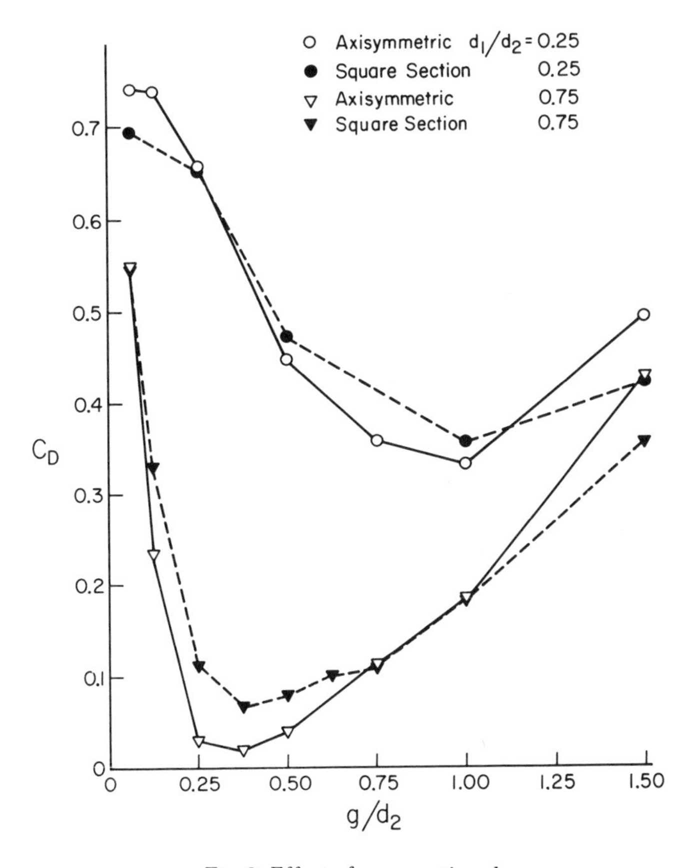

Fig. 8. Effect of cross-section shape.

Side view flow pictures obtained with dye injection are similar to those for the axisymmetric case shown in Fig. 3 and are not included here. More interesting views were obtained by observing the face of the rearbody from an oblique angle. By introducing air into the gap a cavity flow could be created which, while not precisely the same as the fully wetted flow, gave a good impression of the shape of the free

References p. 273.

surface and verified the main features indicated by dye injection in the fully wetted flow. A sketch of the observed flow at an optimum condition (i.e. a minimum value of C_D) is shown in Fig. 9. The most notable feature is the form of that part of the separation surface which springs from a corner of the front plate. In the cavity flow, the wedge-shaped portion of surface is very much like that sketched; it reattaches on the rearbody face inside the corner and is deflected inward. It is probably this reattachment region near the corner that makes the major contribution to the increase in drag as compared to the axisymmetric case, and one might expect some reduction of C_D by cutting away those corners appropriately.

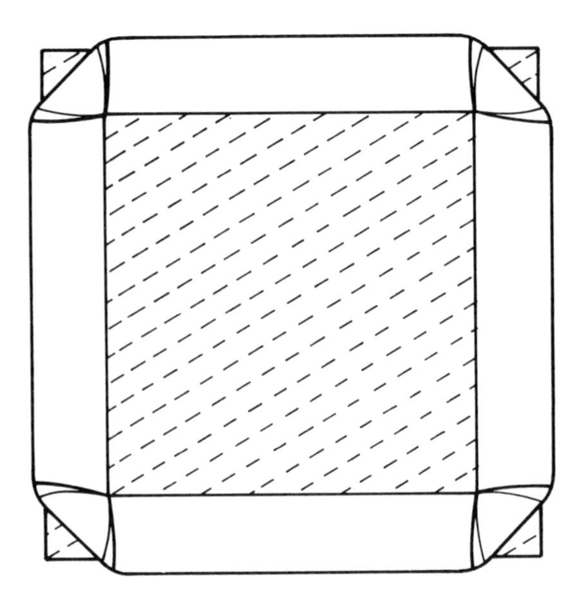

Fig. 9. Front view of cavity separation surface of square cross-section at optimum gap. Cross-hatched areas denote the exposed faces of the front and rearbodies.

ROUNDED CORNERS ON THE AXISYMMETRIC AFTERBODY

The high drag on the rearbody alone could, of course, be drastically reduced, without assistance from a frontbody, by simply rounding its edges sufficiently. A rounding radius equal to one-eighth the body diameter is quite sufficient to reduce the drag of the front face of the rearbody to zero, in the absence of any frontbody, provided the Reynolds number is large enough that premature laminar separation does not occur. In Fig. 10, the rearbody-alone point at $g/d_2 = 0$ and $Re = 500{,}000$ shows the effect of such separation. However, placing a small disc against the face of the rearbody provides sufficient tripping action to avert laminar separation, and the drag coefficient goes to zero.

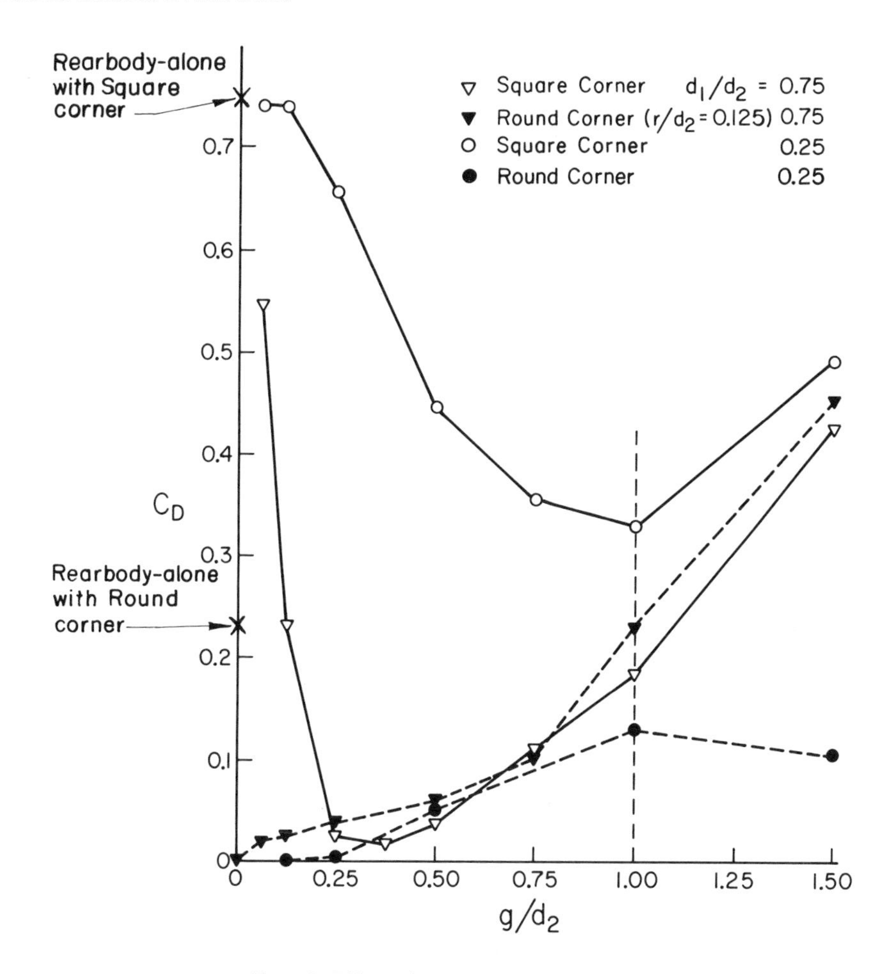

Fig. 10. Effect of corner radius on rearbody.

It is of interest to see what effect a frontbody would have when the rearbody has rounded edges. Some results are shown in Fig. 10 for axisymmetric bodies. Again, the comparison is made for two cases, $d_1/d_2 = 0.75$ and 0.25, respectively, with and without rounded edges. For $d_1/d_2 = 0.75$ and small values of g/d_2 (less than 0.2) rounding of the rearbody results in a very large reduction of drag, because the flow which has reattached onto the face of the rearbody can develop suction on the rounded edges. For large gap ratios ($g/d_2 > 0.2$), on the other hand, the rounding is not helpful. In fact, for intermediate gap ratios, where the square-edged body system has very low drag, rounding of the edges is actually detrimental. For $d_1/d_2 = 0.25$ there is always a very beneficial effect of rounding. This is because the wake of the smaller frontbody is probably already closed at $g/d_2 = 1$ while for the large frontbody ($d_1/d_2 = 0.75$) this development would be delayed to larger values of g/d_2.

References p. 273.

It is clear that for all cases, rounding of the rearbody must be beneficial for $g/d_2 \rightarrow \infty$ since it becomes independent of the frontbody. The vertical line drawn at $g/d_2 = 1.00$ refers to the case $d_1/d_2 = 0.25$; it indicates that for gap ratios greater than 1.00, the two bodies are tending to become independent (see eq. 3), with the system drag coefficient tending to the following values:

$$C_D \rightarrow 0.75 + 1.2\left(\frac{1}{4}\right)^2 = 0.83\,, \quad \text{square corner}$$

$$C_D \rightarrow 0 + 1.2\left(\frac{1}{4}\right)^2 = 0.08\,, \quad \text{rounded corner}$$

For $d_1/d_2 = 0.75$, the asymptotic values are

$$C_D \rightarrow 0.75 + 1.2\left(\frac{3}{4}\right)^2 = 1.43\,, \quad \text{square corner}$$

$$C_D \rightarrow 0 + 1.2\left(\frac{3}{4}\right)^2 = 0.68\,, \quad \text{rounded corner}$$

A most interesting result of the rounding experiment is that for some gap ratios (subcritical and near optimum) square edges are better than rounded ones. This may be important in those instances where some frontbody is required. A closer investigation of the reattaching flow for these conditions needs to be made, however.

CONCLUDING REMARKS

From these results, it appears that there are three flow regimes for the system with a square-edged rearbody that was investigated: (1) If the shielding frontbody is absent or not optimized, the drag coefficient for the system has the ordinary, bluff-body value of order unity. (2) For a well designed system and for optimum values of gap ratio less than critical, the drag coefficient can be reduced almost two orders of magnitude below bluff-body values. (3) For optimum gap ratios larger than critical, the drag coefficients are of intermediate magnitude.

Although the investigation is not complete, it appears that these changes may be characterized by the scale of the nonsteady motions that occur. In the low drag range (2), the scales of the eddies in the free turbulent shear layer spanning the gap are small, i.e. no larger than about 0.3g, and the eddy motion is independent of the gap geometry; in the intermediate regime (3), fluctuations of larger scale occur, possibly because of a cavity oscillation; in the high drag, bluff-body regime (1), the oscillations are of a scale comparable to the body diameter.

A better understanding of these drag related mechanisms and their relation to the system geometry would be helpful in designing for the substantial drag reductions that appear to be technically feasible.

ACKNOWLEDGEMENTS

The work described here was supported with funds from the Ford-Exxon Energy Research Program of the California Institute of Technology and is being continued under a National Science Foundation Grant. For assistance in the early stages of the research, we are grateful to Professor F. Clauser, who designed and built the force balance for the Merrill Wind Tunnel, and to Mr. R. Breidenthal, who did some early measurements on the drag of a half body. Thanks are also extended to Mr. W. Bettes for various helpful suggestions and to Mr. Till Liepmann for assistance with the experiments.

REFERENCES

Hoerner, S. F. (1965), Fluid-Dynamic Drag, published by the author, Brick Town, N.J.

Liepmann, H. W. & Laufer, J. (1947), Investigations of Free Turbulent Mixing, NACA TN 1257.

Plessel, M. S. & Perry, B. (1954), On the Application of Free Streamline Theory to Cavity Flows, California Institute of Technology.

Roshko, A. (1954), A New Hodograph for Free Streamline Theory, NACA TN 3168.

Roshko, A. (1955), Some Measurements of Flow in a Rectangular Cutout, NACA TN 3488.

Saunders, W. S. (1966), Apparatus for Reducing Linear and Lateral Wind Resistance in a Tractor-Trailer Combination Vehicle., U. S. Patent Office 3, 241, 876.

Strück, H. G. (1970), Discontinuous Flows and Free Streamline Solutions for Axisymmetric Bodies at Zero and Small Angles of Attack, NASA TN D-5634.

Ward, T.M. (1976), The Hydrodynamics Laboratory at the California Institute of Technology, Graduate Aeronautical Laboratories, California Institute of Technology.

DISCUSSION

W. W. Willmarth *(University of Michigan)*

I have some comments about the unsteady flow in the separated shear layer between your front and rearbodies. We did some related work at the University of Michigan involving the development of an open window for an aircraft optical device. And, in talking to Dr. Hucho, it sounds as though there are similar ground vehicle problems with sunroofs. Years ago, when I was at Cal Tech, Karamcheti and Roshko looked at the flow over a cavity and found a tremendous radiation of sound, on the order of 140-160 dB, even at 200-300 ft/sec. What happens is that the shear layer flaps across the back edge of the cavity and pumps air in and out. The ingredients of the problem are the cavity volume, a lot of turbulence, and a very thin shear layer compared to the cavity length.

The first thing we did in trying to solve the problem with our open optical window was to look at a two-dimensional configuration (see Fig. 11). We tried to hop the shear layer over the cavity with a small, upstream ramp so that reattachment definitely occurred downstream of the rear edge. That cut the sound radiation quite a bit, but not entirely. At that point we hadn't done anything about the thin shear layer, so we took a crack at that. Practically, it wasn't feasible to increase the upstream body length to get a thicker boundary layer, so we came up with the scheme of using a series of high drag, porous fences just upstream of the cavity. The fences

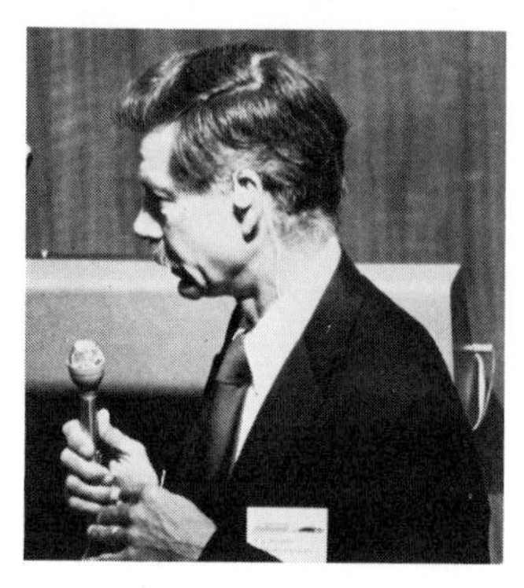

W. W. Willmarth

had to be porous. We tried all combinations, but the best seemed to be with the biggest one first and the next ones succeedingly smaller in a linear fashion. Three or four are all you need. As I see it, the idea was to make some big eddies with the leading fence and then chop them up by having the flow go through and around the smaller screens. The shear layer over the cavity becomes very thick and, as a result, the reattachment is reasonably steady, velocities in the cavity are quite low, and you can kill the resonance.

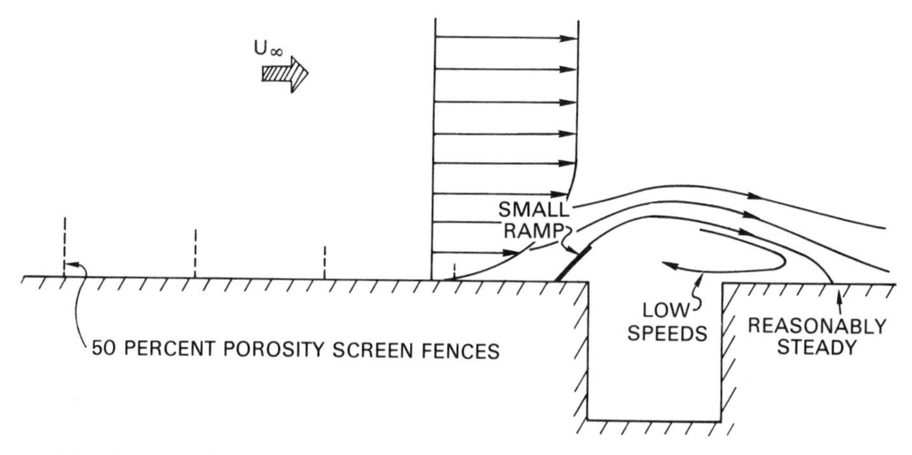

Fig. 11. Control of flow over a two-dimensional cavity using porous fences and a ramp.

The final configuration of the optical device (see Fig. 12) was axially symmetric and more complicated than our two-dimensional configuration. It also had to be insensitive to angle of attack over a range of ±12 degrees at speeds of 100 to 200 m/s. We measured the pressure in the cavity as a function of time, and the odd thing we found was that the smallest fluctuations occurred at a tilt of about 4 degrees rather than at an angle of zero. The rms pressure in the cavity was comparable to that found on the wall beneath a flat plate turbulent boundary layer; the rms velocity was only 0.006 U_∞ with maximum fluctuations of only 0.04 U_∞. It was really quiet. The solution to the cavity problem we worked out here might work on any cavity where you have a very short run upstream of the separated shear layer.

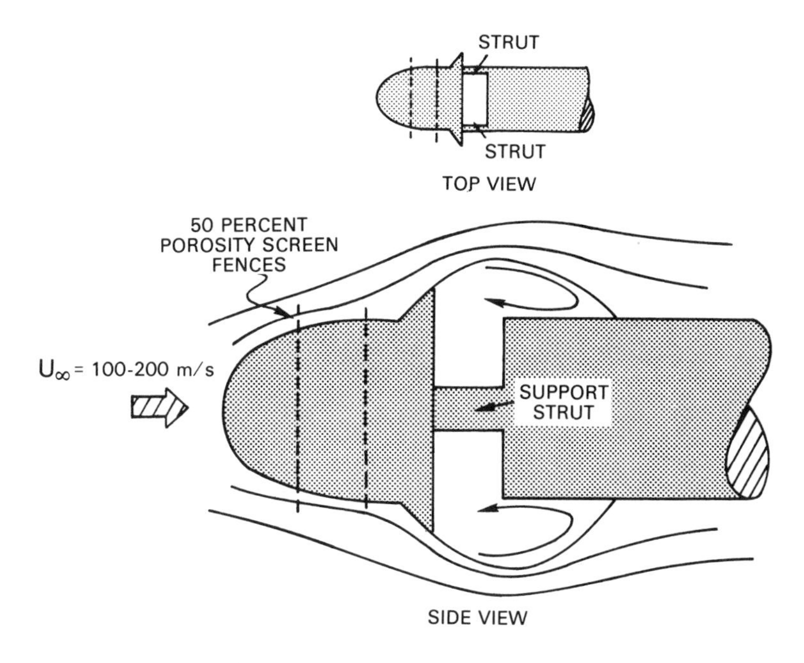

Fig. 12. Practical application of fences and ramp to control flow over a cavity in an axially symmetric body. Additional side support strut detail is shown in smaller top view sketch.

A. Roshko

I think that's a nice piece of work and very relevant to our two-body problem. It seems quite likely to me that a changeover from a highly oscillating to a very quiet cavity flow may correspond to what we called our critical gap length. I would interpret your results as follows, which is, I think, basically the way you looked at it. If you have a free shear layer, even though it's turbulent, you can think of that layer as having an instability problem itself. That is, it might build up large oscillations. The distance required to build up such oscillations is measured in terms of the initial thickness of the layer. I would think that thickening the boundary layer increases the distance required for a large oscillation to build up, and that's perhaps what solved your problem. Similarly, in terms of our critical gap, it would seem that by increasing the boundary layer thickness ahead of the gap, you could extend the critical gap length to a larger value. That's certainly something we'll want to try.

R. T. Jones *(NASA-Ames Research Center)*

Don't you think you could suppress the cavity oscillation by putting a screen just inside the cavity, parallel to the mean flow? I'm sure that if you put a little screen inside the mouth hole of a flute, the player would never be able to make a sound. So you shouldn't need things like fences and ramps on the *outside*, just some damping of the motion *inside.*

W. W. Willmarth

I think the key thing with our oscillating cavity flow *and* with Roshko & Koenig's supercritical and subcritical gap ratios, is the pumping action of the thin shear layer as it leaves the upstream edge of the cavity. If you can do something upstream that will reduce the gradient of the shear, then there will be much less entrainment or pumping. It's the pumping that sucks the cavity down to low pressures and starts the oscillation.

D. J. Maull *(Cambridge University, England)*

I have a comment on Professor Willmarth's work. The cavity problem has been looked at almost ad nauseum, starting with the problem of bomb bay buffeting. His precise problem was almost completely cured by rounding the downstream edge of the bomb bay cavity. I think it makes sense, intuitively, that flows don't like to reattach, nor do humans, on sharp corners. I think it would be better to treat the downstream edge of the cavity rather than trying to modify the approach flow.

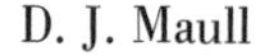

D. J. Maull A. Roshko

A. Roshko

I don't think you can always depend on treating the downstream edge. There's a remarkable counter-example that we saw in the case of supersonic flow over a cavity. With a square downstream corner the flow was rather quiet – there was very weak external radiation of noise. However, when we rounded that corner, which we hoped would even further reduce the radiated noise, we got an incredible whistling – a very strong external noise radiation.*

R. Sedney *(U.S. Army Ballistic Research Laboratories)*

I have another addition to Professor Willmarth's work. About eight years ago we worked on the same problem at the Martin Company. The solution was a ramp which came up at about 45 degrees and bled off about 1/4 of the boundary layer (see Fig. 13). There was a porous plate at the downstream end of the ramp. It was a tremendously efficient solution for cutting down cavity oscillation and external noise radiation.

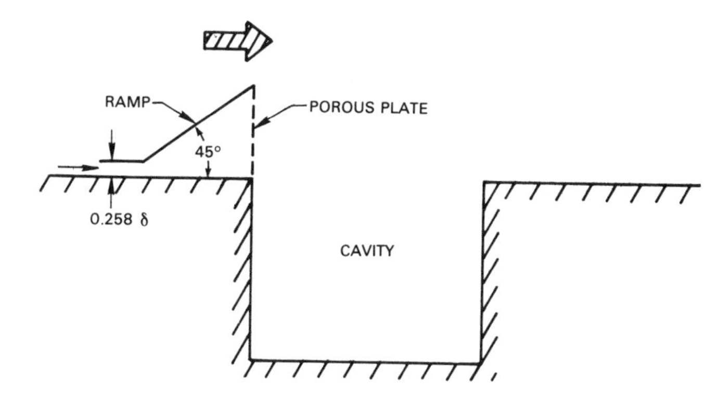

Fig. 13. Combination of ramp, boundary layer bleed, and porous fence to control flow over a cavity.

P.B.S. Lissaman *(Aerovironment, Inc.)*

We've developed a porous air deflector for trucks that seems to exhibit some of the features that have been discussed in conjunction with these cavity flows. Because of our deflector's intentionally selected porosity, the buffeting transmitted to the tractor is enormously reduced. It also seems to work very well in cross winds, and we believe this has something to do with the small scale of turbulence coming through the perforated screen, as well as with the lip-like protrusions provided on its side edges.

R. T. Jones

In all this discussion of the gap flow details have we forgotten that one of the main points of your paper is that you succeeded in *eliminating* the drag of the forebody?

---

*Thomke, G. (1964), *Separation and Reattachment of a Turbulent Boundary Layer Behind Downstream Facing Steps and Cavities, Douglas Aircraft Company Report SM-43062.*

A. Roshko

I think the questions connected with the critical gap are perhaps pertinent in trying to understand the mechanism that destroyed the very low drag levels at subcritical gaps.

R. T. Jones

Since your test body was semi-infinite, its drag did not include that of the base. What final drag coefficient can you predict for us if you include the base drag?

A. Roshko

The base drag coefficient for one of these bodies is about 0.2. Since the forebody drag would be zero, the total drag would also be 0.2.

R. T. Jones

This would be the minimum total drag at zero yaw angle. Crosswinds would create non-zero yaw angles. Wouldn't you have to also put dams in somewhere else to prevent a mismatch between the gap shear layer and the leading edges of the rearbody?

A. Roshko

I think you might have to have *adjustable* flaps, or dams, on the frontbody if you really wanted to do it right, that is to *always* guide the flow precisely onto the leading edges of the rearbody at all angles of yaw. It would get rather complex.

H. H. Korst *(University of Illinois)*

Why have you chosen the dissipation integral from minus ∞ to plus ∞ as the significant force contribution? It would appear that it might be a good idea to use the momentum integral from minus ∞ to the reattaching streamline instead. Did you try that?

A. Roshko

No, we did not. That would be another way to do it. In fact, with the momentum integral it is easier to keep the various terms under control, because in applying the dissipation integral to the gap flow you really are neglecting some dissipation which is occurring farther downstream.

Editors' Comment: In preparing their paper for publication in the Proceedings

Roshko & Koenig did in fact reformulate their calculations according to a momentum balance.

H. H. Korst

My second question concerns the hodograph theory. You probably considered the change in the pressure distribution on the front face of the plate in the sense of Jakob. When Jakob, in 1926, started the modification of the free streamline theory with parametric pressures in the wake, he also found a modification of the pressure distribution on the front face of 2-D plates. That provided a very attractive way of obtaining simple things, such as plate drag, in a more realistic manner. If you take the actual pressure distribution that results from such a parametric definition of the free streamline, you come up with values that are quite close to those we expect from experiment, instead of the high values that you get. Have you considered that in advance?

A. Roshko

Jakob's method, or that of Parkinson & Jandali which is pretty much the same thing, or other free streamline methods, all have the base pressure as a parameter. My impression for something like a plate or a disk is that once you have chosen the base pressure coefficient, that fixes the pressure at the edges of the plate or the disk, and also the pressure distribution over the whole front face of the body. So the force calculated from the method comes out very accurately. So, I don't think that this is going to produce very much difference.

D. J. Maull

Several years ago we had a look at the inverse of Roshko & Koenig's problem – a cavity in an axisymmetric pipe. The reason we were looking at the axisymmetric case was that when we had first tried a two-dimensional cavity, the flow broke down into cells across the width of the cavity. Unfortunately, we found that the axisymmetric case also broke down into cells. These rotated very slowly around the cavity, unless they were fixed in circumferential position by a small, local asymmetry in the cavity geometry. Did you find any cellular structures in your cavities?

A. Roshko

I must admit that we haven't looked at it closely enough to notice. But I think it's interesting to know that a cellular breakdown in a two-dimensional cavity might actually correspond to a circumferential instability.

S. Saunders *(Rudkin-Wiley Corporation)*

Professor Roshko has attributed to me a great deal more understanding of the phenomenon of low drag achievable with bluff bodies in tandem than I've ever had. I originally had the notion of trying to guide the flow over the rearbody with turning vanes. It wasn't until I was preparing a course using Von Mises book* and saw an illustration of the interference effect between two disks, that I experienced a flash of an idea that led to the Airshield™. I realized that the combination of a disk and a cylinder was applicable to the problem of tractor-trailer drag; I think the whole process was strictly serendipity.

When I did the original work, I was concerned with the problem of turbulence and I tested the disk-cylinder combination under three approach flow conditions. The first was in typical wind-tunnel-type low turbulence, and the second and third were in higher turbulence, 4 or 5 percent, with two different scales. One scale was of the order of the diameter of the body; the other was several times the diameter. The tests were quite crude and the results were never published. The net effect appeared to be that you *do* have a resonant system. If the oncoming turbulence contains energy in the range of resonance, you can degrade the performance of the device. When the disk-cylinder combination is close to optimum, the degradation in performance is very low. However, under off-optimum conditions the situation is more critical and the degradation increases.

W. A. Mair *(Cambridge University, England)*

If I understood what Professor Roshko said correctly, I am amazed to find that for the subcritical diameter ratio the only advantage of rounding the edges of the rearbody occurred at the less-than-optimum gap ratios, and that around the optimum condition rounding those edges actually *increased* the drag. Do you think that's because there was a trapped vortex in the square-edged case that wouldn't sit there as quietly when the edge was rounded? And extending that argument, if it's correct, what one really wants to do, perhaps, is to find a way of making that vortex sit there with a rounded edge. I don't know how you would do that, but you might then be able to get the square-edged drag minimum and the round-edged drag advantage at less-than-optimum gaps with the same configuration.

A. Roshko

I believe your arguments are correct. I think that our flow visualization studies actually showed that the vortex *was* a little bit more unsteady with a rounded edge. But why that's so, I don't know. One's intuition would almost say that a rounded edge, where the vortex doesn't have to precisely position itself as on a sharp corner, would be a more stable situation. It seems to be quite the opposite. That example I mentioned earlier involving supersonic flow over a cavity is another example of this same situation; I really don't understand either one.

**Von Mises, R. (1945), Theory of Flight, McGraw-Hill Book Company, Inc., pp. 95-98.*

T. M. Barrows *(Department of Transportation)*

In the axisymmetric case there's evidently a ring vortex between the disk and the cylindrical rearbody; in your square case there must be something similar to a ring vortex between the plate and square cross-section rearbody. When a free-standing ring vortex gets out of round, it tends to be unstable. Do you have any feeling for what the shape of the vortex was in the square case? Was it closer to round, or square?

K. Koenig *(California Institute of Technology)*

Based on our flow visualization studies you could infer that the general shape was slightly rounded on each of the four basic sides with diagonal-like corners. In the optimum case the vortex seemed to wander a little bit inside the cavity, compared to the axisymmetric optimum condition in which the vortex was tightly locked in the cavity. Even though the square case's vortex wandered a little and had this unusual

K. Koenig

shape about half-way between axisymmetric and square, it wasn't unstable. You could perturb the model or the airstream rather severely and it always stayed there.

A. Roshko

When the vortex does move or oscillate a little bit, it's not clear whether the instability originates with the vortex itself or with the reattachment area. Either instability would perturb the other.

A. Leonard *(NASA-Ames Research Center)*

On Fig. 2 you gave us a hint of some results where the frontbody was larger than the rearbody; in fact, you showed that the drag coefficient was less than 1.0 and decreasing with increasing gap. Eventually, at some large gap the coefficient must reach 2.0 so there has to be a minimum at some intermediate gap. Did you study that minimum?

A. Roshko

We didn't study any cases where the frontbody was larger than the rearbody. The data you are referring to in Fig. 2 was for equal diameters. It's hard to say *why* the drag initially decreases with increasing gap but, as you said, a minimum *will* be reached. And when they get far enough apart, the total drag *has* to be the sum of the two.

For the case where the frontbody *is* larger than the rearbody I have thought of the following possibility. If the frontbody is properly shaped such that separation is avoided, and if the free shear layer is then directed onto the leading edges of the rearbody (see Fig. 14), you might end up with an optimum configuration whose drag coefficient, based on the diameter of the rearbody, would be substantially lower than the 0.8, which is the nominal value for the rearbody alone.

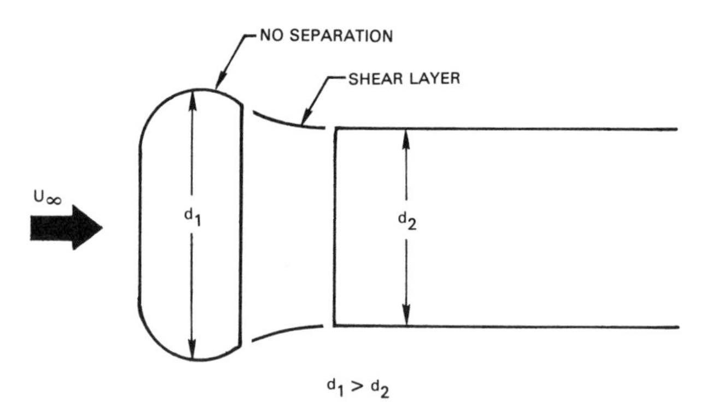

Fig. 14. Possible minimum drag configuration for case where frontbody is larger than rearbody.

F. T. Buckley, Jr. *(University of Maryland)*

Professor Mair commented on the stabilization of the wake flow behind an axisymmetric cylinder by extending a rod from the base. He showed that the average base pressure can be increased depending on the length of the rod. In your experimental setup you had such a rod in the cavity. Could you give us your opinion as to the stability of the system *with* the rod compared to what it would be *without* the rod?

A. Roshko

Our support rod was very small – about 1/8 the diameter of the rearbody. My feeling is that it would make no difference if the rod's diameter were made smaller, or even a little bit larger – especially at these small gaps where there's no question of closure of the separated flow from the frontbody. Do you have any opinion about that Professor Mair?

W. A. Mair

I agree with your comment.

K. R. Cooper *(National Research Council, Canada)*

I just wanted to comment that some of the features of the drag data you presented in Fig. 2 for a body away from the ground with a disk ahead of it, are similar to what you find with deflectors mounted on the top of truck cabs in front of trailers. For example, for your subcritical gap ratios, the drag decreased very rapidly with increasing gap and then rose more slowly away from the minimum. This suggests that when a disk of a given size is too close, the separating shear layer strikes the front face of the rearbody and the drag departs from the minimum faster than for the case where the shear layer misses the body. This behavior is exactly what you find with trucks. However, in contrast to your results with corner rounding on the rearbody, we found that with trucks you get lower drag with any given deflector if you round the corresponding trailer corners. This may be due to the fact that in ground proximity you only have to get the flow smoothly onto the top and upper sides of the trailer, rather than all the way around. In spite of this, I think your overall results are remarkably similar to those with tractor-trailers.

W. W. Willmarth

When Dr. Morel was talking about critical flow regimes, it struck me then, and again later, that when you get a transition from one regime to another, sometimes it's steady and sometimes it's unsteady. It depends on whether the vorticity in the new regime can find a home, i.e. does it enter into the mean flow freely, and so on and so forth. So, you really *should* look at the vorticity. In this connection, I have a question for Professor Roshko. In Fig. 2 you showed that the minimum drag was achieved with a diameter ratio of 0.75. You showed data for diameter ratios less than that, but I don't think there was any data between 0.75 and 1.0. What happens between those two ratios?

A. Roshko A. M. O. Smith W. W. Willmarth

A. Roshko

At a diameter ratio of 0.875 the drag decreases even more rapidly with increasing gap, and the minimum, which is very sharp compared to 0.75, is just a little bit higher and occurs at a smaller gap. After you've passed the minimum it looks as though you get a jump to another flow regime. It's a complex behavior; we didn't put the data on Fig. 2 partly because it would have cluttered up the other curves, and partly because it would have taken a little more explanation than we were prepared for.

A. Roshko

C. Dalton *(University of Houston)*

I have some data* for a group of circular cylinders that might be of interest in the context of Professor Roshko & Mr. Koenig's work. We looked at three cylinders in a line parallel to the freestream direction, all of equal diameter, d, with the face-to-face gap spacing specified as x. If we plot the drag coefficient for the upstream cylinder as a function of x/d, we get the distribution shown in Fig. 15. Near a gap of zero the coefficient is about 1; away from zero a slight minimum occurs, and at large gaps the coefficient becomes about 1.1. For the middle cylinder the drag coefficient starts at about 0.2 or 0.3 and increases to 0.4 or 0.5. For the downstream cylinder the drag is initially *negative*, anywhere between -0.4 and -1.0 depending on the Reynolds number. As the gap increases, the drag of the downstream cylinder increases and eventually reaches the same level as the middle cylinder at the large gaps. These data were obtained in a small wind tunnel with a turbulence level around 2 percent. The maximum Reynolds number was about 8×10^4 so the flows were laminar.

A.M.O. Smith

I've got a few closing remarks combined with some ideas that occurred to me. First of all, at the outset I said this was *funny* aerodynamics. Dr. A.L. Klein at Cal Tech came up with that term, and from what I've seen during this session it's appropriate. When I talked to him one time, I was looking for the type of information that's been

Dalton, C., & Szabo, J. M. (1976), Drag on a Group of Cylinders, Journal of Pressure Vessel Technology, ASME Transactions, Vol. 99, Jan.-Feb., pp. 152-157.

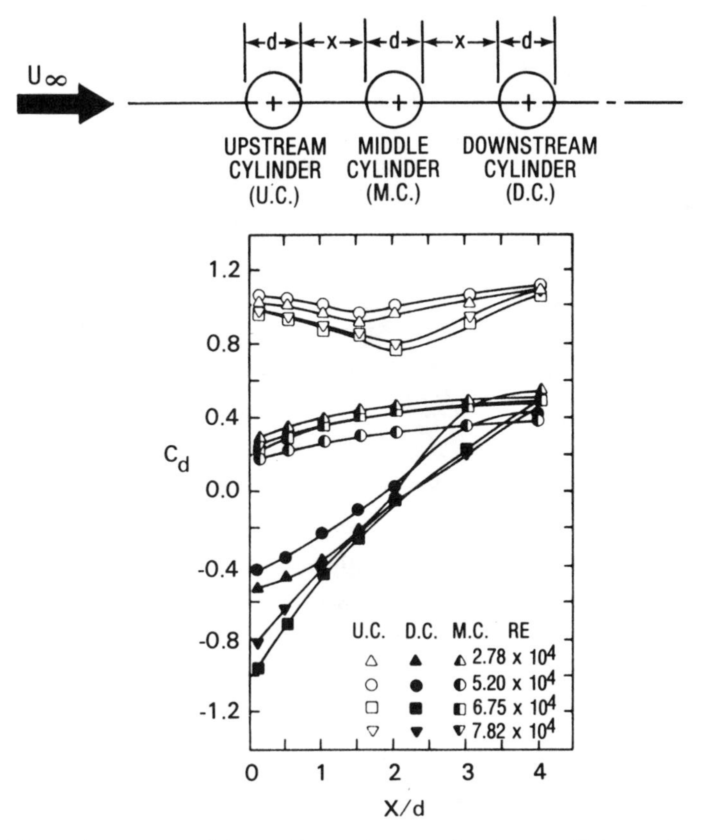

Fig. 15. Effect of spacing and Reynolds number on the drag coefficients of three cylinders aligned in a row parallel to freestream.

discussed in this session and I found that it is mighty rare, so a compendium on this subject will be very useful. The second thing that occurred to me is that this session has been a little more academic compared to the others, which have been a little more applied.

It seems to me we're dealing with only a few types of flows with a lot of variations. One type is that discussed by Roshko & Koenig where there's a "windbreaker" of some sort that has a second body immersed in its separated wake. There are obviously two significant parameters for such a flow – the diameter ratio of the two bodies and the distance between them. So it's not surprising that you find all kinds of drag variations when these parameters are varied, including minimums at optimum combinations. One real world question, of course, for the optimum combinations, has to do with the sharpness of the minimums. If the minimum is real sharp, there would be a problem of running under all conditions and maintaining that minimum. Maybe that defines an area where some work is needed – trying to spread out the conditions at which minimum drag is reached.

The second type of flow we've seen is the swept wing, leading-edge vortex type of thing discussed by Dr. Morel. This is certainly an important flow category. Other important categories include the ordinary wake itself, the flow around sharp leading edges, and the flow over a cavity.

A. M. O. Smith

The unsteadiness of the cavity flow problem seems to come up again and again, as it has during the discussion of Roshko & Koenig's paper. It's quite evident from the discussion that there's been a lot of varied experience with it over the years in conjunction with work on the aerodynamics of aircraft bomb bays. I was involved once with a bomb bay buffeting problem. One of the things we did was to make a high-speed movie of the visualized flow using a water channel. It was almost hypnotic to watch that movie. The flow approached the cavity and tried to make it around the upstream corner. Actually, it made it at first, but then a vortex formed, grew, became too big, and was shed toward the downstream corner of the cavity. The vortex banged into the downstream corner and tended to split in some fashion. There appeared to be a pulse that was felt upstream – a feedback mechanism. The message which I think this adds is that the process of shedding is involved in these cavity flows. It might help to remember this when the effect of a cavity's corner geometry or screens and fences are considered. If you look at averages, these flows may appear to be rather ordinary, but when you look at the details, as we did, you get the strong impression of an unsteady type of flow.

SESSION IV

Session Chairperson
M. V. MORKOVIN

Illinois Institute of Technology
Chicago, Illinois

NUMERICAL MODELING OF BLUNT-BODY FLOWS–PROBLEMS AND PROSPECTS

M. T. LANDAHL

Massachusetts Institute of Technology, Cambridge, Massachusetts

ABSTRACT

The problems associated with numerical modeling of blunt-body flows are discussed. An efficient modeling technique should ideally incorporate the interactions between the turbulent boundary layer near the body, the unsteady, highly vortical wake flow behind the body, and the potential-flow regions outside these. The incomplete understanding of vortical unsteady flow fields, in particular, turbulent boundary layers and their separation behavior, will for the foreseeable future preclude accurate modeling; but even coarse modeling methods could serve an important role in establishing cause-and-effect relationships. In particular, one should aim at finding methods which can be used to predict, at least qualitatively, the effect of small local body changes on local flow patterns and on the overall drag. A case is made for flow-field calculation methods based on the vorticity equations. Such methods have proved successful in aeronautical and meteorological applications. The overall drag and lift can be calculated in terms of the vorticity shed into the wake; in particular, the vortex drag associated with longitudinal vortices due to aerodynamic lift can be analyzed.

NOTATION

a	vortex core radius
A, $\underline{B}$	0(1) parameters depending on vorticity distribution in vortex core
D	drag

References pp. 301-302.

L	lift
$\underline{n}$	unit binormal
$\underline{r}$	radius vector
R	local radius of curvature of a vortex filament
Re	Reynolds number
$\underline{u}$	velocity vector field
U	velocity of body
$\underline{u}_t$	velocity just outside boundary layer
$\underline{v} = \nabla\phi$	potential flow field
Γ	circulation around vortex filament
ρ	density
$\underline{\omega} = \nabla \times \underline{u}$	vorticity, $\underline{\omega} = (\omega_1, \omega_2, \omega_3)$

INTRODUCTION

Flows around blunt bodies (such as automobiles) at high Reynolds numbers present severe difficulties for the analyst since such flows are usually unsteady and turbulent. The basic flow mechanisms are only partially understood, with viscosity playing a role only in localized regions of the flow; this role is crucial for the development of the flow, however. The continuing improvements in the speed and capacity of computers have made direct numerical modeling of increasingly complicated flow fields feasible, a development getting much of its impetus from aerospace engineering where the computer now has become a standard design tool for the calculation of aerodynamic loads on complicated configurations, both for steady and unsteady flows; the latter are of interest in flutter predictions. My aim here is to try to extract from and boldly extrapolate what has been learned in the aerodynamics field that may be of use for the difficult blunt-body flow problem.

Characteristic of most aerodynamic problems is a flow field which is only slightly perturbed from a uniform flow. Also, it is usually irrotational, except in very localized regions such as in the vortex wake extending back from a sharp trailing edge. Numerical lifting-surface theories giving continuous loadings over steady or unsteadily deforming thin wings have been developed into standard aerodynamic design tools (see, e.g., Landahl & Stark, 1968). With proper consideration to detailed behavior near wing and control surface edges these are able to predict load distributions within very few percent of measured values.

For more complicated three-dimensional bodies, discretized rather than continuous load distributions have been used to model the flow, either over a network of straight lines, the so-called vortex-lattice method, or over flat or curved panels, the so-called panel method. The vortex-lattice method was first proposed by Falkner (1948), and then substantially developed and improved in more recent times by, among others, Rubbert (1964), Hedman (1966), and Belotserkovskii (1969). The method employs a suitably selected lattice of straight-line vortex segments over the wing, fuselage and tail, and then determines the strength of each vortex by requiring the flow-tangency condition to be satisfied at preselected control points, usually at the center of each lattice. This leads to a system of linear equations of large order, a problem efficiently tackled by modern computer codes. The vortex-lattice method has the advantage that extremely complicated body shapes may be treated in a reasonably simple manner. As an example of the complicated flight vehicle shapes that may be analyzed in this fashion we reproduce in Fig. 1 the lattice network employed in a paper by Palko (1976). The lattice method was in this case employed to determine flow inclinations induced in the (steady) flow field. This example may be typical of the low-speed aerodynamic problems for whole aircraft configurations at small angles of attack with the flow separating only at a sharp trailing edge that may be fairly routinely, and efficiently, analyzed with such a numerical technique.

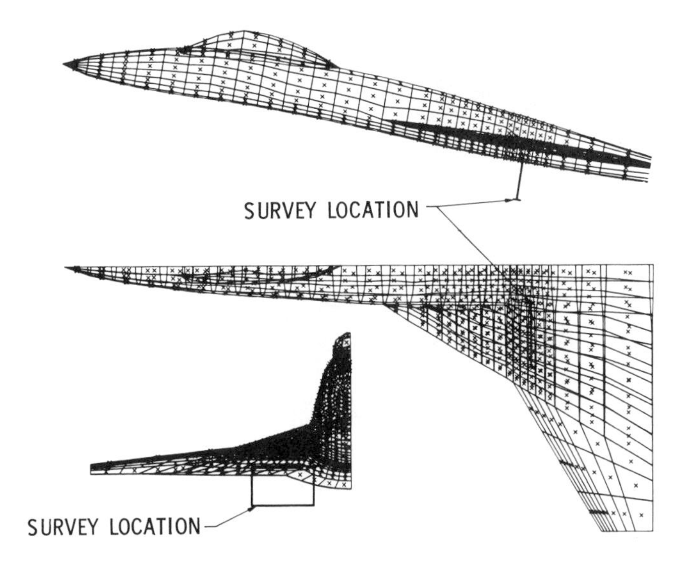

Fig. 1. Mathematical model of the fuselage-wing configuration. From Palko (1976).

References pp. 301-302.

The problem becomes considerably more complicated when one tries to take into account flow separation at other places than the trailing edge, or when the angle of attack is so large that the deformation of the wake (wake rollup) becomes large too close behind the wing, thereby substantially influencing the lift distribution. Wings of small aspect ratio, with highly swept leading edges, often become susceptible to such effects; in particular, leading-edge separation often occurs for such wings producing, somewhat paradoxically, a considerable augmentation of the lift (as well as of the drag). A few examples have been presented in the literature on the use of vortex-lattice and similar methods to model such flows. In Fig. 2 is shown the vortex-lattice scheme employed by Kandil, Mook & Nayfeh (1976) in their calculation of the aerodynamic properties of a low-aspect-ratio delta wing at high angles of attack. To model the leading-edge separation, free vortices coming off the leading edge were added and allowed to interact with the bound ones in the wing, and with the free ones coming off the trailing edge. The positions of these free vortices, composed of straight-line segments, were determined by requiring each segment to be parallel to the local velocity at a control point on the segment. Convergence of the calculation procedure, which is iterative, was speeded up by replacing the system of discrete leading-edge vortices with an equivalent single concentrated core, and the calculation was stopped when the strength and position of this core did not change any more. The vortex wake coming off the trailing edge was, in the calculations shown, assumed to deform only in the immediate neighborhood of the trailing edge; it was taken to be flat and parallel to the free stream further downstream. This approximation was reported to have only a marginal effect on the final results for the

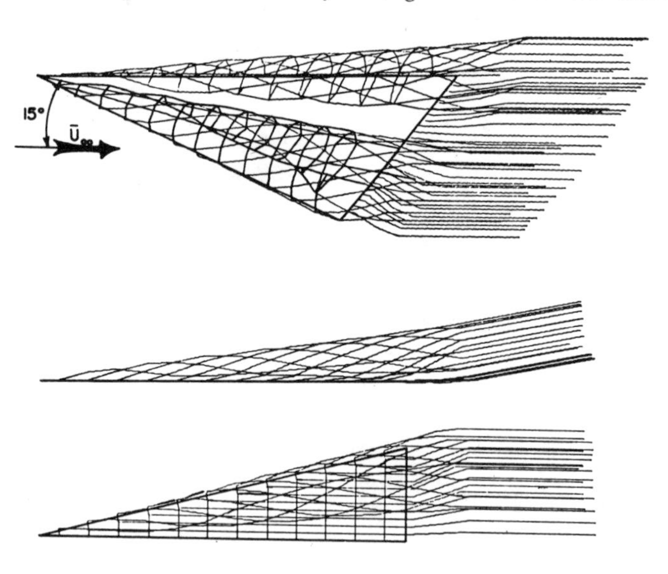

Fig. 2. A typical solution of the wake shape for a delta wing with AR = 1. 12x12 lattice. From Kandil, Mook & Nayfeh (1976).

aerodynamic forces. Clearly, the addition of more vortex segments to achieve a more-detailed representation of the separated wakes would be possible, a refinement primarily limited by computer capacity and cost. A more sophisticated modeling of the wake flow with vortices which consist of curved segments rather than straight ones, and/or which possess a continuous distribution of vorticity, is considerably more difficult and requires a deeper understanding of flows with concentrated regions of vorticity. Some of the properties of such flows will be discussed later.

The vortex-lattice method represents the continuous load distribution on an aerodynamic body by discrete loadings carried along the vortex lines. In interpreting the result for the loading distribution, the discrete line loadings must be considered spread out over the surface in some fashion. Panel methods use less-singular load distributions, but are somewhat more complicated. A widely used one is that developed at McDonnell-Douglas by Hess (1972). This uses a formulation based on sources and vortex sheets distributed over a body surface, which is divided up into flat panels over each of which the source and vortex-sheet strengths are assumed constant. Such a method leads to a surface velocity distribution less singular than that produced by a vortex lattice. An even more refined panel method has been presented recently by Ehlers, Johnson & Rubbert (1975), which employs linearly-varying source and quadratically-varying doublet distributions on curved panels.

PROPERTIES OF VORTICAL FLOWS OF LOW VISCOSITY

It is an important characteristic of flows of small viscosity around solid bodies that the vorticity produced by the body tends to be highly concentrated along sheets and lines. This behavior can be understood from the equation for the vorticity $\underline{\omega} = \nabla \times \underline{u}$ in a fluid of constant density,

$$\frac{D\underline{\omega}}{Dt} = (\underline{\omega} \cdot \nabla)\,\underline{u} + \nu\,\nabla^2 \underline{\omega} \tag{1}$$

the first term on the right-hand side representing the change in vorticity due to deformation (stretching or shortening) of a fluid element, and the second, diffusion by viscosity. An inviscid fluid initially without vorticity will remain irrotational; for a fluid of low viscosity, substantial vorticity production will take place only in the thin boundary layer surrounding the body. For moderately high Reynolds numbers in the laminar regime, the boundary layer thickness divided by the body reference length is proportional to $Re^{-1/2}$ (Re being the Reynolds number), i.e., very small; for higher Reynolds numbers, the boundary layer goes turbulent, usually ahead of or at the point of minimum surface pressure. A turbulent boundary layer is considerably thicker than a laminar one at the same Re, but usually still remains thin whenever attached to the body surface.

References pp. 301-302.

The boundary layer may be looked upon as a thin sheet of vorticity enveloping the body. The strength of the sheet per unit planform area, i.e., the integral of the vorticity across the sheet, is easily seen to be equal to the velocity u_t just outside the boundary layer. The vorticity in the sheet can be shown to move at an equivalent velocity of $u_t/2$ (Lighthill, 1963). The boundary layer will separate from the body surface if subjected to sufficiently strong positive ("adverse") pressure gradients. A separated boundary layer may remain thin a short distance downstream from the line of separation and proceed into the wake as a free shear layer (Fig. 3). Such a free layer of concentrated vorticity is highly unstable and quickly develops wave-like corrugations, which tend to concentrate the vorticity in localized regions and thereby cause the formation of a row of vortices with fairly distinct vortex cores (Fig. 4).

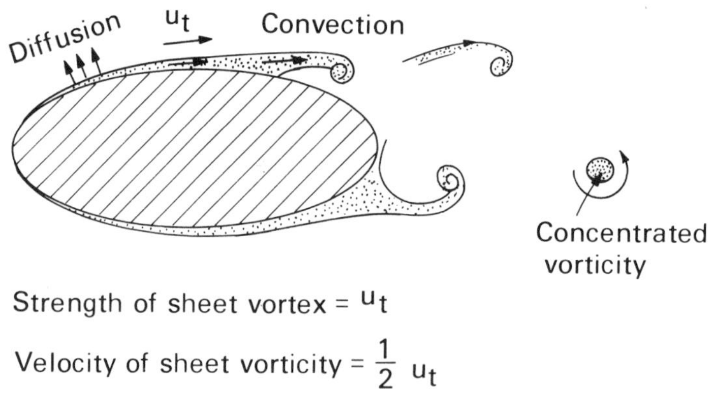

Fig. 3. Vortical flow field around blunt body.

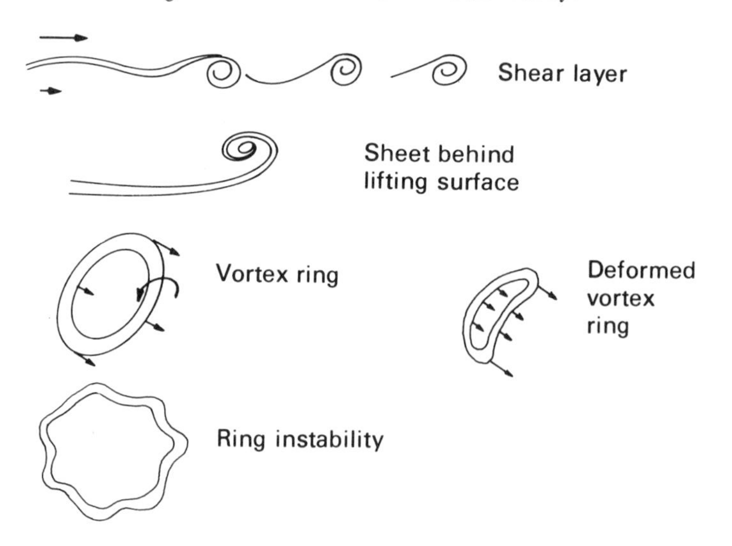

Fig. 4. Behavior of concentrated-vorticity fields in flows of small viscosity.

Because the vorticity field is solenoidal, the vortices must end on boundaries of the flow, or close onto themselves, the later being the usual situation for free-flying three-dimensional bodies. A closed vortex loop cannot maintain a steady shape. This can be shown to follow from the solution for its velocity field in terms of the vorticity,

$$\underline{u} = -\frac{1}{4\pi}\int \frac{(\underline{r} - \underline{r}') \times \underline{\omega}(\underline{r}')\, d\underline{r}'}{|\underline{r} - \underline{r}'|^3} + \underline{v} \tag{2}$$

where $\underline{r}$ = is the position vector, $\underline{r}' = (x', y', z')$ the dummy (integration) variable, $d\underline{r}' = dx'dy'dz'$, and

$$\underline{v} = \nabla\phi \tag{3}$$

is an arbitrary irrotational velocity field. When $\underline{\omega}$ is concentrated along a line, the expression for the velocity field induced by the line vortex simplifies to (Biot-Savart's law)

$$\underline{u}_\Gamma = \frac{\Gamma}{4\pi}\int \frac{(\underline{r} - \underline{r}') \times d\underline{s}'}{|\underline{r} - \underline{r}'|^3} \tag{4}$$

where $d\underline{s}' = \partial\underline{r}/\partial\underline{s}'\, d\underline{s}'$ is the element of the arc, $\underline{r}(\underline{s}')$, defining the location of the line vortex, and Γ is its circulation. By considering the neighborhood $\underline{r} \simeq \underline{r}'$ one can show that for points approaching a *curved* line vortex a logarithmically infinite velocity results. The vorticity must convect with the local flow velocity, hence a *concentrated* curved line-vortex cannot exist in the flow since it would be convected away with infinite velocity. This dilemma is resolved by letting the vortex have a core of small but finite size. Widnall *et al.* (1971) showed that the velocity induced near a vortex, whose core radius, a, is small compared to the local radius of curvature R(s) of the filament, is given approximately by

$$\underline{u}_\Gamma = -\frac{\Gamma}{4\pi R(s)}\left[\underline{n}\ln\frac{a}{R(s)} + \underline{B}\right], \tag{5}$$

where $\underline{n}$ is the unit binormal and $\underline{B}$ is an 0(1) vector that depends on the actual shape of the filament and the vorticity distribution in the core. The Biot-Savart formula for a concentrated line-vortex will produce this result if a small segment ℓ on each side of the point on the line-vortex is excluded from the integration, where

$$\ell = \frac{a}{2}\, e^A, \tag{6}$$

and where A is an 0(1) constant (usually positive) that depends on the detailed

References pp. 301-302.

vorticity distribution. With the use of the cut-off distance ℓ calculated in this manner, one can determine directly from the Biot-Savart law (4) the velocity field induced by any vortex with a small core radius.

From (6) and (5) it follows that a vortex initially lying in one plane but with variable curvature *cannot remain* in one plane, since the portions with a smaller radius of curvature will receive a higher induced velocity in the direction of the binormal and therefore will race ahead of portions with a larger radius of curvature, (Fig. 4). Only one line configuration will remain plane, namely that of constant curvature, i.e., a vortex ring. However, at high Reynolds numbers a vortex ring develops instabilities (Fig. 4) and eventually breaks up (Widnall & Sullivan, 1973; see also review article by Widnall, 1975). Thus, the free vorticity shed by a separated boundary layer behind a three-dimensional blunt body will first tend to concentrate into filament loops which will deform due to the induced velocity field; these will then become increasingly more contorted and finely structured as they move downstream. Occasionally, neighboring vortex cores may interact in a viscous manner and thereby combine to a bigger vortex (or annihilate each other, depending on the signs of the vorticity in each core) and hence lead to an increase in the scale of the eddy motion. However, the main effect of a turbulent flow field on the vorticity field is to drive it into increasingly finer scales until viscous diffusion becomes sufficiently strong to counteract the effects of stretching of the vortex filaments (Fig. 5). The stretching of a vortex with a concentrated core will tend to make the core radius shrink in the same ratio as the fluid element is elongated. This will cause the tangential velocity component in the core to increase but will not change the velocity field at large distances, since the total circulation of the vortex remains unaltered. Therefore, in the modeling of a separated three-dimensional flow with vortices of thin cores, vortex stretching needs only be taken into account in the calculation of the self-induction velocity of the vortex. Hence, the flow field calculated this way may possibly not be crucially dependent on the actual core size used in modeling the separated flow.

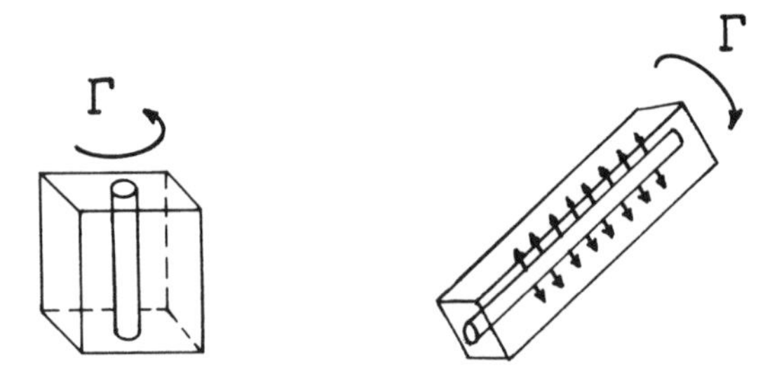

Fig. 5. Convection, stretching and viscous diffusion of vortex filament.

VORTEX MODELS FOR SEPARATED BLUNT-BODY FLOWS

It is in principle possible to attack the vorticity equation (1) directly as an initial-value problem, using any suitable finite-difference method, and then determine the velocity field from (2). The additional potential flow $\underline{v}$ must be determined at each step to be such that the flow-tangency condition is satisfied. Such a method was tried with success by Payne (1958) for the two-dimensional steady flow around a cylinder at a moderately high Reynolds number, and was also recently used by Kinney & Cielak (1975) to calculate the viscous flow around an airfoil in impulsive motion. The advantage in using vorticity instead of the primitive variables $\underline{u}$ and p in a numerical treatment of high Reynolds number flows lies primarily in the limited extent of the regions of high vorticity in such flows. Therefore, the diffusion-type equation (1) need be solved numerically only in a region of *limited* spatial extent, a region, however, that is constantly changing because of convection of the vorticity.

Since in a fluid of low viscosity the vorticity changes only relatively slowly in a convected frame, it would be computationally more efficient to use a Lagrangian formulation based on a discrete representation of the vorticity; this would be especially natural for a separated-flow region in view of the tendency of vorticity to concentrate along lines, as discussed above. Methods based on modeling of a separated wake by discrete vortices have been successfully developed by Chorin (1973) for the two-dimensional flow around a cylinder, and by Leonard (1975) for three-dimensional flow around a sphere. Both employ a conceptually simple model for the shedding of vorticity at the separation line (Fig. 6). At this line, the downstream end of the vortex sheet moves with the average velocity $\underline{u}_t/2$. When the front has moved a preselected distance, a ribbon of the sheet is removed from the downstream edge and formed into a vortex filament. For a ribbon segment of length h, the circulation of the vortex is thus $\Gamma = h\,u_t$.

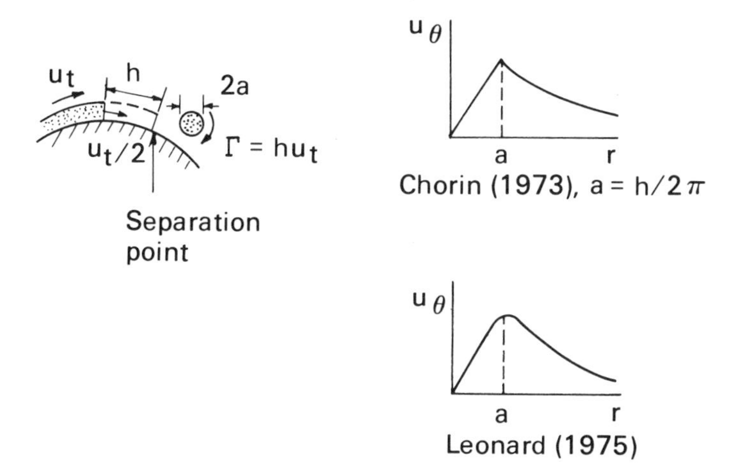

Fig. 6. Vortex-Shedding models used by Chorin (1973) and by Leonard (1975).

References pp. 301-302.

Chorin (1973) uses a Rankine vortex as his core model and selects the core thickness such that the vortex has a velocity increase across the core equal to that across the boundary layer. Leonard (1975) employs a slightly more sophisticated core model with a Gaussian distribution of vorticity in the core, and the height of the center of the vortex above the surface chosen such as to ensure a translational velocity of $\underline{u}_t/2$ for the vortex. The initial core size is allowed to vary within a range, with the upper limit given by the core touching the surface and the lower by assuming the volume of the vortex core to be equal to the fluid volume of the ribbon of boundary layer it replaces. Leonard's (1975) results for the positions of the separated vortices in the wake of a sphere at different times after the start of the motion are reproduced in Fig. 7. A small initial rigid translation of the vortices was introduced to start the formation of an unsymmetrical and unsteady vortex pattern. The separation was here assumed to take place at the point of maximum sphere diameter.

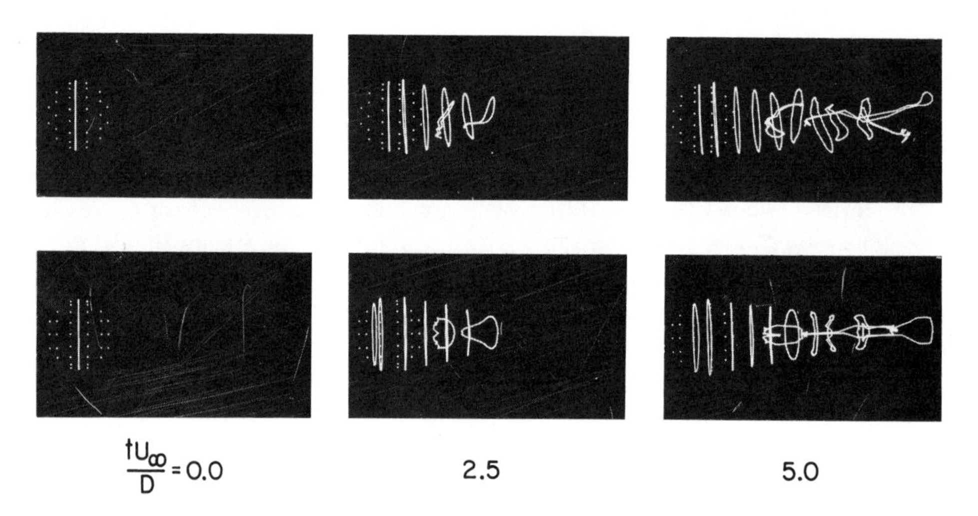

Fig. 7. Simulation of flow past a sphere with an impulsive start as observed from two directions perpendicular to one another. From Leonard (1975).

In principle, there seem to be no obstacles to also carrying out such calculations for more complicated three-dimensional bodies, *provided* one knows where the separation line is located. At each step, one would need to determine the flow field due to the attached vortex sheet enwrapping the body, the velocity field due to the wake vortices, and the potential-flow field required to satisfy the flow-tangency condition at the body surface. The latter problem could presumably be handled by a vortex-lattice method, replacing the sheet of traveling vorticity in the boundary layer by a lattice of bound discrete vortices at each instant. The main weakness of such a model lies in the lack of *a priori* knowledge of the separation-line location, as this depends on the pressure distribution along the boundary layer which is itself influenced by the separation. For bodies with a sharp downstream edge, the problem is considerably simpler since separation usually occurs at such edges. But for general smooth bodies, the determination of the separation line would in principle require calculation of the boundary layer at each time step in order to determine where it would separate. This is an extremely difficult problem even for the simplest case of a laminar and two-dimensional steady boundary layer, and almost impossible in the three-dimensional and unsteady cases. For the turbulent boundary layer in particular, the lack of understanding of even the fundamental mechanisms controlling the flow is very great indeed, and any realistic calculation scheme would have to be based on empirical models. Nevertheless, by incorporating experimental data into such calculations, one might be able to obtain useful information for aerodynamic properties of interest.

The total lift and drag, which are the quantities of primary interest to the engineer, may possibly not be overly sensitive to good detailed modeling of the flow field. This is so because lift and pressure drag can be shown to be directly related to integrals of the vorticity distribution far downstream in the wake. The total lift on a three-dimensional body in an unbounded, inviscid and steady flow subjected to small disturbances is given by

$$L = \rho U \iint_S y\omega_1 \, dS \tag{7}$$

where ρ is the fluid density, U the velocity of the body, y the lateral coordinate of the flow, and ω_1 the streamwise component of the (mean) vorticity ($\omega_1(y,z) = \frac{\partial w}{\partial y} - \frac{\partial v}{\partial z}$) far downstream in the wake (the so-called Trefftz plane in wing theory) and $dS = dydz$. This result can easily be seen to follow from integration of the vertical momentum transported through the Trefftz plane. The integration is to be carried out over the cross-section S of the far wake.

The drag associated with the vortex distribution may be most readily found by determining the rotational kinetic energy shed into the wake per unit time, and then equating this with the work done by the body on the flow. This procedure gives (Cone, 1962)

$$D = -\frac{\rho}{4\pi}\iiiint \omega_1 \, \omega_1' \ln \sqrt{(y-y')^2 + (z-z')^2} \, dS \, dS'$$

where $dS' = dy'dz'$ and $\omega_1' \equiv \omega_1(y',z')$

For a planar vortex-sheet distribution, this reduces to the well-known formula for the vortex drag of a lifting wing. The drag formula indicates that a high vortex drag is experienced when the vorticity is *concentrated.* For a single concentrated vortex, the drag will tend logarithmically to infinity as the vortex-core diameter shrinks to zero,

References pp. 301-302.

as can be seen by substituting a delta function for ω_1. A body surface shape that tends to produce concentrated longitudinal vortices in the wake (for example, one with sharp inclined side edges) would be expected to have a high vortex drag. Also, from a combination of (7) and (8) it follows that for a prescribed lift distribution the vortex drag is inversely proportional to the square of the spanwise dimension of the body carrying the lift, so that in order to achieve low drag the lift should be spread out in the spanwise direction as much as possible. Body modifications that remove local high lift concentrations would therefore be expected to lower the drag.

CONCLUSIONS

The extension of numerical aerodynamic prediction methods, such as the vortex-lattice method and the like, to separated three-dimensional blunt-body flows employing a discrete-vortex representation of the wake appears to be feasible in principle *provided* information is available on the location of the separation line and how it is affected by pressure gradients. In comparison with a standard finite-difference method working directly with velocity and pressure, a vortex method is probably computationally less efficient for the same number of mesh points as discrete vortices used. However, a vortex method would presumably improve in accuracy for high Reynolds number flows (Chorin, 1973) in contrast to the behavior of standard finite-difference methods, and would require fewer mesh points for a given accuracy since the velocity field is irrotational in most of the flow field surrounding the body, even for a blunt body with extensive separation. Furthermore, and not less importantly, the vorticity method deals directly with an essential fluid mechanical aspect of a separated flow, namely the shedding and convection of vorticity in the wake. The overall aerodynamic properties such as lift and drag can be readily obtained from the distribution of vorticity shed into the wake. Hence, one may possibly be able to infer from the character of the vorticity distribution in the wake what qualitative effects modifications to the body surface might have on the aerodynamic properties.

The weakest part of a method of this kind is the determination of the separation line. A smooth body with unsteady vortex shedding will have a separation line which will move unsteadily over the body surface. The knowledge of the mechanism of separation of three-dimensional boundary layers in an *unsteady* pressure field is rather incomplete, to say the least, and a reliable prediction method is not available. This is, of course, especially true in the case of turbulent boundary layers. For these, one would have to hope for improvements in the approximate methods based on semi-empirical turbulence models of the kind now widely used for *steady* boundary layers. A direct numerical simulation of the detailed flow in a turbulent boundary layer, whether by finite differences or vorticity, is out of the realm of the possible because of the large range of eddy scales that have been found to play an important role in the dynamical processes in the viscous and buffer layers near the wall (see recent survey article by Willmarth, 1975).

Finally, it should be emphasized that aerodynamic theory and numerical flow modeling could never realistically replace wind tunnel testing, but should primarily be used as a complement to it with the aid of which basic cause-and-effect relationships may be studied in more-or-less idealized flow situations. What anyone having any experience in applied aerodynamics would testify to is the infinite variety of curious phenomena appearing in flows of real fluids, and their sometimes extreme sensitivity to minute changes in body shape or flow conditions, in particular when flow separation is involved. It will therefore be a long time, if ever, before flows around complicated bodies such as automobiles can be computed directly from a numerical solution of the equations of motion.

REFERENCES

Belotserkovskii, S. M. (1969), Calculation of the Flow Around Wings of Arbitrary Planform in a Wide Range of Angles of Attack, NASA TT F-12, 291.

Chorin, A. J. (1973), Numerical Study of Slightly Viscous Flow, J. Fluid Mech., Vol. 57, pp. 785-796.

Ehlers, F. E., Johnson, F. T., & Rubbert, P. E., (1975) Advanced Panel-Type Influence Coefficient Methods Applied to Subsonic and Supersonic Flows, **Aerodynamic Analyses Requiring Advanced Computers,** *Part II, NASA SP-347, pp. 939-984.*

Falkner, V. M. (1948), The Solution of Lifting Plane Problems by Vortex Lattice Theory, British A.R.C., R & M 2591.

Hedman, S. G. (1966), Vortex Lattice Method for Calculation of Quasi Steady State Loadings on Thin Elastic Wings in Subsonic Flow, FFA Rep. 105, Aeronautical Res. Inst. of Sweden.

Hess, J. L. (1972). Calculation of Potential Flow about Arbitrary Three-Dimensional Lifting Bodies, Rep. No. MDC J5679-01 (Contract N00019-71-C-0524), McDonnell Douglas Corp.

Kandil, O. A., Mook, D. T. & Nayfeh, A. H., (1976), New Convergence Criteria for the Vortex-Lattice Models of the Leading-Edge Separation, NASA SP-405, pp. 285-292.

Kinney, R. B. & Cielak, Z. M. (1975), Impulsive Motion of an Airfoil in a Viscous Fluid. Proceedings of Symposium on Unsteady Aerodynamics (March 8-20, 1975) Vol II, ppp. 487-512. Editor R. B. Kinney. U. S. Air Force and University of Arizona, Tucson, Arizona.

Landahl, M. T. & Stark, V. J. E. (1968), Numerical Lifting-Surface Theory - Problems and Progress, AIAA J., Vol. 6, pp. 2049-2060.

Leonard, A. (1975), Simulation of Unsteady Three-Dimensional Separated Flows with Interacting Vortex Filaments, in NASA SP-347, **Aerodynamic Analyses Requiring Advanced Computers,** *pp. 925-37. (A more accessible reference is Simulation of Three-Dimensional Separated Flows with Vortex Filaments, in* **Lecture Notes in Physics,** *Springer-Verlag, 1977).*

Lighthill, M. J., (1963), Introduction, Boundary Layer Theory, in **Laminar Boundary Layers,** *(L. Rosenhead, ed.) Ch. II, Oxford University Press.*

Palko, R. L. (1976), Utilization of the AEDC Three-Dimensional Potential Flow Computer Program, NASA SP-405, pp. 127-143.

Payne, R. B. (1958), Calculations of Unsteady Viscous Flow past a Circular Cylinder, J. Fluid Mech., Vol. 4, pp. 81-86.

Rubbert, P. E. (1964), Theoretical Characteristics of Arbitrary Wings by a Non-planar Vortex Lattice Method. Boeing Report D6-9244. Boeing Commercial Airplane Division, Renton, Washington.

Widnall, S. E. (1975), The Structure and Dynamics of Vortex Filaments, in **Annual Review of Fluid Mechanics,** *Vol. 7, pp. 141-165, Annual Reviews, Palo Alto, California.*

Widnall, S. E., Bliss, D. & Zalay, A., (1971) Theoretical and Experimental Study of the Stability of a Vortex Pair, in **Aircraft Wake Turbulence and Its Detection,** *(ed. J. H. Olsen, A. Goldberg and M. Rogers) Plenum Press, New York.*

Widnall, S. E. & Sullivan, J. P. (1973), On the Stability of Vortex Rings, Proc. Roy. Soc., Ser. A332, pp. 335-353.

Willmarth, W. W. (1975), Structure of Turbulence in Boundary Layers, in **Advances in Applied Mechanics,** *Vol. 15, pp. 159-254.*

DISCUSSION

Prepared Discussion

A. Leonard *(NASA-Ames Research Center)*

Professor Landahl has already introduced the calculation method I want to describe. Thanks to Helmholtz, Kelvin, Widnall, Moore and Saffman we have a numerical method that is quite relevant to all these bluff body flows. The idea is to model a wake region in terms of a finite number of vortex filaments or loops (Fig. 8). The number of grid points needed for each filament is only that required to adequately describe the vortex shape in three dimensions. None are required out in the potential flow region.

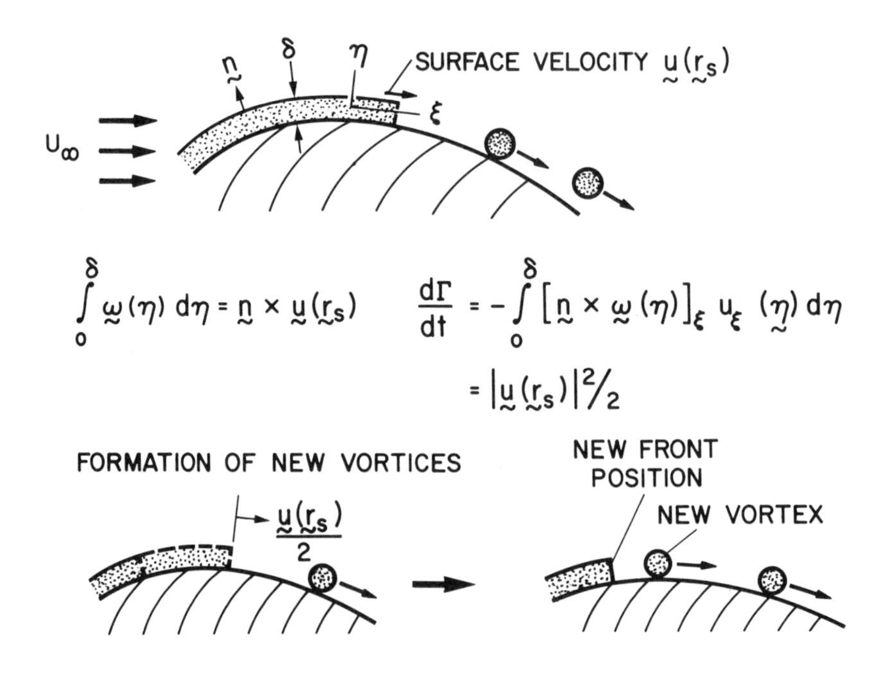

Fig. 8. Boundary layer model.

Fig. 9 shows what our results look like for the flow past a sphere. The line on the sphere is the downstream edge of a sheet of vorticity that is being tracked. The vorticity sheet moves forward with a velocity $U_{edge}/2$, and as it passes this line it is peeled off in pieces to create individual vortices, which are not necessarily axisymmetric. The other loops are closed vortex filaments described by, say, 20 to 100 grid points which have peeled off at earlier times and have moved with the flow field. Each piece of a vortex filament has a core size associated with it, and each filament has a circulation.

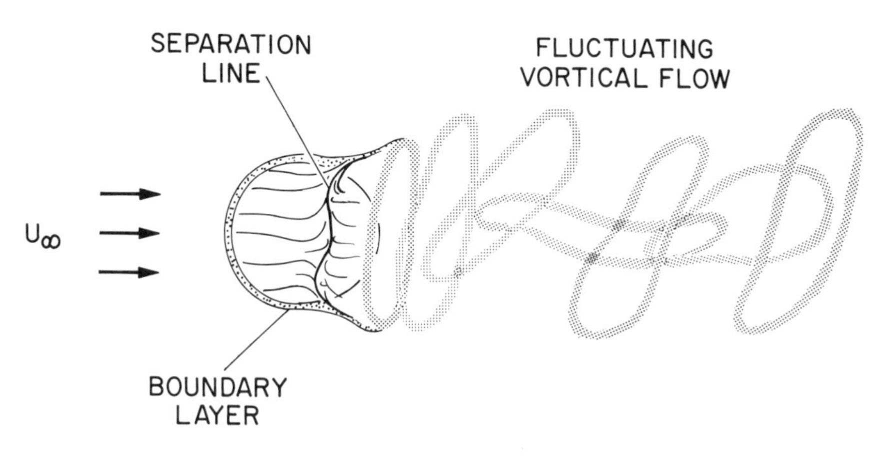

ESSENTIAL INGREDIENTS:

1. UNSTEADY BOUNDARY-LAYER SIMULATION
2. TRACKING OF VORTEX FILAMENTS
3. SATISFY INVISCID BOUNDARY CONDITIONS

Fig. 9. Simulation of 3-D unsteady separated flow.

The experimental application of the calculation technique to a bluff body flow is shown in Fig. 10. The calculated drag agrees quite well with the measured value in the subcritical Reynolds number range.

The next results in Fig. 11 are the side force as a function of time. To kick the flow off its axisymmetric pattern it was given an initial asymmetric perturbation in the time span from 0 to 2-1/2 time units and then let go. By spinning it, I tried to induce a rotational type of disturbance like the one that Dr. Achenbach sees with his hot wires. The computed results seem to agree with that during an initial time period, but

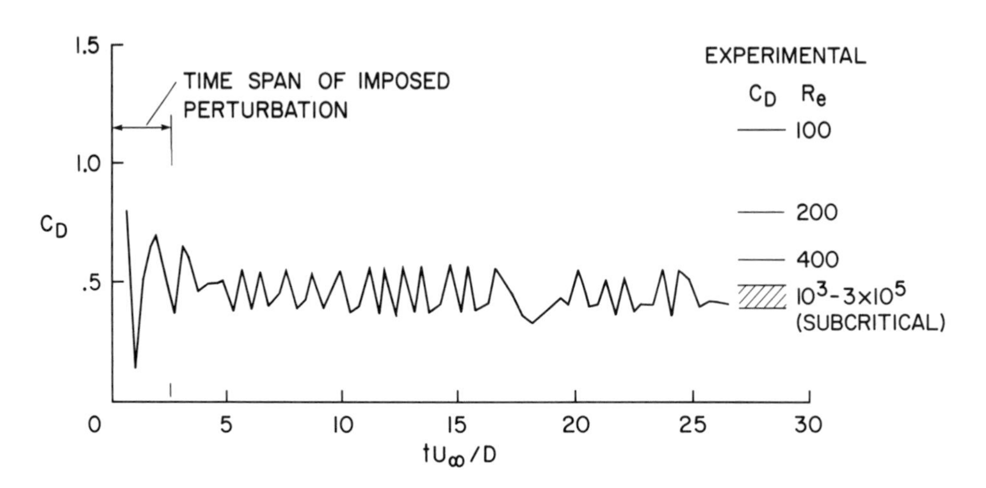

Fig. 10. Sphere drag coefficient.

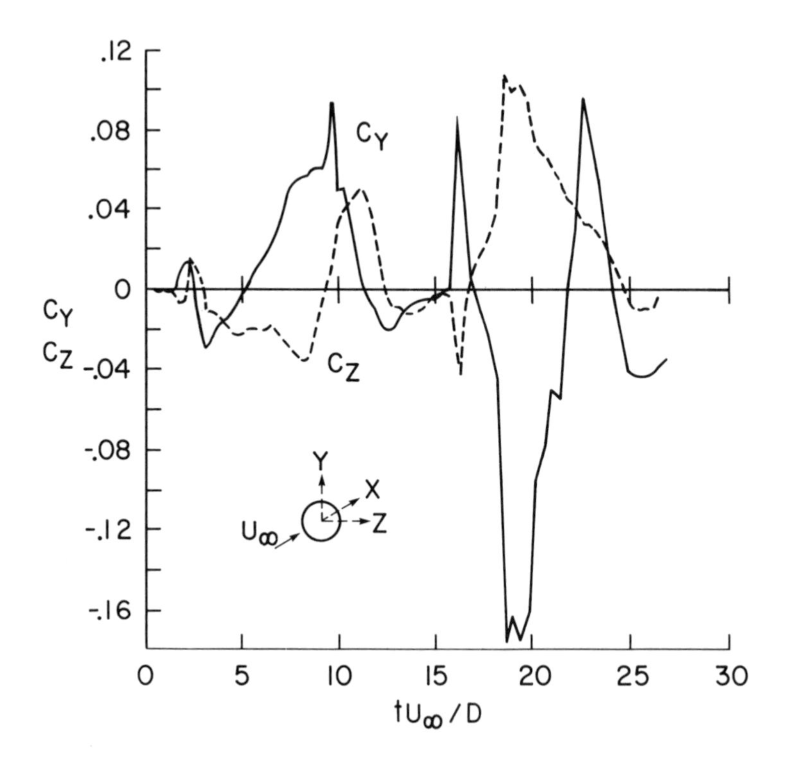

Fig. 11. Side force coefficient.

then a zig-zag type disturbance seems to develop. This can be seen better in Fig. 12 which is a polar plot of the side force. Starting from zero time, the side force rotates around in a spiral trajectory, and then goes into a large flip-flop or zig-zag type motion. Both of these types of motion are observed experimentally for free-falling spheres. There is a helical path under certain conditions, and a zig-zag one under others.

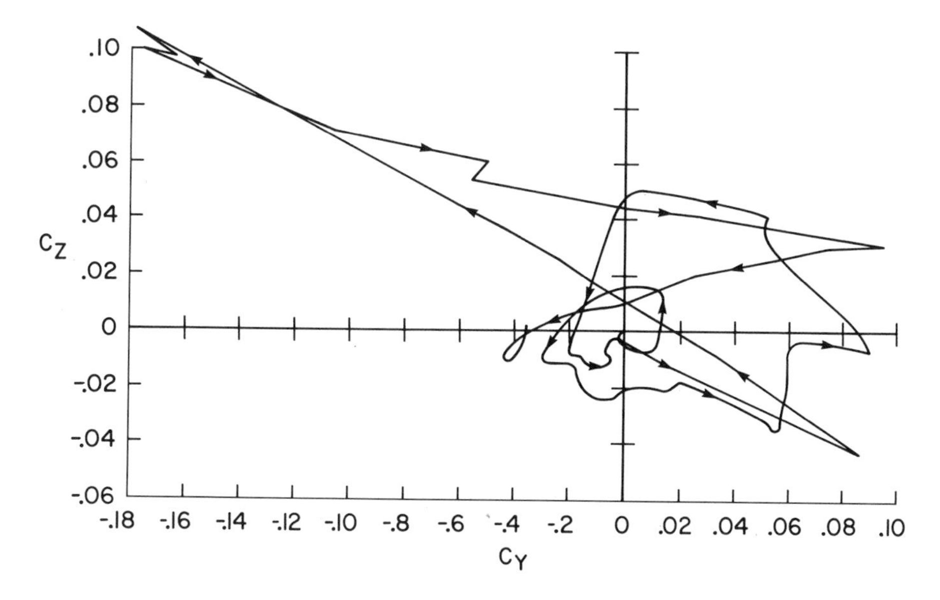

Fig. 12. Side force coefficient.

Editors' Comment – Dr. Leonard then showed a movie of the cathode ray tube output for the computed flow past a sphere. There were 4 sequences: 3 side views and 1 oblique view from the rear.

Although it was not easy to detect the large-scale flow and waviness which could produce the time-varying side force, some swirling and back-and-forth gyration in the wake could be seen in the first sequence. Just downstream of the separation point the free shear surface acted like a tubular sheet that moved around a bit. In the next sequence, for flow visualization purposes, every few time steps a line of passive particles was introduced across the stream one-half radius upstream of the sphere. With no vortex filaments being marked, these lines of "hydrogen bubbles" clearly showed the demarcation between the potential flow and the edge of the wake, with some of the particles becoming entrained into the wake. In the third side view sequence the lines connecting the vortex filament grid points of the first sequence were eliminated so the points tended to act like dye particles injected into the boundary layer. The final sequence was a rear three-quarter view of this arrangement.

Dr. Leonard pointed out that the method treats boundary conditions very well, except the downstream one where all procedures have a problem. The vortices have to be annihilated sooner or later because it becomes too expensive to follow them for very long distances.

R. T. Jones *(NASA-Ames Research Center)*

This technique seems to have promise but there is one thing that seems to be missing, and I think it is probably very important. You seem to employ no criterion for separation. How do you determine where the vortices are going to be inserted?

A. Leonard

That is a very good question. I arbitrarily release them ahead of the point of laminar separation. I don't pretend to do turbulent boundary layers at all. I release the vortex elements right on the separation line, and then the exterior flow just carries them away.

R. T. Jones

Is this a very low Reynolds number flow?

A. Leonard

The method doesn't have a Reynolds number that can be pinned down. That is one of the problems with the vortex methods. To reproduce Reynolds number effects, I think one could combine a conventional boundary layer calculation having a grid on the sphere surface with the vortex method for the outer flow. You should then follow the Reynolds number trends more accurately.

A. E. Perry *(University of Melbourne, Australia)*

You released vortex rings at an equator of the sphere, more or less. There is a lot of photographic evidence that the wake behind a sphere forms elongated loops of limited circumferential extent, so it seems to me that the vorticity should be coming off like horseshoes. If you had to release it as horseshoes, then it might form into structures similar to the daisy-chain-type loops that Dr. Achenbach showed yesterday. Your ring vortices appeared to want to form elongated loops but just couldn't.

A. Leonard

The Achenbach loops have a much lower Strouhal number than my calculations. Therefore, if I do laminar-type wakes (and I believe what you are talking about is the laminar-type wake) the only thing I can hope for is to have 10 or 12 of my higher frequency ring vortices act in concert and look like one of Achenbach's chain-type lower frequency loops; I haven't seen that yet.

A. Leonard and M.V. Morkovin

M. V. Morkovin

We are not here to judge Dr. Leonard's technique per se. It was only offered as an illustration of the capabilities of one of the things Professor Landahl discussed. We should now get back to the main points of Professor Landahl's presentation.

I had the opportunity to read your paper in advance, and you do comment about limitations with regard to the separation line. In light of what Dr. Leonard has presented do you want to add anything?

M. T. Landahl

The real hooker in a method like this is the assumption that the separation line is given. In principle, I presume, we could combine this calculation with a boundary layer calculation along the lines that have been indicated and try to compute the separation line. Now there is the difficulty. To use an understatement, the problem of unsteady separation isn't particularly well known or well understood. There is a lot of quarrel over what is meant by unsteady separation in two dimensions. In three dimensions, I'm not sure the problem has even been attempted.

When you come to turbulent boundary layers the problem gets even worse. Any hope that you could simulate the flow directly by finite-difference calculations, or the like, from the Navier-Stokes equations is practically zero because we know now-a-days that the range of scales in a turbulent boundary layer is so enormous. Furthermore, the small scales are very important to the overall dynamics of turbulent boundary layers so that the capacity problem and running time limitations of any foreseeable computer make it completely impossible. One would have to hope for improvements in the kind of turbulence models that are currently fashionable. On the other hand, it is possible that one could work with some simpler separation criterion.

I am not so sure that the flow field is that overly sensitive to the precise location of separation. For a two-dimensional flow field, yes; but the three-dimensional flow field around a more-or-less slender body is perhaps not quite as sensitive as far as the vorticity shed into the wake is concerned. And after all, that is all that matters for the drag.

M. V. Morkovin

Isn't the drag formula you presented basically Reynolds number independent? What are the limitations on it? It is not linearized is it? Secondly, how would you attach a Reynolds number to Dr. Leonard's calculations?

M. T. Landahl

No, the formula is not linearized; there is no limitation to small perturbations. A method like this is inherently nonlinear, because nonlinearity enters in through the relationship between the velocity and vorticity fields. You can build up any velocity field by vortex elements.

As for any limitations in Reynolds number in the vortex-type calculations, I find by reading Chorin's paper that the accuracy should improve with increasing Reynolds number. There is, however, a limitation in the method, namely the assumption about the nature of the vortex elements that are released. I think it can be removed by first sizing the vortex core so that it contains the same kinetic energy as the strip of boundary layer being represented. Secondly, the distance between vortices should perhaps be related to the typical distance between vortices shed into a free shear layer. Whether that's important or not, I don't know.

S. J. Kline *(Stanford University)*

Your presentation gave a very nice explanation of vorticity and of some of the things one can understand from it. But maybe I can play the Devil's advocate. If you were Dr. Hucho or Dr. Sovran, or someone else working in the automobile industry, what would you do with your equations that would tell you what to do about road vehicle drag, right now or in the next five years.

M. T. Landahl

What I would do is study the vorticity shed into the wake – trace the vorticity from the bodies; that is where the secret lies. Then with a numerical method like the one described you could presumably get an understanding of how a local change of body geometry would change the vorticity pattern in the wake. You would then try to modify the surface of the vehicle so as to cancel, as much as possible, any strong concentrations of vorticity, both unsteadily shed ones and longitudinal ones.

There is one thing that should be pointed out. For at least some time to come, numerical methods should never be used as a replacement for wind tunnels. They should be used to study simplified flows so that you can gain an understanding of cause and effect.

M.T. Landahl

M. V. Morkovin

I would like to make clear what Professor Landahl was asked to do in his paper. His task was to survey the current prospects for this type of analysis and computation and to project what might be possible in the future. He was not asked to advocate a method to us. If he concludes that the prospects are rather dim right now then that is also very important because there are technical managers who have been reading stories in the media to the effect that wind tunnels are no longer needed because adequate results can be computed. It is in this sense, then, that we are discussing the issues here.

H. M. Nagib *(Illinois Institute of Technology)*

You say that getting an indication of the vorticity in a wake is important. What kinds of things are you advocating in wind tunnel measurements, i.e. the type of wake survey, the type of measurement? In particular, do you expect people to make vorticity measurements?

M. T. Landahl

Yes, I think that would be a good idea. You can either make vorticity measurements or you can measure cross-flow velocity distributions and from that also get the wake. I think the unsteady effects can be worked into the formula. The nice thing about the vorticity formula is that you only have to integrate over the region occupied by vorticity. If you do it in terms of velocity you can also integrate the kinetic energy shed into the wake; that's where the drag formula comes from, basically. But then you have to integrate all the way from infinity.

A. Roshko *(California Institute of Technology)*

I want to follow up on your comment that one might look in the wake for the effect of modifications in body geometry. I think the difficulty here is with separation. In aeronautical practice the wake *is* connected to the details of a body's overall flow pattern, but that flow is essentially completely attached. With the amount of flow separation that exists on most cars it may not be possible to discriminate local problem areas in a vehicle's flow field on the basis of observations in the wake. Since small changes in geometry at the front of a car can make large changes in the separation pattern, I think the connection has to be thought through again.

M. T. Landahl

Well, there are some cases where you know where separation will occur; for example, if a body has sharp edges. You also know that if you have sharp edges you will have strong concentrations of vorticity. What you want to do is to shed vorticity where the velocity outside the boundary layer is low because that will only shed a small amount of kinetic energy into the flow. That kind of an intuitive idea might be helpful.

R. T. Jones

I want to amplify the point that was just made. The thing really lacking in this technique, which is otherwise very promising, is the knowledge of where the separation point is. In the case of a sphere you know that when it separates around the equator you get a very large drag; when the boundary layer becomes turbulent and it closes in more at the rear, the drag coefficient drops by an order of magnitude. So everything depends on your ability to determine where the vortices are shed.

M. T. Landahl

Yes. On the other hand, to understand it the other way around you could say to the engineer, please design me a car that has a separation point right there.

W. W. Willmarth *(University of Michigan)*

I support Professor Landahl's idea of looking at the vorticity in the wake of bluff bodies. We measured the forces on a 22 in. diameter sphere in a 100 ft/sec air stream. The wake was asymmetric and was randomly oriented around the periphery of the sphere. This generated a side force which randomly varied in direction and magnitude. However, for short periods of time (up to a minute in some cases) it stayed fixed in circumferential position, producing a steady side force. There has to be a trailing vortex pair associated with this force. By observing and measuring it, you should be able to determine both the magnitude and orientation of the instantaneous side force, and possibly also learn something about its basic nature.

F. T. Buckley, Jr. *(University of Maryland)*

Could you tell us about the possibility of using such a method for calculating flows which reattach after they have separated? It seems like most of your applications were at the base of vehicles. As sketches used in this symposium have shown, local separations that occur near the front of vehicles usually reattach.

M. T. Landahl

I haven't thought much about that.

M. V. Morkovin

Those separation bubbles represent a change from laminar to turbulent flow and that is beyond the capability of this study at the moment.

J. E. Hackett *(Lockheed-Georgia Company)*

I just want to mention that we are in fact making wake measurements of the type that Professor Landahl is suggesting. We use rakes of 5-hole probes to measure the pitch and yaw aft of wings and bodies, making measurements only in regions where there is vorticity. These are identified by a stagnation pressure defect and/or by flow visualization. With appropriate software that we have written we convert these measurements to vorticity and stream function distributions. The product of these two is then integrated over the cross-flow plane to yield vortex drag. Our particular inspiration for this approach was derived from a study of Eric Maskell's.*

In the future we plan to extend the technique by using an LDA system to make the velocity measurements.

**Maskell, E. C. (1973), Progress Towards a Method for the Measurement of the Components of the Drag of a Wing of Finite Span, RAE Tech. Rept. 72232.*

PROSPECTS FOR NUMERICAL SIMULATION OF BLUFF-BODY AERODYNAMICS

C. W. HIRT and J. D. RAMSHAW

University of California Los Alamos Scientific Laboratory

Los Alamos, New Mexico

ABSTRACT

An improved understanding of the aerodynamics of bluff bodies such as road vehicles can lead to significant reductions in gasoline consumption and to increased safety and comfort. To achieve these goals improved theoretical and experimental techniques are urgently needed. This paper explores the potential of using numerical-simulation methods for predicting and interpreting aerodynamic phenomena affecting bluff bodies. As a basis for discussion, a prototype finite-difference method is described, and illustrated with sample calculations of air flow about simple bluff bodies. The limitations of this scheme are then discussed in detail, together with some suggestions for extensions that could be realized in the immediate future. The paper concludes with speculations on what could be achieved in the next five to ten years to produce a generally useful research tool for bluff-body aerodynamics.

I. INTRODUCTION

An improved understanding of the aerodynamics of bluff bodies such as road vehicles can have many important consequences. For example, aerodynamically designed vehicles can exhibit significant energy savings, increased safety, reduced noise levels, and other desirable features. Because of the complex structure of air flows about road vehicles, however, detailed investigations generally require the gathering and interpretation of extensive amounts of experimental data. This is particularly true considering the wide variety of vehicle shapes and air flow configurations of interest. Experimental investigations covering only a narrow range of parameters still require nontrivial expenditures of time and money to construct apparatus, run tests, record data, and to draw meaningful conclusions. For this reason

References pp. 348-350.

it is desirable to seek additional means of improving and extending the theory of bluff-body aerodynamics.

In this paper we explore the possibility of utilizing numerical-simulation techniques to help in the gathering, interpretation, and presentation of data on the aerodynamic processes affecting road vehicles. Although the numerical methods to be discussed can be used in much the same manner as an experimental apparatus, they are by no means meant to, nor are they capable of, replacing experiments. Numerical simulations, however, can play an important role in interpreting and extending experimentally obtained results. For example, a numerical solution that agrees with measured data at *isolated* instrument stations can then be used to give a *complete* picture of the flow structure in the region computed. Furthermore, once a correlation has been established between numerical and experimental results for a given configuration, then the effects produced by varying flow parameters, altering boundary conditions, or making other changes can often be more readily accomplished and interpreted numerically than experimentally. Thus, even when a numerical method is not so highly developed as to allow complete *a priori* predictions, it may be very useful as a means of interpolating and extrapolating from a small number of carefully selected experimental results.

Although numerical methods have proven highly useful in many areas of fluid dynamics, the complexities associated with air flows about bluff bodies have previously discouraged extensive numerical work in this area. Nevertheless, it is the contention of this paper that the three-dimensional numerical simulation of air flows about bluff bodies is now feasible. The extent of what can be accomplished today, the costs involved, and the limitations and possibilities for the next decade are topics that will be discussed in the remaining sections of this paper. The emphasis throughout will be on three-dimensional problems.

No attempt will be made to review all the excellent numerical schemes currently available. Instead, a rather simple three-dimensional finite-difference method is described and used as a prototype upon which to base a discussion of the important features and limitations common to essentially *all* numerical methods. This prototype method is outlined in Sec. II, and its use is illustrated with calculations of three-dimensional air flows about simple bluff bodies. The examples reveal a number of limitations in the prototype code that must be overcome before a generally useful research tool can be developed. These limitations, and suggestions for their removal, are discussed in Sec. III. Some of them are associated with specific simplifications contained in the prototype code that can be readily overcome by using modifications based on currently available techniques. Others are common to all existing numerical methods and will require new innovations in methodology for their removal. Section IV contains some long-range projections of the advances likely to be made in computing road-vehicle aerodynamics in the next ten years.

There has been no effort to include a comprehensive list of references relating to topics discussed in this paper. Where references are given they are either meant to be representative, or are those with which the authors are most familiar. Any omissions or errors in priority are unintentional.

II. A PROTOTYPE NUMERICAL METHOD FOR BLUFF-BODY AERODYNAMICS

To encourage the utilization of numerical methods in fluid dynamics, a series of simple, well-documented computer codes is being developed and installed in the Argonne Code Center (Hirt *et al.* 1975). This family of codes, given the generic name SOLA (an acronym for solution algorithm), presently contains three members. All are two-dimensional, time-dependent, finite-difference techniques for solving the Navier-Stokes equations. The first two codes, SOLA and SOLA-SURF, are for incompressible fluids. SOLA-SURF is an extended form of SOLA that can handle free-surface problems. The third code, SOLA-ICE, is a partially implicit solution method for compressible as well as incompressible fluids.

For the purposes of this paper a three-dimensional version of SOLA, referred to as SOLA-3D, will be adopted as the prototype code. When it has been fully documented, SOLA-3D will also be submitted to the Argonne Code Center.

A. The Prototype Code – The SOLA-3D code utilizes a simplified formulation of the well known Marker-and-Cell (MAC) numerical solution method (Harlow *et al.* 1965). However, it differs from the original MAC method in several respects. The Lagrangian marker particles used in MAC to identify free surfaces are not needed for the flows treated by SOLA. The uniform mesh used in MAC has been extended to a variable rectangular mesh in SOLA-3D to give improved resolution in regions of rapid spatial variations. The SOLA codes also have improved finite-difference approximations that can be easily modified to have an optional second-order-accurate formulation (Hirt & Stein 1976a). In addition, the simple structure of all the SOLA codes permits a wide varietv of boundary conditions and other variations to be easily inserted.

Details of SOLA-3D will only be sketched here to the extent needed for later discussions. A more complete description of an early experimental version of this code is available (Hirt & Cook 1972), although it does not contain the improved differencing and variable-mesh features of SOLA-3D.

The differential equations to be solved are,

$$\frac{\partial u}{\partial t} + \frac{\partial u^2}{\partial x} + \frac{\partial uv}{\partial y} + \frac{\partial uw}{\partial z} = -\frac{\partial p}{\partial x} + g_x + \nu\left(\frac{\partial^2 u}{\partial x^2} + \frac{\partial^2 u}{\partial y^2} + \frac{\partial^2 u}{\partial z^2}\right)$$

References pp. 348-350.

$$\frac{\partial v}{\partial t} + \frac{\partial uv}{\partial x} + \frac{\partial v^2}{\partial y} + \frac{\partial vw}{\partial z} = -\frac{\partial p}{\partial y} + g_y + \nu\left(\frac{\partial^2 v}{\partial x^2} + \frac{\partial^2 v}{\partial y^2} + \frac{\partial^2 v}{\partial z^2}\right) \tag{1}$$

$$\frac{\partial w}{\partial t} + \frac{\partial uw}{\partial x} + \frac{\partial vw}{\partial y} + \frac{\partial w^2}{\partial z} = -\frac{\partial p}{\partial z} + g_z + \nu\left(\frac{\partial^2 w}{\partial x^2} + \frac{\partial^2 w}{\partial y^2} + \frac{\partial^2 w}{\partial z^2}\right)$$

where (u,v,w) are velocity components in the respective coordinate directions (x,y,z), p is the fluid pressure divided by the constant fluid density, ν is the kinematic viscosity, and (g_x, g_y, g_z) are body accelerations in the (x,y,z) coordinate directions. These momentum equations are to be supplemented with the incompressibility condition

$$\frac{\partial u}{\partial x} + \frac{\partial v}{\partial y} + \frac{\partial w}{\partial z} = 0\,. \tag{2}$$

The numerical solution of these equations utilizes a mesh of rectangular cells with edge lengths δx_i, δy_j, δz_k, where subscripts refer to the ith cell in the x-direction, the jth cell in the y-direction, and the kth cell in the z-direction. The location of variables in a typical cell are shown in Fig. 1. The entire mesh consists of IMAX cells in the x-direction, JMAX cells in the y-direction, and KMAX cells in the z-direction. A single layer of cells around the mesh perimeter is reserved for the setting of boundary conditions, so that the fluid-containing region of the mesh consists of IMAX-2 by JMAX-2 by KMAX-2 cells. The proper use of these boundary cells eliminates the need for special finite-difference equations at the boundaries, i.e., equations used in the mesh interior are also used unchanged at the boundaries of the mesh.

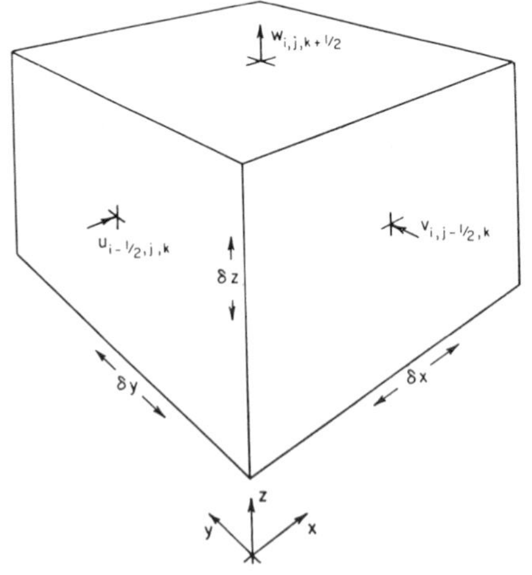

Fig. 1. Location of velocity components with respect to a typical mesh cell (i,j,k). Pressure is located at cell center.

A cycle of calculation to advance the flow configuration through a time interval δt consists of the following two major steps:

1. Finite-difference approximations of the momentum equations, equation (1), are used to obtain guesses for the new time-level velocities using the previous time-level quantities in all convective, viscous, and pressure gradient terms. The new velocities thus obtained will not necessarily satisfy the incompressibility condition, equation (2). Therefore,

2. the pressure is adjusted in each mesh cell to insure that the finite-difference approximation of equation (2) for each cell is satisfied. These pressure adjustments must be done iteratively, because a change of pressure in one cell will upset the balance in neighboring cells. The number of iteration sweeps through the mesh, necessary to obtain a desired level of convergence to equation (2) in all cells, varies with each problem. However, a typical number for problems described in this paper is 5 to 20. Typically, more iterations are required to get a problem started, because large initial flow transients require large pressure adjustments. The lower iteration number is more typical as nearly steady flow conditions are approached. Of course, the iteration number drops to unity when steady conditions are reached.

From a mathematical point of view, the iteration process is used to obtain the solution of a Poisson equation for pressure, although this equation is not explicitly written out in the code. From a physical point of view, the pressure iteration is necessary to account for the long-range influence of rapidly propagating acoustic pressure waves that maintain a uniform density.

The particular finite-difference approximations used in SOLA-3D are not essential for purposes of this paper and for this reason are not provided here. In principle, any set of finite-difference approximations can be inserted into the code without altering its basic structure. The only requirements are that the chosen approximations do not lead to numerical instabilities, and that they reduce to the original differential equations in the limit of vanishingly small space and time increments. These requirements are met by the approximations used in all SOLA codes.

B. Application to Flow Over a Cube – To illustrate the use of SOLA-3D for bluff-body aerodynamics a calculation has been made of the steady flow of air over a cube sitting on a ground plane inside a wind tunnel with model area blockage of 6.25%. The air flow is directed normal to a face of the cube. Fig. 2 schematically illustrates the type of mesh arrangement used for the calculations. Because the flow is normal to a face, symmetry is assumed about a vertical mid-plane and only half the flow field is computed, as indicated in Fig. 2. It should also be noted that mesh intervals are crowded up near the edges of the obstacle where more flow detail is expected. The obstacle is defined within the mesh by setting to zero all velocity components in cells defining the obstacle. Possibilities for defining more general body shapes will be taken up in the next section (III.A.1).

References pp. 348-350.

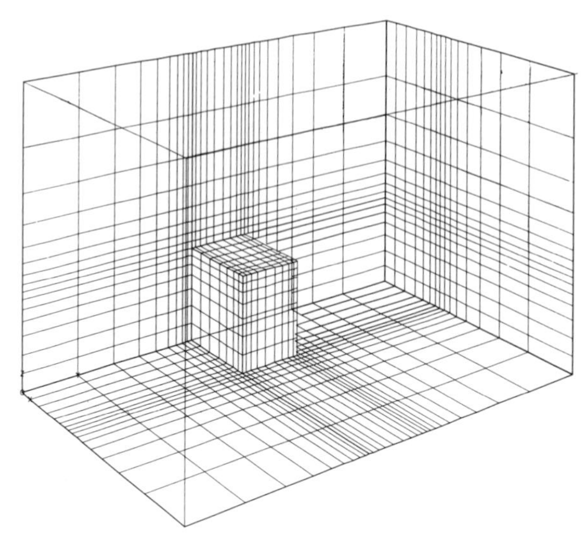

Fig. 2. Schematic of variable mesh arrangement for flow about a prismatic body.

The incident wind is defined by specifying velocities over the upstream face of the computing mesh. These velocities may be constant or may have prescribed space and time variations. For example, a power law profile, $u = u_o(z/z_o)^n$, representing a turbulent boundary layer is useful when investigating wind flows about large buildings. In the present example the calculations are compared with wind tunnel data (Chien *et al.* 1951), in which a uniform wind was used.

Zero normal velocities and zero shear stresses were maintained at the side walls of the mesh to represent the wind tunnel walls and the plane of symmetry. At the downstream boundary, zero velocity gradients in the flow direction are imposed at the beginning of each calculational cycle, but the boundary velocities are then allowed to change during the pressure iteration. This outflow, or continuative, boundary condition has been found in many cases to have little adverse upstream influence, provided the boundary is far enough downstream from the region of interest. In general, the choice of boundary conditions at the perimeter of the mesh is an important and difficult problem area. This is discussed in more detail in the next section (III.A.3).

The cube was assigned a unit edge length and the incident velocity was uniform with magnitude $u_o = \sqrt{2}$, so that pressure coefficients, $c_p = (p - p_o)/(u_o^2/2)$, on the cube surface were simply equal to the computed pressure minus the free-stream reference pressure p_o. Because some turbulence was assumed to exist in the incident flow, and certainly some is generated by the cube, a kinematic viscosity of .001 (representing an effective Reynolds number, based on the cube dimension, of 1400) was used in the calculation to represent a small constant eddy viscosity. Better ways to represent the effects of turbulence are discussed later (III.C). The corresponding molecular viscosity of air in the same units is approximately 3 x 10^{-5}.

The finite-difference mesh, including boundary cells, consists of 32 cells in the downstream (y) direction, 19 cells in the vertical (z) direction, and 19 cells in the horizontal cross-stream (x) direction (similar to, but not the same as, shown in Fig. 2). The physical dimensions of the mesh in units of the cube edge length are: 7 downstream, 4 vertical, and 2 cross stream (one-half of symmetric problem). Approximately 1 hour of CDC 7600 computer time was required to reach a steady flow pattern starting from an initially uniform flow. A brief description of the accuracy and numerical stability limits imposed by choices of space and time increments is given in the next section (III.B.1).

Sample results computed with SOLA-3D are shown in Figs. 3 to 7. The velocity vectors in Figs. 3 to 6 are in planes containing, or near, the faces of the cube. Each vector is drawn from the center of a calculational cell with a direction and length proportional to the average cell velocity (a plus sign is at the *foot* of each vector). From the vector plots containing the top and side faces it is seen that flow separations have developed at the leading edges of these faces. The wake flow is very complex, containing several rotating flow regions. In this example, flow separation occurs at the sharp leading edges of the cube as expected, but for more general boundary shapes special care must be taken as described in the next section (III.A.2).

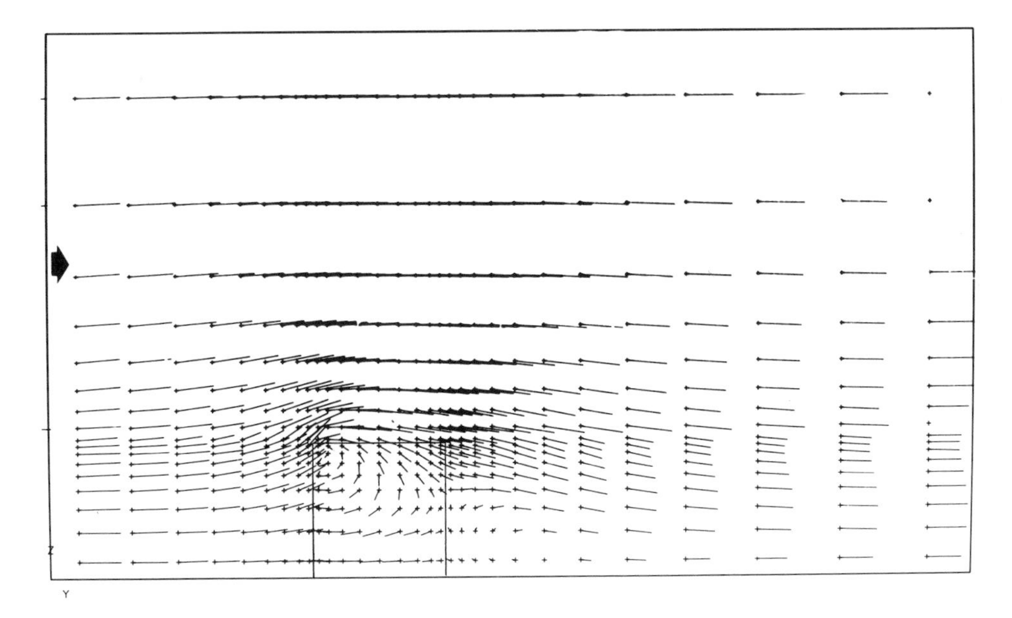

Fig. 3. Computed velocity vectors in plane containing side face of cube. Each vector starts at the mesh point (+) to which it corresponds.

References pp. 348-350.

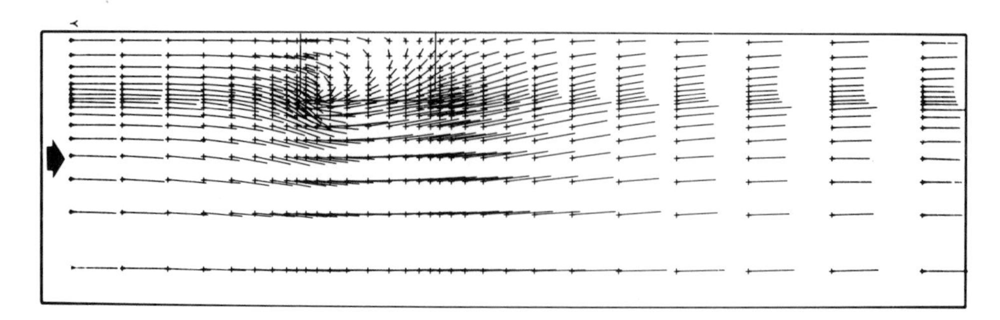

Fig. 4. Computed velocity vectors in plane containing top face of cube.

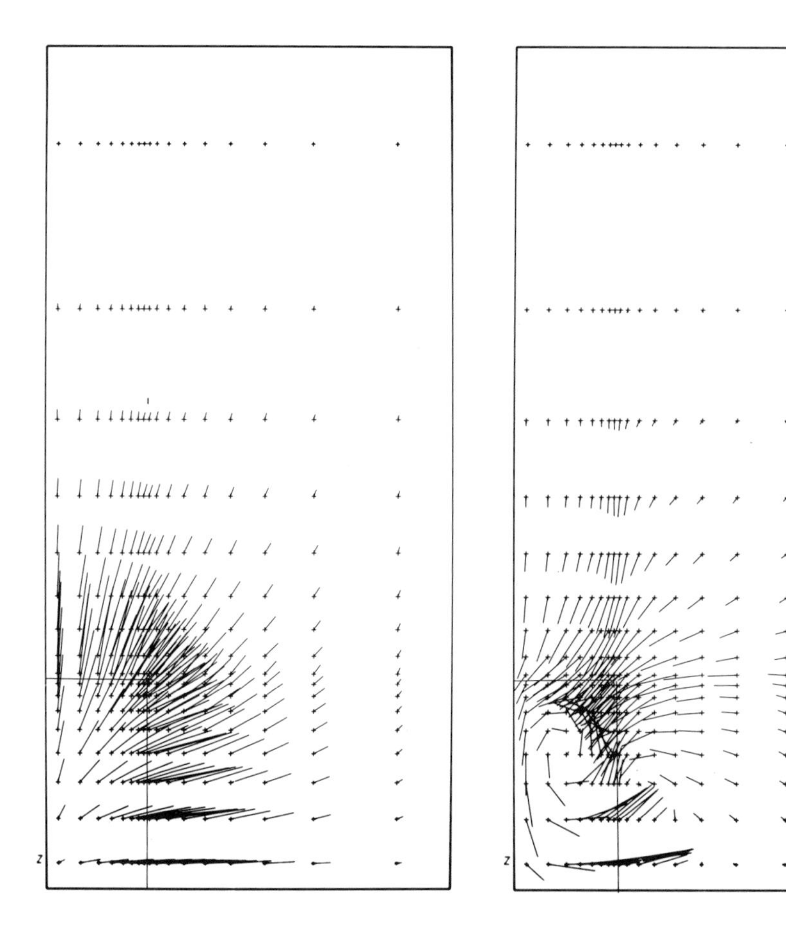

Fig. 5. Computed velocity vectors in plane containing front face of cube (only right symmetric half is shown).

Fig. 6. Computed velocity vectors in plane near rear face of cube.

Pressure-coefficient contours on the cube faces are shown in Fig. 7, which also contains the experimentally obtained data (Chien *et al.* 1951). The best agreement is on the front face where the pressure distribution is a direct result of the incident flow. The slight tendency of the experimentally obtained data to form closed contours near the bottom of the front face indicates that a small boundary layer has developed on the ground plate on which the cube is mounted in the wind tunnel. No ground boundary layer develops in the calculations because of the free-slip boundary condition.

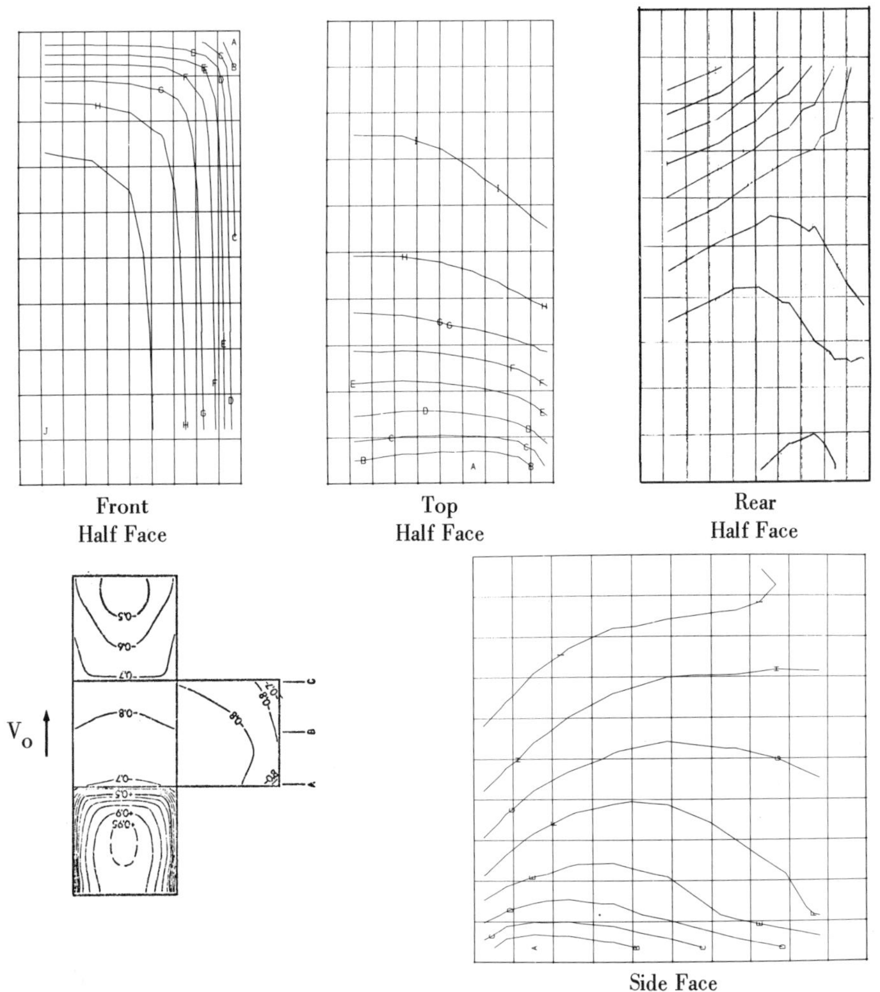

Fig. 7. Calculated pressure coefficient contours on cube faces. Insert shows experimentally measured pressures. Calculated contours correspond to equally spaced values between minimum values (A) and maximum values (J). The minimum and maximum values for each face are: Front A = -.156, J = 1.071; Side A = -1.505, J = -.502; Top A = -1.474, J = -.502; Rear A = -.558, J = -.347. (Pressure coefficients exceeding unity occur because the reference pressure does not exactly correspond to the undisturbed upstream flow. The reference pressure is taken from the upstream face of the computing mesh, only a finite distance from the body.)

References pp. 348-350.

Pressures computed on the top and side faces show somewhat more structure near the leading edges than observed in the experiments. However, the average pressure coefficient away from the leading edges is -0.72 on both faces, which is in good agreement with the data. Similar features are seen on the rear face where a weak pressure gradient is computed at the base of the rear face, and the average of the pressure coefficient over the upper half of the face is -0.55.

In comparing the magnitude of the computed and measured pressures, it should be borne in mind that measurements made by different researchers typically show variations in pressure coefficients on side and rear faces of approximately 10 to 20 per cent. Of greater concern in the comparisons, however, is the appearance of more structure in the computed results. An explanation for this can be traced to an inaccurate representation of the effects of molecular and turbulent viscosity. To see this, refer to Table I which contains the maximum and minimum contour values computed on each face of the cube using different values of kinematic viscosity. The values used were .01, .001, and 0.0. From the table it is seen that larger viscosities produce a greater spread in pressures, that is, stronger gradients on the side and rear faces, but have relatively little influence on the front face. Also the differences observed between the .001 and 0.0 cases are less than the differences between the .01 and .001 cases. At the lower viscosity the pressure coefficients are approaching the experimental results. Thus, these results indicate that the computations are probably too "viscous", even with $\nu = 0.0$, and that better results would be obtained by increasing numerical resolution (or accuracy) in the vicinity of the cube. It may also be that some of the differences observed between various physical experiments are the result of different levels of ambient turbulence, surface roughness, and the extent of the approach boundary layer development.

Because of its simple shape and the availability of experimental data, the cube example has been used here to illustrate the possibilities for numerical simulation of bluff-body aerodynamics. Examples of other SOLA-3D calculations involving uniform and boundary-layer winds over various prismatic structures are described by Hirt & Stein (1976b). In each case the calculations have been compared with wind tunnel data. Taken together, these examples establish a good basis for proceeding to more complicated aerodynamic problems.

C. Application to Flow Over a Tractor-Trailer – In an effort to explore the possibilities of using SOLA-3D for road-vehicle aerodynamics, a calculation has been performed for steady head-on flow about a simplified tractor-trailer configuration. A cross section of the truck and its placement in the finite-difference mesh is shown in Fig. 8. The incident air flow is uniform and the truck is situated above the ground plane to permit air flow underneath. Wheels and other under-carriage structure are not modeled. All corners of the vehicle, which is represented by three rectangular blocks, are sharp. The basic vehicle shape was scaled from a toy model and is clearly not representative of all tractor-trailer configurations.

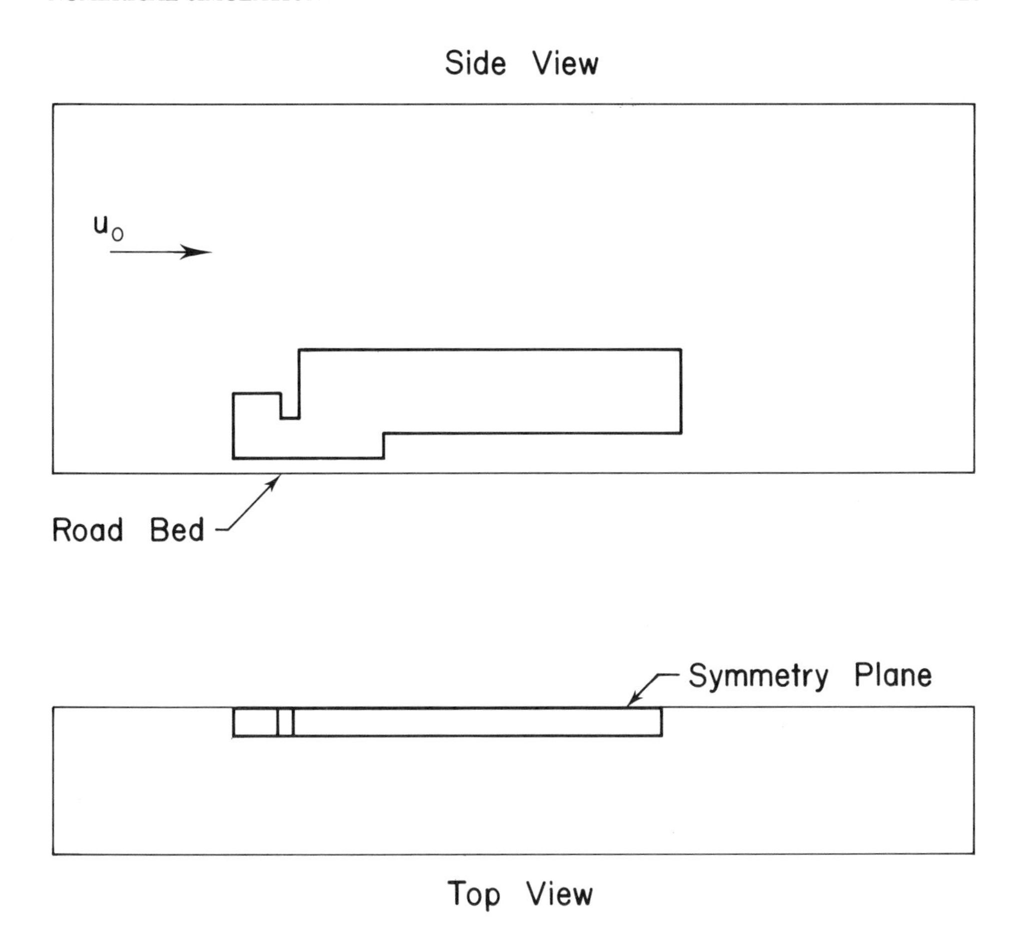

Fig. 8. Schematic of tractor-trailer geometry and mesh boundaries.

The calculational mesh employed for this problem contained 11 cells in the cross stream, or x-direction, 30 cells in the vertical, or y-direction, and 38 cells in the flow or z-direction. The calculation took approximately 90 minutes on a CDC 7600 computer, using the second-order-accurate finite difference approximation available in SOLA-3D. Preliminary calculations using a viscosity of .001 (representing an effective Reynolds number, based on trailer width, of 1400), although stable, produced somewhat hashy looking velocity distributions, indicating that numerical errors were significantly influencing the results with the limited resolution available. For this reason, a viscosity of .01 (an effective Reynolds number of 140) was used to obtain the results shown in Figs. 9 to 12.

References pp. 348-350.

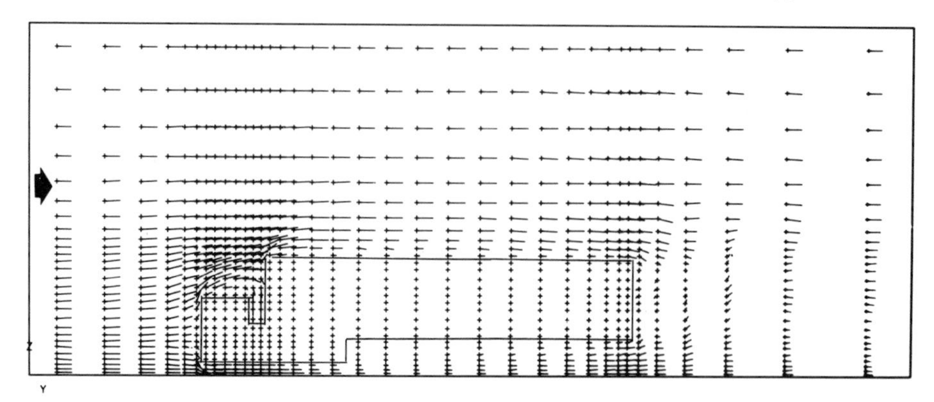

Fig. 9. Computed velocity vectors in symmetry plane of flow around tractor-trailer.

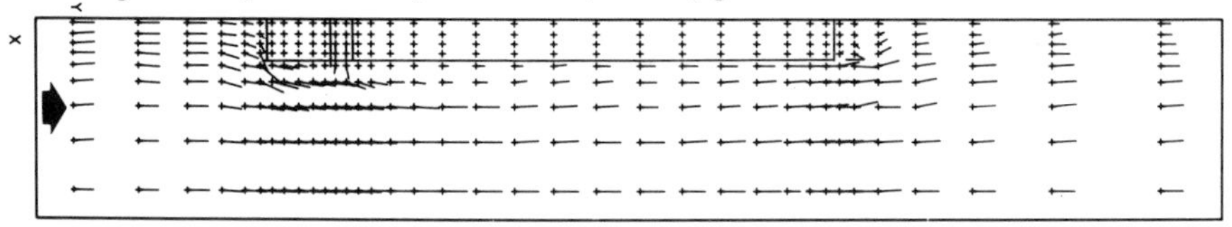

Fig. 10. Computed velocity vectors in horizontal plane intersecting tractor cab and trailer at mid cab height. Top edge of figure is plane of symmetry of flow.

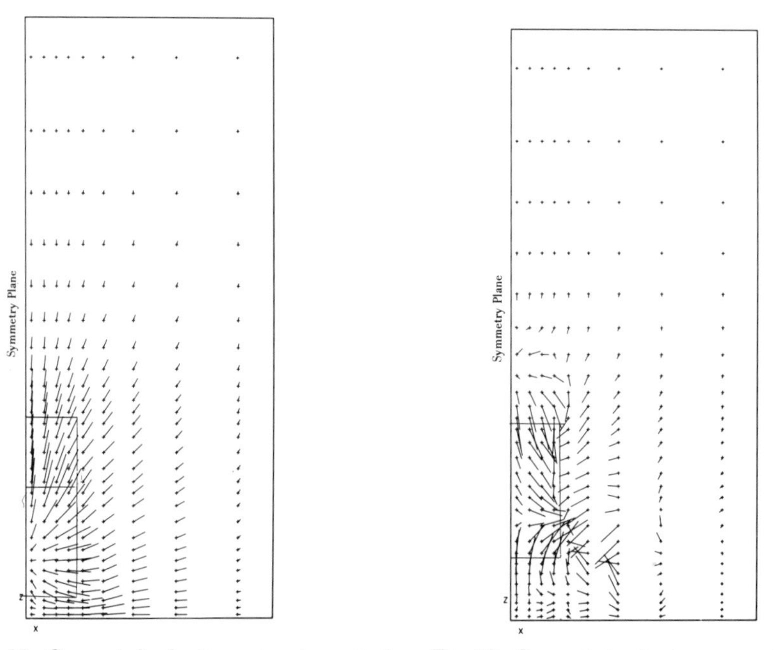

Fig. 11. Computed velocity vectors in vertical plane near front face of cab.

Fig. 12. Computed velocity vectors in vertical plane near rear face of trailer.

Unfortunately, the larger viscosity largely suppresses the reverse-flow details on the top leading edges of the cab and trailer that are observed in the small-viscosity calculations (compare Fig. 13 with Fig. 9).

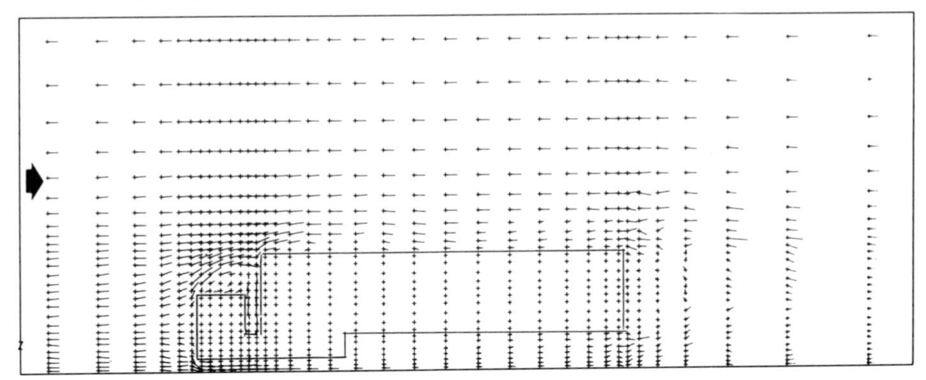

Fig. 13. Computed velocity vectors in symmetry plane of flow around tractor-trailer as obtained with lower (.001) kinematic viscosity. More flow structure is seen near leading edges of cab and trailer, but flow is very irregular in wake region.

No large reverse flow is observed at the leading side edge of the cab. It is believed that this is a result of the jet-like lateral flow entering the side region from the open space between the cab and trailer (Fig. 10).

A flow stagnation does develop on the face of the cab, with downward directed flow below the stagnation point (Fig. 11). This is not, however, associated with an eddy in front of the cab, but only reflects the presence of flow into the open space between the under-carriage and the ground. From Fig. 12 it can be seen that the wake region is highly complex, and contains a variety of interacting vortices. Pressure distribution, lift and drag coefficients, and other details can be obtained from the calculated results, but are not included here.

No attempt has been made to compare these calculations with experimentally obtained data. In retrospect, however, it is likely that the lateral boundaries of the computing mesh could have been placed somewhat closer to the vehicle and the mesh refined to increase resolution around the leading edges of the cab and trailer. Nevertheless, these limited results illustrate that numerical simulations are a potentially powerful tool for investigating aerodynamic phenomena associated with road vehicles. These results also show that many extensions and improvements are necessary before a code like SOLA-3D can be used for extensive practical applications.

In the next section the limitations of numerical methods are put in more quantitative terms and the possibilities for relatively short term improvements are explored.

References pp. 348-350.

III. LIMITATIONS AND POSSIBILITIES FOR IMPROVEMENT

In the previous section we identified a variety of limitations and special assumptions contained in the prototype code SOLA-3D. These limitations fall into three basic problem areas. First, there are problems associated with boundary conditions; for example, how to define arbitrarily shaped bodies, how to resolve boundary layers, how to calculate or define separation points, and how to prescribe inflow and outflow boundary conditions at the mesh perimeter. Second, there are problems associated with the accuracy of the numerical approximations; for example, what constitutes an adequate discretization of time and space, what is the importance of higher-order approximation schemes, would schemes based on finite-element or Galerkin methods be superior to finite-difference methods, and what are the limitations imposed on a solution by numerical stability requirements? Finally, there are miscellaneous problems, such as how to represent the effects of fluid turbulence and how to optimize the solution algorithms to get the best results for a given expenditure of computer resources.

It is obvious that most of these problems are not uniquely tied to SOLA-3D but are common to all approximation methods. Thus, in the following paragraphs the discussion of the above problems will be kept as general as possible.

A. Boundary Conditions – The specification of numerical boundary conditions is a somewhat tricky, and sometimes subtle, business that requires considerably more care and attention than one might naively suspect from the ratio of boundary cells to interior cells. For convenience of discussion, the general problem of boundary conditions may be subdivided into three areas: 1. geometrical specification of boundary shapes; 2. boundary conditions on the surface of solid bodies in the interior of the mesh; and 3. boundary conditions at the external boundaries of the computing mesh.

1. Geometrical Specification – Many bluff bodies of interest may be regarded as approximately rectangular in shape, and hence can easily be represented in a rectangular computing mesh. In most cases, however, the body of interest is not rectangular and the capability to represent curved boundaries of reasonably arbitrary shape is needed. It is not always realized that sufficiently simple curved boundaries can often be represented quite well within a rectangular mesh through the use of variable spatial increments, provided that the details of the boundary layer flow are not required. For example, a quonset hut of semicircular cross section was successfully modeled with the SOLA-3D code by choosing the spatial increments in such a way that the mesh points fall on the boundary surface, Fig. 14. A more general and flexible method of representing curved boundaries is provided by the ALE (Arbitrary Lagrangian-Eulerian) method (Trulio 1966; Hirt *et al.* 1974; Pracht 1975), which is embodied in the two-dimensional Los Alamos codes YAQUI and CHOLLA and in the three-dimensional code BAAL. This method allows the spatial coordinates of each mesh point to be separately specified, and in addition provides for an

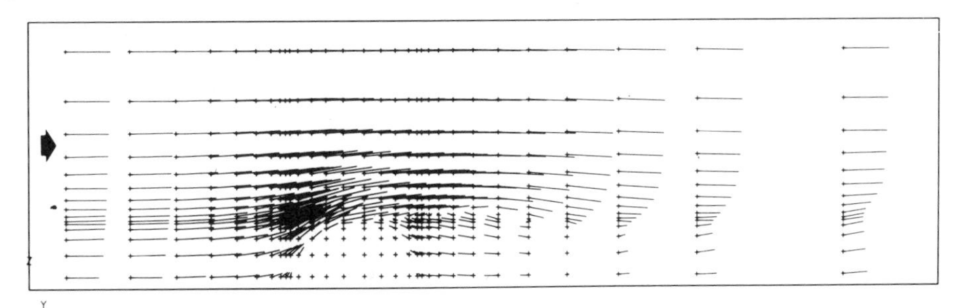

Fig. 14. Computed velocity vectors in symmetry plane of flow over a quonset hut. Variable mesh permits some curved boundary shapes to be modeled.

arbitrary time-dependence of the mesh-point coordinates to permit simulation of moving boundaries. As an example, the YAQUI mesh used for a calculation of flow past a circular cylinder is shown in Fig. 15, together with the calculated velocity-vector plot and pressure and vorticity contour plots. Although somewhat more complicated in structure than rectangular-mesh codes, ALE-like codes are conceptually no more difficult to comprehend, and their difference equations can readily be written in conservative form, which is usually, but not invariably, desirable for accuracy. Other methods are also available for handling curved boundaries. Viecelli (1969) has successfully treated curved boundaries by a modification of the method commonly used to represent free surfaces in MAC-type codes (Harlow *et al.* 1965). A simplified form of Viecelli's method (Hirt *et al.* 1975) is used in

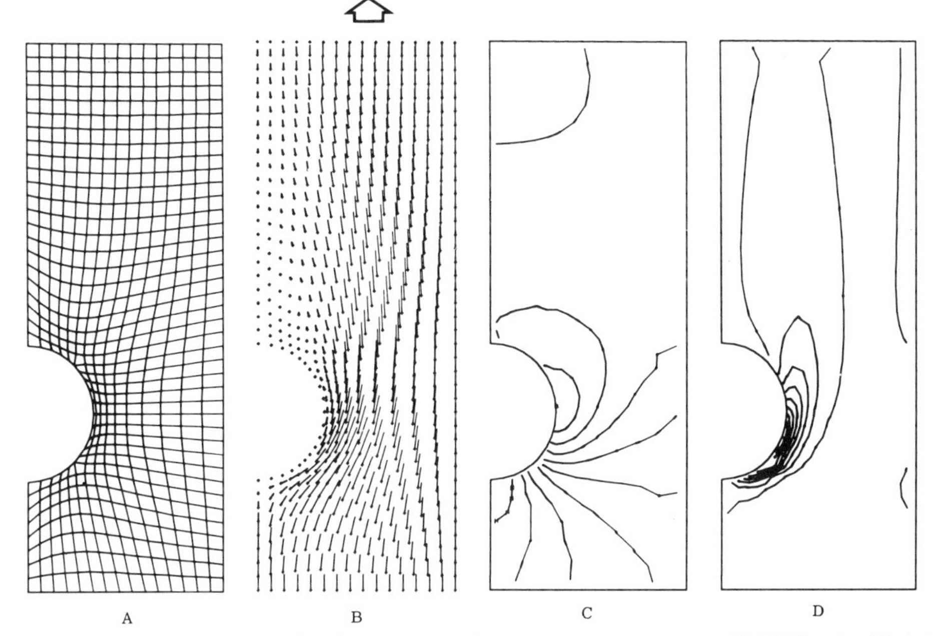

Fig. 15. Calculated results for flow over a cylinder using variable-mesh YAQUI code. Mesh is shown in A, velocities in B, pressure contours in C, and vorticity contours in D.

References pp. 348-350.

SOLA-SURF and in SOLA-3D to describe a curved surface with elevation $z=H(x,y)$. It would not be difficult to generalize this single-valued function of (x,y) to arbitrarily curved surfaces. The major effort needed would be to program the logic necessary to determine how the boundary velocity components must be set to satisfy the proper wall boundary conditions.

In contrast to most finite-difference codes, finite-element codes are normally written in such a way as to allow for arbitrary boundary shapes. This fact is sometimes regarded as an advantage that finite-element methods enjoy over the more conventional finite-difference methods. However, ALE-type finite-difference codes have essentially the same capability and have the advantage of greater simplicity.

Finally, we mention the use of coordinate transformations to represent curved boundaries by defining a new system of coordinates in which the boundaries become rectangular. This method can also be used to increase resolution in regions of rapid spatial variation. Typically the differential equations are transformed into the new coordinate system and finite-difference analogs are then written for the transformed equations. Some workers feel that this procedure provides increased accuracy over the more straightforward procedure of using a variable mesh, either rectangular or ALE-like. We disagree with this view for several reasons. First, such statements appear to be based on the observation that a deterioration in the formal order of accuracy (typically from the second to first order) results from the use of a variable mesh, while by differencing the transformed equations in a regular mesh second-order accuracy can apparently be preserved. This advantage is illusory, however, because the order of accuracy of a given finite-difference expression is not invariant to coordinate transformations. That is, a finite-difference expression that is second-order accurate in the transformed coordinates will still be only first-order accurate in the original coordinates (see Appendix). Moreover, even if this were not the case it is known that conclusions about accuracy based on the formal order of the truncation errors can be very misleading, especially in problems where relatively coarse zoning is dictated by economic and/or storage constraints, as is virtually always the case in three-dimensional calculations. In addition, the coordinate-transformation approach as usually applied has certain undesirable characteristics (Roache 1972), among them the loss of conservation properties. This difficulty can be avoided by constructing the difference equations from integral balances over elementary cells in transformed-coordinate space, but when this is done one obtains difference equations essentially equivalent to those of the ALE method (Stein *et al.* 1976).

2. *Solid-Boundary Conditions* – We now discuss the question of the boundary conditions to be applied at the surface of solid bodies within the computing mesh. If computer time and storage were no object, the answer to this question would be simple: one would simply use no-slip velocity boundary conditions at solid walls in conjunction with very fine zoning, and let boundary layers, separation points, etc.

emerge naturally from the calculation. Even if this were feasible, turbulence would be a complicating factor, but one that could be taken into account either by the use of known eddy-viscosity formulas or by the use of more sophisticated turbulence transport models (see Sec. III.C). In practice, however, computer time and storage constraints preclude the accurate representation of boundary layers (except perhaps by an iterative coupling of a boundary-layer solution algorithm to a coarsely-zoned "outer" calculation using free-slip boundary conditions; see Sec. IV). Fortunately, in connection with the prediction of drag and the pressure distribution on bluff bodies, we are largely indifferent to the details of the boundary layer flow except in so far as they determine the separation point. The reason is that form drag will be the dominant drag mechanism in almost all cases, and it is reasonable to expect that if the location of separation is predicted correctly one will obtain reasonable results for pressure distributions and hence for form drag. In the case of bodies with sharp corners, it is known in advance that separation will occur at these corners, and we might then expect to get reasonable results without doing anything special about boundary conditions. The results of the first calculation discussed in Sec. II partially bear out this expectation, although the variations produced by varying the kinematic viscosity show that the boundary conditions do have a non-negligible, although fairly small, effect even when the separation point is known in advance.

In the case of bodies without sharp corners the location of separation is not known in advance and something more must be done. In some cases the location of separation may be known fairly well on empirical grounds. If so, this knowledge can be put into the calculation by setting the tangential velocity closest to the wall at that location equal to zero. Continuity, equation (2), will then force the flow to separate at that location, and again reasonably accurate predictions of drag and pressure distributions can be expected. If the location of separation is completely unknown *a priori* then it is necessary to devise a reasonable approximate procedure for estimating the wall shear stress, and then impose this shear stress as a boundary condition. This is necessary because the location of separation is primarily determined by the combined effects of the wall shear stress and the adverse pressure gradient. (In addition, one can obtain in this way an approximation to the friction drag by integrating the approximate wall shear stress.) One possible way of doing this is to use the wall shear stress obtained by matching the tangential velocity nearest the wall to the turbulent logarithmic law-of-the-wall velocity profile (Launder & Spalding 1972). It is not clear that this will always be a good approximation, however, because the velocity nearest the wall may be located outside the law-of-the-wall region. Furthermore, the law of the wall may not be applicable in regions of recirculating flow, although errors made in such regions would presumably have only minor effects on the location of separation. At any rate, it is clear that the capability of this or any other shear-stress prescription to correctly predict the location of separation must be determined by comparisons of numerical calculations with experimental data. It would be desirable to perform a systematic comparison of a variety of approximate

References pp. 348-350.

shear-stress prescriptions with experimental data to determine which of them are likely to give the best results in different situations. Such studies could readily be carried out, at reasonable expense, with existing codes such as SOLA and SOLA-3D. We believe that this is one of the next steps that should be taken in an effort to simulate road-vehicle drag and aerodynamic effects numerically.

3. Mesh-Perimeter Conditions - Finally, we discuss the problem of specifying boundary conditions at the external boundaries of the computing mesh. This problem has several aspects. One difficulty is that the difference equations are almost invariably of a higher order than the differential equations they represent, in the sense that they require additional boundary conditions that would be redundant or inconsistent in the differential problem. Ideally, these additional boundary conditions should be chosen to be redundant rather than inconsistent, but this would require prior knowledge of the solution and hence is not generally possible. Fortunately, in practice it appears possible in many cases to violate the consistency condition without contaminating the entire solution, the effects of the inconsistency being confined to a thin layer at the boundary.

A frequently more serious difficulty, although not one of a purely numerical nature, is that the physical boundary conditions (such as pressure distribution over an outflow boundary, or incoming velocity profile over an inflow boundary) are often not known precisely. Outflow boundaries behind bluff bodies are particularly troublesome. since irregular nonsteady wake regions may persist for many body-diameters downstream (Hansen & Cermak 1975). Fortunately, in many cases, the details of the flow near the body appear to be relatively insensitive to the details of the boundary distributions, provided that the boundary is not "too close" to the body and that a numerical outflow boundary condition can be found that allows the flow to exit smoothly from the computational region with minimum adverse upstream influence. This is the purpose of the continuative outflow boundary condition described in connection with the calculations of Sec. II. This prescription works well in some problems but not in others. Orszag (1976) recently described a new continuative outflow procedure for incompressible flow based upon the introduction of an artificial compressibility near the boundary. Many other continuative outflow prescriptions may be imagined (Orlanski 1976; Hanson & Petschek 1976). In general, the best prescription for a given problem, and the precise distance from the body at which the boundary becomes "too close," are likely to be problem-dependent and must be determined by numerical experimentation. Thus, we recommend that further work be done to identify the best methods of treating outflow boundary conditions for bluff-body aerodynamic calculations.

B. Accuracy Considerations — Probably the most important and most difficult problem associated with a numerical approximation is how to assess its accuracy. A related question is, what prescriptions can be used to determine *a priori* the numerical resolution necessary to achieve a desired level of accuracy. Although no hard and fast rules can be established to give a satisfying answer to these questions for all cases, there are some guidelines based on simple order-of-magnitude estimates.

1. Reynolds Number Restrictions – Let us first determine the highest flow Reynolds number that a numerical method can resolve. There are two ways to do this. Using our physical intuition, we argue that the flow processes on a scale comparable to the mesh spacing δx should be smoothly varying. This assumption is the basis on which finite-difference approximations are made. It would require strong viscous effects, and hence a Reynolds number for flow processes on the scale δx that is equal to or less than unity. Thus, if δu is the velocity variation over distance δx, the condition for an accurate approximation is

$$R_\delta \equiv \frac{\delta x\ \delta u}{\nu} \lesssim 1 ,$$

where R_δ is referred to as a cell Reynolds number. This result may be translated into a condition on the problem Reynolds number, R, by defining $N\delta x = L$ and $N\delta u = u$, so that

$$R \equiv \frac{Lu}{\nu} \lesssim N^2 ,$$

where N is the number of mesh intervals used to define the characteristic length L and characteristic velocity u.

An alternative way to arrive at this result is to estimate the influence of the truncation errors introduced by a finite-difference approximation. That is, when the discrete equations are expanded in Taylor series in terms of the space and time increments, the resulting differential equations will differ from the original differential equations by higher-order-derivative terms having coefficients proportional to powers of the space and time increments (Richtmyer & Morton 1967). Usually the most significant members of these residual, truncation-error terms are those that contribute diffusive-like effects in competition with the actual viscous dissipation (Hirt 1968). For a difference approximation or order (p+1), where $p \geq 1$, the lowest-order truncation error will modify the kinematic viscosity according to

$$\nu + a\ \delta x^{p+1} \left(\frac{\partial^p u}{\partial x^p} \right)$$

where a is a numerical coefficient of order unity. Thus, for an accurate representation of viscous effects the second term must be small compared to the first. In the finite-difference approximation the second term will be approximated as $a'\ \delta x\ \delta u$, where a' is another numerical coefficient. Consequently, we are led to the conclusion

References pp. 348-350.

that for an accurate approximation we must satisfy

$$\frac{a' \, \delta x \, \delta u}{\nu} < 1 ,$$

which, except for the a′, is the result previously obtained.

In the case of first-order methods the kinematic viscosity is typically augmented by an error term of the form (Hirt 1968)

$$a \, \delta x \, |u| .$$

and the Reynolds number restriction becomes $R \lesssim N$. This is qualitatively different from second and higher-order schemes, because the error involves $|u|$ rather than its derivatives.

An example of the increased accuracy that can be achieved by a second-order method in contrast to a first-order one is shown in Fig. 16. The example concerns the unstable growth and coalescence of eddies in a shear layer. With the first-order method numerical errors were observed to smooth the shear layer to such an extent

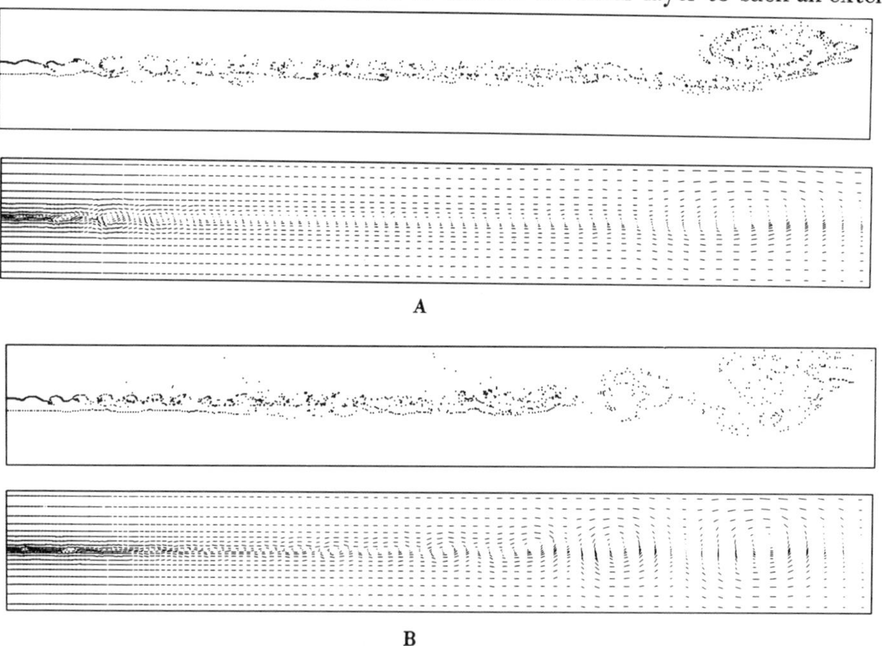

Fig. 16. Calculated results for velocities and marker-particle streams caused by instability of a shear layer. Results obtained by first-order-accurate method (A) and second-order-accurate method (B). The plate separating the two uniform-velocity streams ends at the left edge of the figures. The higher velocity is below the plate. In the velocity-vector plots the mean velocity of the two streams has been subtracted.

that it eventually became stable. In contrast, with the second-order method the true physical instabilities are much less strongly damped. Of course, this is an extreme example, because the shear layer is initially a discontinuity. In other problems where large gradients of this sort are not present, first-order methods may yield perfectly acceptable answers. Incidentally, the fact that the error estimate for second and higher-order schemes is independent of the order suggests that little is gained by using schemes of order greater than two.

For three-dimensional calculations that are currently feasible, N is of order 10 to 20, which according to the above discussion limits calculations to Reynolds number less than about 400. The use of a variable mesh that distributes the N points such that more resolution is provided in regions of rapid spatial variations is equivalent to increasing N. The amount of increase can be estimated as follows: Let δx_0 represent the smallest mesh spacing, and assume a geometric rate of increase in successive cell sizes such that the Nth cell has width $\delta x_0\ (1+a)^N$. The percentage change at each step, a, is typically chosen to be 0.1. If a is much larger, the truncation errors introduced by the non-uniformity of the mesh (see Appendix) can become unacceptably large. Because the N cell widths must sum to L it is not hard to show that

$$N_{eff} = L/\delta x_0 = [(1+a)^{N+1} - 1]/a\ .$$

Thus, for N = 10 and a = 0.1 the effective N could be approximately 20 instead of 10 and accurate calculations at Reynolds numbers of order 400 should be possible. A similar estimate for N = 20 and a = 0.1 shows that Reynolds numbers of a few thousand are possible.

On the basis of these estimates, three-dimensional numerical calculations of aerodynamic phenomena for which R is typically 10^5 or larger would appear to be out of the question with the present generation of computers. Fortunately, it isn't necessary to reproduce the exact Reynolds number of a flow. If it were, wind tunnels, wave tanks, and other experimental research tools would not be as useful as they are. Experimentally this fact rests on the observation that many fluid flows become independent of Reynolds number once a certain minimum value is exceeded (Schlichting 1960). In some cases, for example flow about rounded bodies, the principal effect of Reynolds number is to change the location of separation points. In such instances, laboratory simulations often resort to trip wires or other means of inducing separation at desired locations without having to attain Reynolds-number similarity (Chien *et al.* 1951). When this is done, however, it becomes necessary to carefully check the validity of the simulated results through comparisons with full-scale results.

Much the same situation exists with numerical simulation methods. The results presented in Sec. II demonstrate that theoretical predictions can be made that are not

References pp. 348-350.

unreasonable. Separation did occur at the sharp edges of the cube as expected, and the pressures on the cube were in fair agreement with wind tunnel data. For more complicated body shapes it will be necessary to insure that separation occurs at the correct locations (see Sec. III.A).

2. Required Spatial Resolution – The principal shortcoming of the results computed for the cube appeared to be associated with poor resolution of the separated flow regions, which affect the average pressure distributions on the faces of the cube. The hardest question, therefore, comes down to a determination of what constitutes "sufficient" resolution for purposes of making useful engineering predictions. In other words, how big must N be?

If it were necessary to correctly compute the details of *all* the turbulent eddies the problem would be several orders of magnitude beyond the capabilities of presently envisioned computers. This may be seen by estimating the ratio of the characteristic macroscopic length scale (L) to the diameter of the smallest eddies (ℓ). Defining the smallest eddy size as characterizing scales on which viscous effects are dominant, and relating the large and small scales through the turbulence - energy dissipation rate, we find (Landau & Lifshitz 1959)

$$\frac{L}{\ell} \sim R^{3/4},$$

where R is the flow Reynolds number. If we identify δx with the smallest length scale (ℓ) and $N\delta x$ with the largest scale (L), then the ratio L/ℓ is equal to N. Thus, even for a relatively low Reynolds 10^5, N would have to be of order 10^4, or three orders of magnitude larger than that used in the examples of Sec. II. Fortunately we know that such resolution is often unnecessary. For engineering purposes we are usually most interested in the dominant flow patterns. These consist of the mean flow and the largest eddy structures, which contain most of the energy of the flow. The smaller eddies provide a means for removing kinetic energy from the larger eddies and dissipating it as heat. The details of this removal and dissipation are often not essential, and in such cases the dissipation mechanisms inherent in numerical methods may alone be sufficient to produce reasonable results. In other cases additional dissipation mechanisms must be provided; for example, a constant eddy viscosity like that used in Sec. II, or through the use of more-sophisticated turbulence models as will be described in Sec. III.C.

In this connection, we may mention the use of the subgrid-scale (SGS) eddy-viscosity method which will be discussed in Sec. III.C. The SGS method in effect defines a wavelength-dependent artificial viscosity that is large for small wavelengths of order δx, but small or negligible for larger wavelengths. Thus the effective Reynolds number for the large-wavelength motions of interest may be much larger than that for the short-wavelength motions where dissipation takes place, and hence much larger than the estimates given above.

Returning to the question of what constitutes sufficient numerical resolution for bluff-body calculations, we next need an estimate for the sizes of the dominant secondary-flow structures likely to be generated. Experience indicates that these structures are roughly the same size as the lateral dimensions of the bodies generating them (Morkovin 1972). In the cube problem, for example, the separated-leading-edge flows and the base recirculation have dimensions equal or nearly equal to that of the cube. Therefore, the value of N chosen to represent the body should be sufficient for all the principal flow features generated by that body.

For simple prismatic bodies like the cube, the above estimate is ideal, but for more complicated bodies like an automobile there arises a major difficulty. Local body features such as bumpers, cowlings, fenders, stabilizer fins, mirrors, etc. that protrude from the body can each induce a secondary-flow field. This requires that a range of scale sizes must be simultaneously resolved. The range would typically be one order of magnitude, corresponding to a minimum-significant-size scale of 6 inches. However, even this modest range of scales, implying a ten fold increase in N cannot be handled by present day computers, because this corresponds to three orders of magnitude increase in computer resources for three-dimensional calculations. Fortunately, there is a way around this difficulty. The flow features to be resolved are generally fixed at the bodies where they are generated (Morkovin 1972). Therefore, the increased resolution needed for their definition can also be localized and need not correspond to a uniform increase in resolution over the entire flow field. It is possible, for example, to locally subdivide coarse mesh cells into successively finer zones (Bradshaw & Kramer 1974; Brandt 1972) or to use ALE-type meshes (Trulio 1966; Hirt *et al.* 1974; Pracht 1975) that are deformed in such a way as to give the desired resolution where it is needed (for other possibilities see Sec. IV). Techniques such as these still require further development and testing for aerodynamic applications, but where they have been used they have generally been successful.

Unfortunately, this is as far as we can go on heuristic grounds alone in setting the limits on the resolution and computer requirements needed for practical bluff-body aerodynamic calculations. The next step requires that we proceed to more-detailed comparisons between calculations and experiments. In particular, the next step is to determine for simple body shapes the minimum resolution necessary to produce acceptable results. The cube calculations presented in Sec. II indicated that $N = 10$ was marginal. Those comparisons, however, were obtained with a first-order finite difference method. Additional calculations are needed to test second-order schemes, to find optimum variable-mesh spacings, and most importantly, to test larger values of N. To support a detailed program such as this, it would be desirable to have a corresponding experimental program with the capability to provide detailed velocity distributions, turbulence levels, and pressures in selected regions of interest.

References pp. 348-350.

3. Higher-Order Methods – We have seen that higher-order finite-difference methods have essentially the same Reynolds number limitations as second-order methods. This conclusion follows from a truncation-error analysis as well as from the requirement that wavelengths of order δx or less be damped out in the calculation. It should also be remarked that the truncation-error estimate of accuracy will only be valid when the dependent variables vary slowly over distances of order p δx, where p is the order of the difference approximation. When the dependent variables vary more rapidly than this the higher-order approximations may actually yield worse results than lower-order schemes (Roache 1972). In addition, higher-order schemes require higher-order boundary conditions, which are not always easy to specify (Roache 1972). Trulio (1964) has also shown that for problems involving shock waves or weak discontinuities, the rate of convergence of higher-order schemes as the mesh is refined is no better than that of second-order schemes. These observations reinforce our earlier conclusion that higher-order schemes are not likely to provide significant improvements over second-order schemes, especially in three-dimensional problems where the zoning must be relative coarse. Hence, higher-order schemes should be approached with reservations. Not only is more effort required to develop a code and obtain a solution, but the solution once obtained may be no more accurate, and in some cases even less accurate, than that obtained with second-order methods.

4. Other Numerical Methods – The above arguments regarding Reynolds-number restrictions and resolution requirements are applicable in a general way to all numerical approximation methods, including finite-element and Galerkin-type techniques. Some persons may object to this last remark on the ground that the proper choice of the weighting and/or basis functions used to represent flow variations within a cell can, in principle, give subcell resolution not possible with a finite-difference method. However, this is true only in the simplest of situations, where something is already known about the subcell flow structure. For the problems of interest here, subcell variations are unknown *a priori* and are undoubtedly different in different regions of the flow. In such cases, the resolution obtained with finite-element methods will be determined primarily by ρ_p, the number of parameters per unit volume available to describe the flow, and cannot be expected to exceed that obtained with a finite-difference method using the same ρ_p. In particular, the smallest length scale that can be represented in the calculation will be comparable to that of a finite-difference method using the same ρ_p. This length scale must be damped for stability reasons and to avoid aliasing errors (Roache 1972). Therefore, the same Reynolds number limitations already obtained for finite-difference methods are likely to hold for finite-element and Galerkin methods, and so little is to be gained by their use.

C. Turbulence – For a road vehicle traveling at 22 m/s (50 mph) through still air, the Reynolds number is approximately

$$\mathrm{Re} = 1.54 \times 10^6\ L\ .$$

where L is some typical linear dimension of the body in meters. Thus, it is clear that the wake flow behind such vehicles will be highly turbulent. In addition, the boundary layer on the surface of such vehicles may be expected to become turbulent at a distance D from the leading edge of the vehicle, where

$$D = \frac{\mathrm{Re}^*}{1.54 \times 10^6}\ \text{meters}$$

and where the critical Reynolds number Re* is on the order of 3×10^5 for smooth bodies, but somewhat lower for rough bodies or if turbulence is present in the free stream. Thus $D \lesssim 20$ cm and the boundary layer on the surface of the vehicle may be expected to be almost entirely turbulent.

The effects of turbulence may be represented in various contexts with varying degrees of sophistication. As emphasized by Launder & Spalding (1972), it is foolish to model turbulence with more sophistication and complexity than is actually needed for the purpose at hand. The question is one of sensitivity; it is not necessary to model turbulence with high accuracy if the effects of primary interest are only weakly dependent on how the turbulence is modeled. In general, the detailed structure of a turbulent flow field is far more sensitive to the turbulence modeling than are the gross characteristics of the flow field. In the present context our objective is to compute the pressure distribution over the surface of the body, and not such fine-scale details of the flow as the structure of the turbulent boundary layer, the viscous sublayer, etc. In fact our principal concern is with integral properties of the pressure distribution, such as drag, lift, and torque. These considerations, together with the encouraging results of first-cut calculations such as those discussed in Sec. II, have led us to the opinion that satisfactory numerical predictions of pressure distributions (and therefore drag, lift, and torque) *a fortiori* can probably be achieved without a great deal of sophistication in the turbulence modeling. However, not enough work has yet been done to permit a firm conclusion in this regard, so the question must remain open for the time being. This is one of the principal areas in which we believe near-term future research should be performed. For the present, we merely discuss a number of possible turbulence modeling approximations that one might contemplate using in the calculation of bluff-body aerodynamic effects.

All of the turbulence representations that we shall discuss proceed by adding to

References pp. 348-350.

the laminar viscosity ν a turbulent or "eddy" viscosity ν_t. The various representations differ in how ν_t is to be determined. More complicated turbulence representations have been proposed, but they are not likely to be practical for routine use in bluff-body aerodynamic calculations. Ordinarily, ν_t is much larger than ν, so that ν can be omitted in turbulent regions of the flow.

The simplest possible representation of turbulence is to use a constant value for ν_t in all parts of the flow field. This constant value may be estimated by

$$\nu_t \sim u' \delta$$

where u' is some measure of the magnitude of the turbulent velocity fluctuations and δ is the dominant length scale of these fluctuations (Launder & Spalding 1972). This approximation is clearly very poor for predicting the detailed structure of a turbulent flow (for example, it predicts a parabolic velocity profile for Poiseuille flow rather than the logarithmic turbulent velocity profile), but the results of the calculations described in Sec. II show that it is not so bad for predicting gross features of the flow. The possibility also remains that the results could be improved by a better choice of wall boundary conditions, while still keeping ν_t constant.

The next level of approximation consists of allowing ν_t to vary with position throughout the flow. The classical model of this type is Prandtl's mixing-length hypothesis (Schlichting 1960), in which $\nu_t = \ell_m^2 \, | \partial u / \partial y |$. The mixing length ℓ_m is not constant throughout the flow; correlations have been developed for its dependence on position in various simple flow fields (Launder & Spalding 1972), and in the absence of other information these correlations can be extrapolated to more complicated flows as well.

A variant of the mixing-length hypothesis that is particularly appealing for use in three-dimensional finite-difference calculations is the subgrid-scale (SGS) eddy viscosity model of Deardorff and others (Deardorff 1971). In this method one sets $\ell_m = C\,\delta x$, where C is a constant which is adjustable in the neighborhood of unity. This choice for ℓ_m has the effect of modeling only the turbulent motions whose length scales are smaller than δx and hence cannot be resolved in the mesh, while allowing turbulent motions of larger scale to occur and be followed during the calculation. More precisely, the SGS form of ν_t artificially enlarges the dissipation length scale (Kolmogorov microscale) of the turbulence until it becomes of order δx, regardless of the local rate of turbulent-energy dissipation. It is thus closely analogous to the von Neumann-Richtmyer artificial viscosity for shock waves (von Neumann & Richtmyer 1950), which artificially enlarges the shock thickness to order δx regardless of the shock strength. Since turbulence is inherently three-dimensional, the use of the SGS method can be justified only in three-dimensional calculations, but in the present context this is not a restriction.

In contrast to many methods of turbulence modeling, the SGS method can represent the transition to turbulence, when this transition begins by a fluid-dynamical instability of the well-resolved longer wavelengths. The method also has the virtue of being extremely simple and easy to implement. Indeed, in many respects the SGS approach seems ideally suited to the type of problems under present consideration, and we feel that it should receive extensive testing within the context of numerical simulations of bluff-body aerodynamics. It may well turn out to be all that is necessary to satisfactorily model turbulence in this context.

Finally, we mention the turbulence-transport modeling approach (Harlow 1973), in which ν_t is obtained as the proper dimensional combination of a small number (usually two) of turbulence parameters, typically the turbulent kinetic energy and dissipation rate, which are themselves obtained by solving additional partial differential equations. These equations model the effects of convection of turbulence by the mean flow, self-diffusion of turbulence, production of turbulence by shear in the mean flow (e.g., at solid walls), and decay of turbulent kinetic energy to heat. Turbulence-transport models have become very popular in recent years, and have enjoyed a great deal of success in modeling the detailed features of many complex turbulent flows. A model of this type has in fact already been used, by Gosman and co-workers at Imperial College, in three-dimensional calculations of flow past a cube (Gosman 1976). However, the superiority of turbulence-transport models over simpler methods has not yet been demonstrated in the context of flow past bluff bodies, where zoning is necessarily considerably coarser than in most of the fine-scale calculations in which these models have been developed and tested. Until and unless such superiority becomes evident, the additional complexity of the turbulence-transport models makes it difficult to justify their routine use in bluff-body calculations. Experimental bluff-body calculations with these models should be continued, however, to form a basis for comparison with the simpler models, the SGS model in particular.

D. Computer Resources – The Limitations section would not be complete without some discussion of computer-time and memory-size requirements. For the most part, memory size is not a controlling factor, because large computers usually have large auxiliary memories. Computing times, on the other hand, can easily get out of bounds and so the discussion here will concentrate on this problem.

The time required to perform a calculation involves several considerations. First, a given solution algorithm will require a fixed amount of time to perform the arithmetic operations needed to advance the flow variables at one mesh point one step forward in time. This time is referred to as a Grind. The Grind times for SOLA-3D on a CDC 7600 computer are approximately 0.23 ms for the first-order method, 0.42 ms for the second-order method, plus 0.03 ms per iteration in either case. Thus, for an average of 10 iterations per time step, the first-order method has a Grind of 0.53 ms and the second-order method a Grind of 0.72 ms. These estimates

References pp. 348-350.

are upper bounds, because they were calculated from total running times that included extensive tape dumps, typed listings, and computer graphic outputs.

Using the Grind time as a basis, the total time needed to perform a calculation can then be estimated from the number of mesh cells and the number of time steps needed for the problem. The number of cells is dictated by accuracy requirements as discussed in preceding sections. The number of time steps depends on numerical stability requirements and on whether the problem of interest involves time-dependent or steady state processes.

The choice of a time step to insure a numerically stable calculation often boils down to the question of accuracy in time (Hirt 1968). For example, in SOLA-3D the time step must be limited to a value that allows fluid to move no further than one cell width in a given step (Hirt & Cook 1972). This is related to numerical accuracy, because the explicit finite-difference approximations used to represent convection and diffusion in SOLA-3D assume interactions between neighboring cells only. Even when the difference approximations are written in more implicit forms, which in principle couple all cells together simultaneously, the resulting coupled nonlinear equations must be solved by iterative processes that cannot allow large changes to occur across more than one cell in an iteration. In this sense, iterative implicit difference methods for convection and diffusion reduce to time-like explicit methods. The experience of many investigators is that implicit methods for convection and diffusion may reduce the number of cycles needed to achieve a steady state solution, but the computer time required is typically the same as or more than that required by explicit calculations to reach the same degree of steady state convergence. Of course, in carefully selected problems one method may be shown superior to another. The point is that, in general, there seems to be no substitute for the physical requirement that body-generated shear layers, regions of separation, wakes, and other flow features associated with bluff bodies must be convected into the flow field from the body surface. The establishment of these features, which depend on local balances of competing processes, seems to require a method that proceeds in an orderly time-like fashion; otherwise unwanted transients and improper balances may be generated that take a long time to eliminate.

Let us assume that the time T necessary to attain a steady state is roughly equal to the time needed for the mean flow u to travel a distance of several, say m, characteristic body lengths L; that is, $T \sim mL/u$. Stability and accuracy limitations, as previously noted, require time steps δt such that

$$\delta t \lesssim \frac{\delta x_0}{u}$$

where δx_0 is the minimum cell dimension. Typically, $\delta t = \frac{1}{3}\frac{\delta x_0}{u}$. The required number of time steps (or iterations in implicit methods) is then

$$N_t = \frac{3mL}{\delta x_o} = 3m\, N_{eff} ,$$

that is, several times the effective number of mesh cells needed to resolve spatial detail of size L. Finally, the time or cost for the calculation of a steady flow is proportional to

$$3m\, N_{eff}(kN)^3 \times \text{Grind} ,$$

where k is the number of L lengths needed in the mesh to span the body plus a suitable distance beyond it to where external boundary conditions can be applied with confidence. Using the previous estimate for N_{eff} in Section III. B.1,

$$\text{COST} \propto \frac{3m}{a} \left[(1+a)^{N+1} - 1\right] (kN)^3 \times \text{Grind} ,$$

where it is recalled that a is the percentage change in the size of neighboring cells in a variable mesh.

This result is a useful guide in setting priorities for future developments. First of all, there is clearly a strong dependence on N, which shows the importance of striving for the minimum resolution that still provides accurate results. Of almost equal importance is k, which is proportional to the distance mesh boundaries must be kept away from the flow region of interest. Therefore, improved boundary conditions could have significant consequences for reducing costs. Finally, a reduction either in Grind time or the time necessary to achieve steady state (as measured by m) has, in comparison, a relatively small influence on the cost of calculations.

IV. LONG RANGE PROJECTIONS

The previous sections have attempted to show that some aerodynamic processes affecting simple bluff bodies are within the capabilities of existing numerical-solution methods and existing computers. Projections of what could be attained in the next decade are the subject of this final section. These projections are subdivided into hardware developments and software developments.

A. Projected Hardware Developments – Although the history of computer development over the past 30 years has been phenomenal, there is evidence that an upper limit is being reached with the present class of super computers. This is illustrated in Fig. 17, which shows the execution bandwidth of the fastest computers

References pp. 348-350.

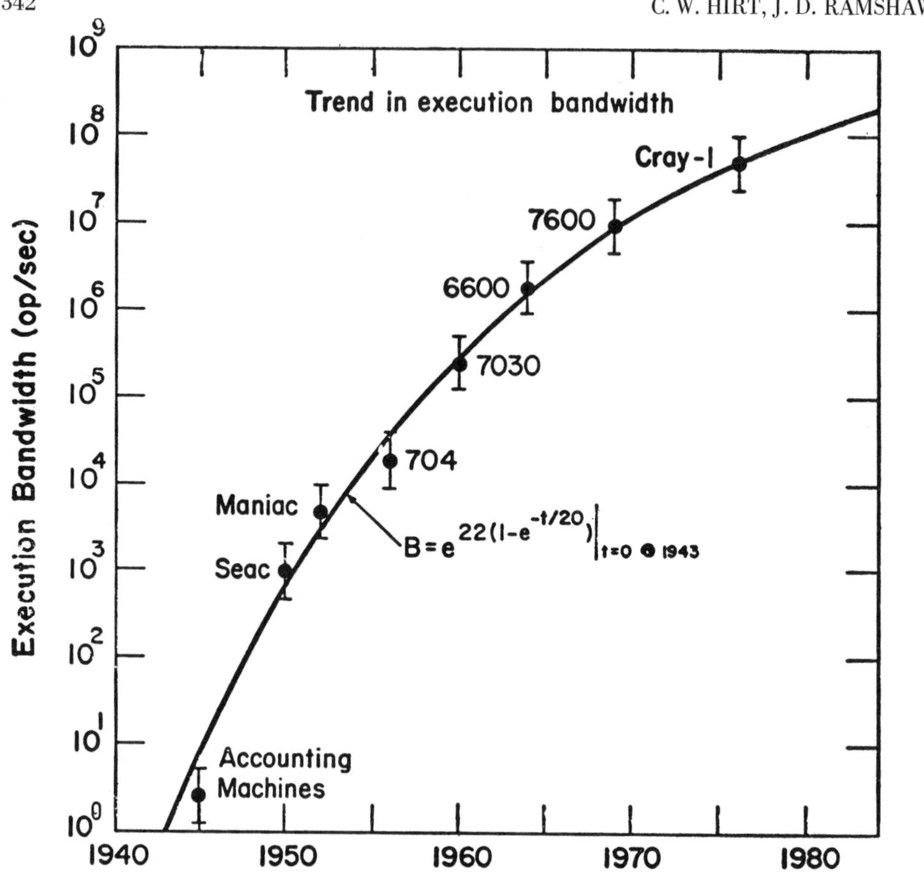

Fig. 17. Extrapolation of growth of capabilities of largest computers.

as a function of time. A few multiples of ten to the ninth operations per second is the apparent limit of the trend. If attained, it would mean more than a hundred-fold increase in speed over the CDC 7600 computer. The CRAY-1 machine is currently undergoing testing and appears to be, at worst, 5 times faster than the 7600, and with careful programming practices it may be as much as 10 times faster. These developments are independent of numerical methodology and will occur as a matter of course.

Perhaps a more interesting area for speculation is the prospect for utilizing special-purpose and mini-computer systems. Although smaller and slower, in some areas these computers have been shown to be cost effective (Orszag 1973) when compared to the super computers. The biggest drawback of mini-computer systems with respect to three-dimensional computations is likely to be the requirement for large amounts of memory that may have to be supplied with slow-access auxiliary memory units. The potential of mini-computers for large-scale fluid dynamics calculations needs to be more fully investigated. Their use could put numerical simulation techniques in the grasp of almost any university or large industry.

B. Projected Software Developments – Several areas where existing computational techniques could be profitably extended have already been mentioned; for example, the use of sub-meshes to give increased local resolution, and new techniques to provide better boundary conditions for outflow boundaries. Both of these areas are likely to produce major short-term savings in the cost of aerodynamic calculations.

There are, however, other possibilities for achieving similar gains. Matched-expansion techniques and integral methods useful in analysis might be combined with numerical methods such that either inner or outer flow regions with relatively simple analytic solutions could be used as boundary conditions for a detailed numerical calculation in the remaining portions of the flow field. One example of this (Buckingham & Birnbaum 1975) has been applied to a three-dimensional problem of a blast wave passing over a simple structure. A coarse grid finite-difference numerical calculation was performed and its output used as an external boundary condition for an integral method that was used near the body surface. Results of the integral solution, however, were not allowed to affect the finite-difference solution, although this could presumably be done with a slightly more elaborate scheme. The purpose of the integral method was to obtain better surface pressures for the computation of blast-loading histories.

Similarly, studies of isolated flow regions around wheel wells, window vents, windshields, radiators, stabilizers, and other areas of interest on road vehicles could be attacked using coarse-mesh calculations of full vehicle configurations to produce boundary conditions for the more detailed flow calculations in isolated regions. This would also be similar to the use of a sub-mesh, except that it might be possible to do the two calculations separately, in which case the computer requirements would be less than that required for a combined calculation. In addition, many refined calculations to investigate variations in local detail could be performed without repeating the coarse-mesh calculation.

Numerical schemes like the Arbitrary-Langrangian-Eulerian (ALE) method used in the YAQUI code (Hirt *et al.* 1974) offer a basis for new techniques with self-adaptive features, that is, schemes that automatically and continuously adjust their space and time increments to optimum values (Trulio 1969). With such schemes the problem of the *a priori* choice of optimum meshes and time-step controls would be eliminated. For example, the ALE mesh can be run in the accurate Lagrangian mode as long as mesh distortions are not too severe. When distortions begin to approach unacceptable limits some rezoning is applied. The amount of rezoning is tied to the degree of mesh distortion and is generally kept as small as possible (Norton & Ruppel 1976). In some stationary flow cases the mesh may be aligned with streamlines but otherwise be Eulerian, which also provides increased accuracy.

In this same vein, it would be useful to have self-testing and correcting programs

References pp. 348-350.

that could detect instabilities or the accumulation of errors and make appropriate corrections without operator intervention. Also, the automated use of sub-meshes that could be added in regions of developing detail, or could be deleted where detail was no longer needed, would be highly useful.

Even without self-testing or adaptive methods, existing hardware developments permit direct operator interaction with running programs. By continuously monitoring calculations while they are in progress, developing difficulties such as instabilities, poor-convergence effects, and boundary-condition problems can be detected and corrected before large amounts of computer time are wasted. This would have been useful for the examples presented in Sec. II, where it was necessary to periodically monitor the calculations in order to choose optimum values of the pressure-iteration convergence level. To insist on too tight a convergence level during the early stages of the calculations would have produced accurate transients, but would also have required more computer time. Because only the steady state results were of interest in these examples, the convergence criterion was initially very loose, and then gradually tightened as the flow evolved toward a steady state. The monitoring of the calculations was done by running the problems in short (5 to 10 min.) pieces. To do this, however, required more computer time for tape dumps and restart calculations than would have been necessary if an interactive system had been used.

A newly evolving capability (McKay *et al.* 1976) to ascertain the sensitivity of computed results to parameter and model variations may find use for aerodynamic studies. This capability, which is being developed to check the sensitivity of nuclear-reactor safety analysis codes, involves special strategies for varying parameters and evaluating their consequences.

Clearly, the possibility of computing time-dependent aerodynamic loads is within the capabilities of a code like SOLA-3D. Some results involving simple wind variations have already been attempted (Hirt & Stein 1976b). More detailed experimental data for simple geometries would be useful for checking and establishing benchmark calculations.

An interesting area for future development is the coupling of statistical techniques with deterministic ones. Techniques in this category would be useful for studying the effects of random wind gusts or large-scale turbulent fluctuations in the wake of one vehicle on another vehicle. Another combination of statistical and deterministic techniques has been proposed (Hirt 1975) for the theoretical treatment of aerodynamic noise propagation and control.

Accurate techniques for computing the aerodynamic dispersal of exhaust gases, dust, and dispersed water droplets have already been partially developed (Sklarew 1970; Hotchkiss & Hirt 1972). As long as the dispersed material is sufficiently rarefied, these schemes can be added to air flow calculations with little difficulty to produce useful auxiliary capabilities.

When the dispersed material has a significant effect on the air flow and can no longer be treated as a passive element, more complicated numerical methods must be used. The recent development of methods for treating multiphase and multicomponent systems for reactor safety applications (Harlow & Amsden 1975) provides the initial technology needed for this type of extension.

Projections for future aerodynamic computations would not be complete without some discussion of coupled fluid-structure interactions. Various two-dimensional calculations have already been performed for the dynamics of rigid bodies floating in a liquid, which involves one kind of fluid-structure coupling (Nichols & Hirt 1976). Calculations have also been performed for deformable structures interacting with fluid phenomena (Dienes *et al.* 1976). It would be natural to extend these techniques to investigate the aerodynamic stability of road vehicles. Perhaps the most useful study to begin with would be the determination of added-mass and damping coefficients for selected, small-amplitude, forced motions of a vehicle in a given wind field.

Although this list of specialized applications and possible future developments could be carried on in many directions, it is perhaps long enough to show that there are many possibilities and considerable potential for achieving useful results in the next few years.

V. ACKNOWLEDGMENTS

The authors particularly wish to acknowledge their appreciation to Leland R. Stein, who wrote the SOLA-3D code and performed the three-dimensional calculations described in this paper. We are also indebted to J. Worlton for providing Fig. 17. This work was performed under the auspices of the U. S. Energy Research and Development Administration Contract W-7405-ENG-36, and was supported in part by the National Science Foundation, Grant No. AG-430.

APPENDIX – EFFECT OF COORDINATE TRANSFORMATIONS ON TRUNCATION ERRORS

There exists a difference of opinion as to whether the use of analytical coordinate transformations prior to the construction of difference equations is generally more accurate than simply differencing the untransformed equations in a variable mesh. To obtain insight into this question we consider a simple one-dimensional example that can be worked out in detail. Let the linear spatial coordinate be x and the transformed coordinate be

$$\xi = \sqrt{x}\,. \tag{A-1}$$

References pp. 348-350.

The coordinate ξ is discretized in equal increments:

$$\xi_i = i\,\Delta\xi = \sqrt{x_i}\ , \qquad \text{(A-2)}$$

which has the effect of providing increased resolution near $x = 0$. From (A-2) we immediately obtain

$$x_i = \xi_i^2 = i^2 (\Delta\xi)^2\ , \qquad \text{(A-3a)}$$

$$\Delta x_p \equiv x_{i+1} - x_i = (2i+1)\,(\Delta\xi)^2 = \Delta\xi(2\xi_i + \Delta\xi)\ , \qquad \text{(A-3b)}$$

$$\Delta x_m \equiv x_i - x_{i-1} = (2i-1)\,(\Delta\xi)^2 = \Delta\xi\,(2\xi_i - \Delta\xi)\ , \qquad \text{(A-3c)}$$

$$x_{i+1} - x_{i-1} = \Delta x_p + \Delta x_m = 4i(\Delta\xi)^2 = 4\xi_i\Delta\xi\ . \qquad \text{(A-3d)}$$

Now consider the physical quantity $Q = \partial\phi/\partial x$, where ϕ is some dependent variable that may be considered as a function of either x or ξ. We will consider two difference approximations to Q. The first, denoted by Q_1, will be obtained by directly differencing Q in the x coordinate system about the point x_i:

$$Q_1 = \frac{\phi_{i+1} - \phi_{i-1}}{x_{i+1} - x_{i-1}}. \qquad \text{(A-4)}$$

The second, denoted by Q_2, will be formed by transforming Q into the ξ coordinate system,

$$Q = \frac{\partial\phi}{\partial x} = \frac{\partial\phi}{\partial\xi}\frac{\partial\xi}{\partial x} = \frac{1}{2\xi}\frac{\partial\phi}{\partial\xi}\ , \qquad \text{(A-5)}$$

and then differencing this form as

$$Q_2 = \frac{1}{2\xi_i}\,\frac{\phi_{i+1} - \phi_{i-1}}{2\Delta\xi}\ . \qquad \text{(A-6)}$$

Next we examine the truncation errors of Q_1 and Q_2. Expanding (A-4) in Taylor series, we find

$$Q_1 = \left.\frac{\partial\phi}{\partial x}\right|_i + \frac{1}{2}\left.\frac{\partial^2\phi}{\partial x^2}\right|_i (\Delta x_p - \Delta x_m) + O(\Delta x_p^2) + O(\Delta x_m^2)\ . \qquad \text{(A-7)}$$

Since $\Delta x_p \neq \Delta x_m$, the difference approximation Q_1 of (A-4) is first-order accurate in Δx. Expanding (A-6) in Taylor series, we obtain

$$Q_2 = \frac{1}{2\xi_i} \frac{\partial \phi}{\partial \xi}\bigg|_i + \frac{(\Delta\xi)^2}{12\,\xi_i} \frac{\partial^3 \phi}{\partial \xi^3}\bigg|_i + O(\Delta\xi)^4 \ . \qquad \text{(A-8)}$$

Thus the difference expression Q_2 of (A-6) is second-order accurate in $\Delta\xi$. Some workers have concluded on the basis of similar considerations that coordinate transformations are preferable to variable meshes. However, combining Eqs. (A-3d) and (A-6) we find

$$Q_2 = \frac{1}{2\xi_i} \frac{(\phi_{i+1} - \phi_{i-1})}{(x_{i+1} - x_{i-1})/2\xi_i} = Q_1 \ . \qquad \text{(A-9)}$$

Thus the difference expressions Q_1 and Q_2 are in fact identical, and we see that the same difference expression can be first order in Δx but second order in $\Delta\xi$. Therefore the order of accuracy of a difference expression is not invariant to coordinate transformations, so it is meaningless to say that a given difference expression is, e.g., second order; one must say second order in what.

Also notice that, since Q_1 and Q_2 are identical they must be *equally accurate,* regardless of whether accuracy is measured by truncation errors or other means.

It is easy to see directly that Q_1 is second-order accurate in $\Delta\xi$. From Eqs. (A-3b) and (A-3c) we see that, although Δx_p and Δx_m are both of order $\Delta\xi$, their difference is of order $(\Delta\xi)^2$, and is in fact simply equal to $2(\Delta\xi)^2$. Thus Eq. (A-7) can be rewritten as

$$Q_1 = \frac{\partial \phi}{\partial x}\bigg|_i + \frac{\partial^2 \phi}{\partial x^2}\bigg|_i (\Delta\xi)^2 + O(\Delta\xi)^2 \ , \qquad \text{(A-10)}$$

which makes it even clearer that Q_1 is simultaneously accurate to order Δx and to order $(\Delta\xi)^2$.

Since Q_1 and Q_2 are identical, there must be a relation between the truncation errors in Eqs. (A-7) and (A-8). This can be explicitly verified by considering the truncation error term

$$T_2 \equiv \frac{(\Delta\xi)^2}{12\,\xi_i} \frac{\partial^3 \phi}{\partial \xi^3}\bigg|_i \qquad \text{(A-11)}$$

References pp. 348-350.

that appears in Eq. (A-8) for Q_2. After several applications of the chain rule to $\partial^3\phi/\partial\xi^3$, one finds

$$T_2 = (\Delta\xi)^2 \left.\frac{\partial^2\phi}{\partial x^2}\right|_i + \frac{2}{3}\xi_i^2 (\Delta\xi)^2 \left.\frac{\partial^3\phi}{\partial x^3}\right|_i , \qquad \text{(A-12)}$$

which contains the truncation error

$$T_1 \equiv \frac{1}{2}\left.\frac{\partial^2\phi}{\partial x^2}\right|_i (\Delta x_p - \Delta x_m) = (\Delta\xi)^2 \left.\frac{\partial^2\phi}{\partial x^2}\right|_i \qquad \text{(A-13)}$$

that appears in Eqs. (A-7) and (A-10) for Q_1. The truncation errors in Q_1 and Q_2 do do not correspond term by term, but in their entirety they must be identical.

TABLE I

	Front		Top		Side		Rear	
Experiment	.95	<.5	–.7	– .8	–.7	– .8	–.5	– .7
Viscosity								
0.0	1.064	–.055	–.562	–1.38	–.565	–1.037	–.412	– .629
.001	1.071	–.156	–.502	–1.474	–.502	–1.505	–.347	– .558
.01	1.074	–.164	–.412	–1.544	–.391	–2.605	–.303	–1.016

REFERENCES

Bradshaw, R. D. and Kramer, J. L. (1974), An Analytical Study of Reduced-Gravity Propellant Settling, National Aeronautics and Space Administration report NASA CR-134593.

Brandt, A. (1972), Multi-Level Adaptive Technique (MLAT) for Fast Numerical Solution to Boundary Value Problems, Proc. Third International Conference on Numerical Methods in Fluid Mechanics, July 3-7, 1972, Paris, 18 *pp. 82-89 Lecture Notes in Physics, Springer-Verlag, NY.*

Buckingham, A. C. and Birnbaum, N. K. (1975), Three-Dimensional Compressible Flow Over a Rigid Structure: Explicit Finite-Difference and Integral Method Coupled at Slip Walls, Proc. 2nd AIAA Computational Fluid Dynamics Conference, Hartford, Conn., June 19-20 (1975).

Chien, N., Feny, Y., Wang, H-J., Siao, T-T. (1951), Wind-Tunnel Studies of Pressure Distribution on Elementary Building Forms, Iowa Institute of Hydraulic Research, State University of Iowa, Iowa City, Iowa.

Deardorff, J. W. (1971), On the Magnitude of the Subgrid Scale Eddy Coefficient, J. Comp. Phys. **7**, *pp. 120-133.*

Dienes, J. K., Hirt, C. W., and Stein, L. R. (1976), Computer Simulation of the Hydroelastic Response of a Pressurized Water Reactor to Sudden Depressurization, Proc. Fourth Water Reactor Safety Research Information Meeting, Washington, DC, September 27-30, 1976.

Gosman, A. D. (1976), private communication.

Hansen, A. C. and Cermak, J. E. (1975), Vortex-Containing Wakes in Surface Obstacles, Project THEMIS Technical Report No. 29, Fluid Dynamics and Diffusion Laboratory, College of Engineering, Colorado State University, Fort Collins, CO.

Hanson, M. E. and Petschek, A. G. (1976), A Boundary Condition for Significantly Reducing Boundary Reflections with a Lagrangian Mesh, J. Comp. Phys. **21**, *pp. 333-339.*

Harlow, F. H. and Welch, J. E. (1965) Numerical Calculations of Time-Dependent Viscous Incompressible Flow, Phys. Fluids **8**, *2182; Welch, J. E., Harlow, F. H., Shannon, J. P., and Daly, B. J. (1966) The MAC Method: A Computing Technique for Solving Viscous, Incompressible, Transient Fluid-Flow Problems Involving Free Surfaces, Los Alamos Scientific Laboratory report LA-3425.*

Harlow, F. H. (1973) Editor, Turbulence Transport Modeling, Vol. XIV of AIAA Selected Reprint Series.

Harlow, F. H. and Amsden, A. A. (1975), Numerical Calculation of Multiphase Fluid Flow, J. Comp. Phys. **17**, *19.*

Hirt, C. W. (1968), Heuristic Stability Theory for Finite-Difference Equations, J. Comp. Phys. **2**, *pp. 339-355.*

Hirt, C. W. and Cook, J. L. (1972), Calculating Three-Dimensional Flows around Structures and over Rough Terrain, J. Comp. Phys. **10**, *pp. 324-340.*

Hirt, C. W., Amsden, A. A., and Cook, J. L. (1974), An Arbitrary Lagrangian-Eulerian Computing Method for All Flow Speeds, J. Comp. Phys. **14**, *pp. 227-253; Amsden, A. A. and Hirt, C. W. (1973), YAQUI: An Arbitrary Lagrangian-Eulerian Computer Program for Fluid Flows at All Speeds, Los Alamos Scientific Laboratory report LA-5100.*

Hirt, C. W. (1975), Numerical Hydrodynamics: Present and Potential, Proc. Workshop on Numerical Hydrodynamics, May 20-21, 1974, National Academy of Sciences, Washington, DC.

Hirt, C. W., Nichols, B. D., and Romero, N. C. (1975), SOLA – A Numerical Solution Algorithm for Transient Fluid Flows, Los Alamos Scientific Laboratory report LA-5852; Cloutman, L. D., Hirt, C. W., and Romero, N. C. (1976), SOLA-ICE: A Numerical Solution Algorithm for Transient Compressible Fluid Flows, Los Alamos Scientific Laboratory report LA-6236.

Hirt, C. W. and Stein, L. R. (1976a), A Simple Scheme for Second Order Accuracy in Marker-and-Cell Codes, unpublished note.

Hirt, C. W. and Stein, L. R. (1976b), Numerical Simulation of Wind Loads on Buildings, paper in preparation.

Hotchkiss, R. S. and Hirt, C. W. (1972), Particulate Transport in Highly Distorted Three-Dimensional Flow Fields, Proc. Computer Simulation Conference, San Diego; also Los Alamos Scientific Laboratory report LA-DC-72-364.

Landau, L. D. and Lifshitz, E. M. (1959), **Fluid Mechanics**, *Pergamon Press, Addison-Wesley Publishing Co., Inc., Reading, Mass.*

Launder, B. E. and Spalding, D. B. (1972), **Mathematical Models of Turbulence**, *Academic Press, New York, NY.*

McKay, M. D., Conover, W. J., and Whiteman, D. E. (1976), Report on the Applications of Statistical Techniques to the Analysis of Computer Codes, Los Alamos Scientific Laboratory report LA-6479-MS.

Morkovin, M. V. (1972), An Approach to Flow Engineering via Functional Flow Modules, Proc. Themis Symposium on Vehicular Dynamics, Rock Island, Illinois, November 1971.

Nichols, B. D. and Hirt, C. W. (1976), Numerical Calculation of Wave Forces on Structures, Proc. Fifteenth Conference on Coastal Engineering, July 11-17, 1976, Honolulu, Hawaii.

Norton, J. L. and Ruppel, H. M. (1976), YAQUI User's Manual for Fireball Calculations, Los Alamos Scientific Laboratory report LA-6261-M.

Orlanski, I. (1976), A Simple Boundary Condition for Unbounded Hyperbolic Flows, J. Comp. Phys. 21, *pp. 251-269.*

Orszag, S. A. (1973), Minicomputers vs Supercomputers: A Study in Cost Effectiveness for Large Numerical Simulation Programs, Flow Research Note No. 38, Flow Research, Inc., Kent, Washington.

Orszag, S. A. (1976), Turbulence and Transition: A Progress Report, to be published.

Pracht, W. E. (1975), Calculating Three-Dimensional Fluid Flows at All Speeds with an Eulerian-Lagrangian Computing Mesh, J. Comp. Phys. 17, *pp. 132-159.*

Richtmyer, R. D. and Morton, K. W. (1967), Difference Methods for Initial-Value Problems, *Second Edition, Interscience Publishers, J. Wiley and Sons, New York, NY.*

Roache, P. J. (1972), Computational Fluid Dynamics, revised printing, Hermosa Publishers, Albuquerque, New Mexico.

Schlichting, H. (1960), Boundary-Layer Theory, *Sixth Edition, McGraw-Hill Book Co., New York, NY.*

Sklarew, R. C. (1970), A New Approach: The Grid Model of Urban Air Pollution, Proc. 63rd Annual Meeting of the Air Pollution Control Association, St. Louis, Missouri, June 14, 1970.

Stein, L. R., Gentry, R. A., and Hirt, C. W. (1976), Computational Simulation of Transient Blast Loading on Three-Dimensional Structures, to be published in Computer Methods in Applied Mechanics and Engineering.

Trulio, J. G. (1964), Studies of Finite Difference Techniques for Continuum Mechanics, Air Force Weapons Laboratory report WL TDR-64-72, Kirtland Air Force Base, Albuquerque, New Mexico.

Trulio, J. G. (1966), Theory and Structure of the AFTON Codes, Air Force Weapons Laboratory, Kirtland Air Force Base Report No. AFWL-TR-66-19.

Trulio, J. G. (1969), Puff Rezone Development, Air Force Weapons Laboratory Report AFWL-TR-69-50, Kirtland Air Force Base, Albuquerque, New Mexico.

Viecelli, J. A. (1969), A Method for Including Arbitrary External Boundaries in the MAC Incompressible Fluid Computing Technique, J. Comp. Phys. 4, *pp. 543-551; A Computing Method for Incompressible Flows Bounded by Moving Walls, J. Comp. Phys.* 8, *pp. 119-143 (1971).*

von Neumann, J. and Richtmyer, R. D. (1950), A Method for the Numerical Calculation of Hydrodynamic Shocks, J. Appl. Phys. 21, *pp. 232-237.*

DISCUSSION

M. V. Morkovin *(Illinois Institute of Technology)*

Since I have had the opportunity to read this paper in advance and to have previous discussions with Drs. Hirt and Ramshaw, I'm going to take the prerogative of making the first comments.

In demonstrating the validity of their computations, Dr. Ramshaw compared their pressure distributions to measured values. He probably doesn't appreciate how restricted the measured pressure distributions may be. I'm sure the experimenters didn't have as many pressure taps as they have mesh points, and that the measured values were inferred from some kind of averaging. The calculations may actually be a better approximation of local pressure than the experimental data they are comparing with. A better means of calibrating the power of the computations is desirable.

M.V. Morkovin

What we agreed on in our previous discussions was that it makes a lot of sense to pose questions for the calculations in those areas where they are particularly good at representing flows. One of these strengths is that they can handle *streamwise* as well as crosswise vorticity. We have seen that in many of the flows of interest to this Symposium there is solid competition and even interference between these two types of vorticity. In order to test the capabilities of the computations with respect to concentrated vorticity, we thought about the possibility of computing two different critical geometries, each of which is associated with configurations that are simple enough to be accommodated relatively easily by the calculations.

The first and main one involves the Nakaguchi drag maximum for the flow past a two-dimensional bar. There is only crosswise vorticity, and a fairly sharp maximum in the drag. If the computed results match the critical behavior, even though there might be a little shift in the critical aspect ratio of the bar, I think we would then have confidence that the basic issue of the length of formation of the vortex eddies which contributes to the maximum *is* being adequately represented.

The second involves streamwise vorticity and the slanted-base results of Tom Morel. Will the computations adequately account for streamwise vorticity, and in particular will they be able to discriminate geometries having the streamwise vorticity of Regime II from those having the crosswise vorticity of Regime I?

These are conceptual tests of the *power* of the calculations, rather than some comparison of detailed pressure distributions which always tends to be favorable so that one is never quite sure about what it proves. I think we more-or-less agreed on all this. Would you care to add any comments?

J. D. Ramshaw

Yes, we talked about this a couple of nights ago. I didn't have a chance to include it into the talk, but I agree with everything Professor Morkovin has said. This is exactly the kind of experimental data we need to have, and to compare the code against. We hope to have the opportunity to compare against the two-dimensional measurements later on.

A. Roshko *(California Institute of Technology)*

I really appreciated this talk. Dr. Ramshaw presented a very lucid exposé of the situation we are in – both its possibilities and limitations.

I'd like to comment on the concept of eddy viscosity. Even in the simplest flows, say a very simple jet, it is a pretty dubious concept at best, and one that is very limited. What you simply have to admit is that, for example, if you choose a viscosity of 0.001 then what you are really computing is a flow at a Reynolds number of 1,000. If you compute it accurately, *that* is the flow you are going to describe. Does that bear any resemblance to a turbulent flow at, say, some very high Reynolds number? Qualitatively the flow patterns might very well be the same, but I don't think the pressure distributions, forces and so on are ever likely to be modeled accurately enough to reproduce separation, especially when the separation is off a smooth surface rather than at an edge.

J. D. Ramshaw

It's not really equivalent to a Reynolds number of 1,000 because that would have a constant viscosity in all parts of the field.

A. Roshko

But that is what you are using!

J. D. Ramshaw

This is what we are using, but I am not advocating it. The particular calculations I showed had a constant viscosity and therefore are equivalent to a lower Reynolds number flow as you say. However, we made this assumption only for the very first calculations to see what it would do. It was the simplest first try we could make to explore the sensitivity of the calculations to the eddy viscosity modeling.

A. Roshko

Then the problem is how you are going to know how to model the viscosity, and that is getting away from numerical analysis itself. Turbulence modeling will improve the situation a bit, but I'm not sure whether the computations incorporate enough physics to permit sufficiently accurate calculations of bluff body flows.

J. D. Ramshaw

People have done research with turbulence models in more finely zoned calculations and have done pretty well. There are things that can be done to let eddy viscosity vary throughout a field. Studies should be made to see how sensitive the results are to various models. It is not obvious to me that a variable eddy viscosity will automatically throw you out of the ball game.

J.D. Ramshaw

A. Roshko

No, it doesn't throw you out of the ball game, but I think it limits your home runs quite a bit.

J. D. Ramshaw

It's another area where we need to compare more with experimental data and see what we can get away with. It is not something you can predict *a priori.*

A. Roshko

Not to be completely negative, I think that testing the computations in the manner Professor Morkovin has suggested is really going to be very worthwhile. In some cases the large-scale motions will be the controlling ones, and these probably can be computed.

J. D. Ramshaw

Yes, we intend to continue making bluff body calculations.

R. T. Jones *(NASA-Ames Research Center)*

It's a heroic effort, but what I have learned from your talk is that you really haven't calculated flows except at low Reynolds numbers. In as much as it is constant, the introduction of a Boussinesq-type viscosity to take account of the effects of turbulence only has the effect of, or aspects similar to, lowering the Reynolds number. However, a laminar boundary layer separates as soon as it encounters any significant adverse pressure rise, say 10% at the most. A turbulent boundary layer, on the other hand, can run against a pressure rise of 60% of the effective q, and that makes a big difference. For example, in the case of a sphere it reduces drag by a factor of five. So I think that if you are going to apply your computational method to road vehicles you should demonstrate its capabilities by showing that you can somehow accommodate the transition at high Reynolds number that occurs on a sphere.

In my opinion, the best chance for doing the sphere problem would be, instead of using the full Navier-Stokes equations, to just use the Euler equations in combination with some separation criterion. I would be willing to bet a large sum that you would be more accurate.

J. D. Ramshaw

That is a possibility I didn't mention. One could take some of the separation criteria that have been developed over the years and use them in conjunction with the calculations.

S. J. Kline *(Stanford University)*

I, also, admired your setting forth all these limts of numerical computations because they are currently under discussion in the technical community and are very important.

As R. T. Jones and Professor Roshko said, it is too simple to say that you can adjust the eddy viscosity and have the "turbulent calculations come out." If I may use the words of Professor Saffman, one of Professor Roshko's colleagues, we ought to distinguish *post*diction from *pre*diction. *Post*diction means that we have analytically fit sets of data that have already been measured. If I understand your objective here, it is a true *pre*diction not *post*diction.

J. D. Ramshaw

That's the ultimate objective, but we have to go in stages.

S. J. Kline

Okay, but if you are going to do the complete calculation it should be a *pre*dictive method. In our experience, for turbulent flows for which large sets of data exist, calculation procedures can be developed which match this data quite well. But when we have tried to extend the calculations beyond the boundaries of the data base we have fallen on our faces. So one has to be a little careful about saying that "turbulent calculations will come out."

J. D. Ramshaw

I don't remember saying exactly that. I didn't mean to imply that turbulence is not a problem in road-vehicle type calculations.

S. J. Kline

As far as eddy viscosity is concerned, I can show you data for a simple channel flow on which Coriolis forces have been imposed. The eddy viscosity starts out at four times the value in a non-rotating channel, drops to zero, goes to negative infinity, comes back to plus infinity and goes to zero. In a case such as this, the limitations that Professor Roshko mentioned are rather severe. I'm not trying to be negative. I'm just trying to point out some of the real difficulties that I see.

Matching a boundary layer calculation to an outer-flow calculation is possible, but the matching problem is severe. The older matching techniques used in boundary layer calculations do not work. It is not that I think matching isn't possible, I just want to sound a warning that it can't be done in the classical way. One has to interact in the first order.

J. D. Ramshaw

That is clearly another order of magnitude beyond what we are presently doing, but we will eventually have to get into it and worry about it.

What you are saying about eddy viscosity applies to all the turbulence models that have attained such popularity in recent years. Most of them are essentially based on a scalar eddy viscosity, and they don't pretend to be able to do everything. But there is an amazing number of things that they can do. In the bluff body context of this symposium we should try to identify those things that we can do that simpler way, so that we can then see what we have to do in a more complicated way.

M. V. Morkovin

The essential thrust of the three discussers was a note of caution about overly simplified treatments of turbulence effects. When you hear about how well the powerful turbulence methods are doing, it is often from people who have a vested interest.

As you have heard from each of the discussers, we really appreciate the effort that has been put into this paper. This was the most honest presentation of any numerical method that I have heard in ages.

GENERAL DISCUSSION AND OUTLOOK FOR THE FUTURE

Moderator: M. V. MORKOVIN

Illinois Institute of Technology

Moderator

I have been asking myself and thinking about the following question. What have we added during this Symposium to the picture painted by Dr. Hucho in his opening paper – the definition of the automobile drag problem as he sees it, and VW's systematic method for approaching the optimization of drag coefficient? I personally have had a difficult time conceiving how one would really improve on his approach, but I think he was asking for suggestions from our collective wisdom. All we have really done during these two days is introduce a few concepts that might *guide* the optimization, but I'm not sure just how helpful it has been. I would like to set the stage once more by asking Dr. Hucho to restate the problem he was trying to excite us about. However, in doing so I would like to make this point to him. The concepts and flow processes we have discussed have a finite characteristic digestion time, and two days is absolutely not enough. Please remember that the impact of what you told us yesterday will be felt for some time, so we are presuming there *will* eventually be feedback.

W.-H. Hucho *(Volkswagenwerk AG, Germany)*

What I would like to get across is sumarized in Fig. 1, which has to be looked at in context with Fig. 33 of my paper. Let me start my comment with the second body of revolution. Having a fineness ratio of $d/\ell \approx 0.3$, which is comparable to the height-to-length ratio of a car, such a body of revolution has a drag coefficient of, say, 0.05. If this body is taken from free air and brought close to ground, its drag goes up due to the fact that the now non-symmetrical flow contains strong separations. If this

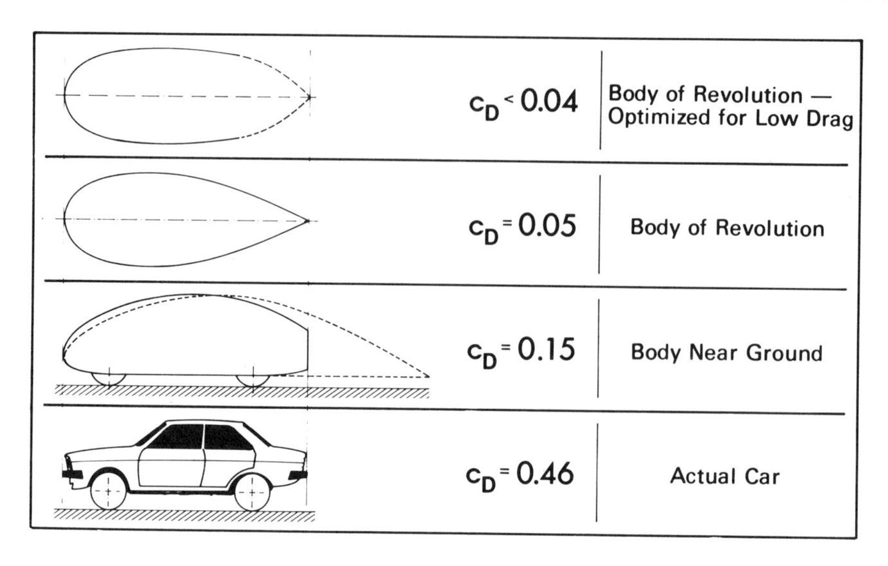

Fig. 1. Drag coefficients, C_D, of bodies and passenger cars.

body is reshaped such that separation is avoided wherever possible, we arrive at W. Klemperer's body shown in dotted outline on the third line of the figure. It has a drag coefficient of 0.15, which is *three* times as high as the value for the body of revolution we started with. Our own work has shown that the $C_D = 0.15$ can almost be maintained when Klemperer's body is modified as shown in solid outline in the figure, with wheel segments added and fitted into the enveloping box of a car. If the drag coefficient of this "ideal" bluff body at a typical car ground clearance is compared with that of an average European car ($C_D \approx 0.46$), we again find that the drag is *three* times higher.

W.-H. Hucho

Now, my interest as a practitioner is mainly directed at the following three questions:

1. What are the reasons for the dramatic drag increase from the "ideal" car-like bluff body to the actual car?
2. What must be done to the actual car to make use of this large, and ostensibly available, drag reduction potential?
3. Can use be made of this potential without going to exotic body shapes that nobody will buy?

Another question, which I rated of second importance in my paper, is the following:

What are the mechanisms causing the large drag increase from the body of revolution in free air, $C_D = 0.05$, to the bluff body in ground proximity with its shape properly matched to the flow field near the ground, $C_D = 0.15$?

Investigating this question is by no means meaningless. In doing so we would hope to arrive at a better understanding of the flow around bluff bodies near ground. Eventually we might be able to reduce the drag of the car-like body below the C_D = 0.15 that was called "ideal" above. Thus we would set a lower starting point for the change from the bluff body to the real car. However, before making the gap between existing cars and the ideal bluff body near ground even larger than it already is today, I recommend investigating how use can be made of the already proven potential that exists.

Finally, the remaining question is, what is the lowest drag coefficient that a body of revolution having a given fineness ratio can be trimmed to? I do not deny that there are broad areas in applied fluid mechanics where an answer to this question is urgently needed. However, I doubt whether it will be of great practical importance for road vehicle aerodynamics.

The only way to find answers to item 2 in my list of primary questions is to carry out systematic research on the flow around cars. We have to define different flow modules and these modules have to be investigated in the same manner as Tom Morel treated the flow around a slanted base. This procedure is made difficult by the fact that a car can't readily be split into different nearly isolated components like an aircraft; we have to be aware of the strong interaction that can occur between the various flow modules.

Moderator

What you are really asking, in another way is, where does the underbody drag that Mason & Beebe talked about come from? Of all the things I've heard, the presence of the ground and what it does has been the most baffling. I would like to ask those among us who have some experience in this area to apply their intuition. What additional things should we be looking for?

T. Morel *(General Motors Research Laboratories)*

I-would like to make a point concerning the effect of ground in increasing drag. In my investigation I had a simple body of car-like proportions which was tested both away from and near to ground. I did not see the three-fold drag increase that Dr. Hucho mentions. The drag coefficient increased only from 0.23 to 0.25 in bringing the body into ground proximity. It had a smooth underbody, and no simulated wheels.

W.-H. Hucho

You are referring to my three-fold increase due to ground effect *per se* (from 0.05 to 0.15). This was for a low-drag body with attached flow throughout. If you have a body which already has a drag coefficient greater than 0.20 in the free stream, implying a significant amount of flow separation, then that is probably a different problem.

A.M.O. Smith *(University of California at Los Angeles)*

I'm not surprised at the increase from 0.05 to 0.15. It suggests several things to me. First of all there is the classical problem of the potential flow over a log on the bottom of a stream that can be found in Milne-Thomson. Since no flow can go underneath, you get much higher velocities over the top of the log than you would when it is far from a ground surface. This will cause more separation and increase drag. Secondly, when you have a body close to ground (say an airfoil) the aft portion of the underbody region becomes an internal-flow type of diffuser. This generates larger adverse pressure gradients on the lower surface of the body than it would have in free air, causing more extensive separation. Putting these two mechanisms together, the increase from 0.05 to 0.15 that Dr. Hucho describes doesn't surprise me. You could more or less identify where the drag comes from by using the model build up technique employed in the aircraft industry. Start with the simplest basic building block of a car, and then add the other components and the geometric detail one element at a time while measuring the drag of each build-up in a wind tunnel.

There are a number of drag reducing methods from aircraft practice that should be practical for cars, e.g. vortex generators, nose flaps. Also, I once devised something that may be of interest here. We cut off the rear third of a hollow airfoil, whose cross-section looked something like that of the body on line 2 of Dr. Hucho's figure, and put an almost semi-circular 2-D cap in the opening that was formed. The edges of the cap were parallel to the airfoil surfaces at the cut-off point, and the height of the cap was a little smaller than the thickness of the airfoil at the opening so that slots were formed top and bottom between the cap and the airfoil. By pressurizing the interior cavity, air was blown out these slots in a direction tangential to the surface of the cap, which was sloped inwards about 40 to 45° at that point. Wind tunnel tests showed that by blowing upstream of the normal point of separation in this way, we were able to eliminate separation at the tail end. We investigated a range of jet velocities and found that this kind of boundary layer blowing worked even with internal pressures lower than what could be obtained with ram air. So I think there really are a number of possibilities for reducing the pressure drag of road vehicles. The problem is to reduce them to acceptable configurations.

E. C. Maskell *(Royal Aircraft Establishment, England)*

Can I suggest that the essential difference between the 0.05 case and the 0.15 is quite simply that you've removed all possibility of having symmetrical flow? In the 0.05 case you can have a flow field which is, in fact, symmetrical about the plane of symmetry of the body. On the other hand, in the second case the only possible symmetry is with the wall, so that the flow around the body itself is essentially asymmetric. Now whether that would account for a factor of three, I wouldn't know.

Moderator

So we could pose the question of whether that alone could account for the factor of three. I think that's amenable to a certain amount of computation.

W. W. Willmarth *(University of Michigan)*

I'd like to follow up on what Eric Maskell said. Suppose you take Dr. Hucho's non-lifting low-drag body in free air and simulate bringing it close to ground by cutting it in half longitudinally and moving the two pieces apart by a distance of twice the ground clearance. Because of the image effect, the velocity over the curved surface of each half will be higher than when the two pieces were joined. This will produce a lift force which, in turn, will generate steamwise vorticity and create drag. This is at least one mechanism by which drag will be larger when a body is brought close to ground.

W.-H. Hucho

Okay, then the question is, how can the streamwise vorticity be avoided?

W. H. Bettes *(California Institute of Technology)*

I think we need to be careful here. The body with $C_D = 0.15$ was more complex than the one in free air in that it had wheels and wasn't at all the same in geometry. If you are really dealing with the free-air body I'd be surprised if there was even a factor of two difference as it is brought into ground proximity.

W.–H. Hucho

The Klemperer body (1922) had no wheels, but we have achieved essentially the same drag coefficient *with* wheels. In any event, whether it is a factor of two or three isn't important. In either case it is much larger than the 10 percent effects that we usually work on.

W. H. Bettes

As I mentioned yesterday, we have tested a streamlined land-speed-record car* that was essentially a long, slender body of revolution. It had wheels, and some internal flow to simulate engine cooling. We measured a drag coefficient of the order of 0.11 to 0.12.

P. W. Bearman *(Imperial College, England)*

What was the drag coefficient when the body was away from the ground?

W. H. Bettes

We had no interest in that situation.

K. R. Cooper *(National Research Council, Canada)*

I have measured a drag coefficient of 0.07 for a streamlined body in ground proximity. This was not far removed from the value for a body of revolution of the same cross-section and length in free air at the same Reynolds number. It had two wheels on the ground but no internal flow. I actually tested it at three different ground clearances. As it got closer to ground the drag decreased, slowly but steadily. So in fact, whether ground proximity increases drag by a factor of two or three or whatever may very well depend on the exact shape of the body being considered.

J. L. Stollery *(Cranfield Institute of Technology, England)*

I don't think Dr. Hucho's comparison between the drag coefficients of 0.05 and 0.15 is valid. We did wind tunnel testing during the development of Donald Campbell's land-speed-record car (Fig. 2). The results pertinent to ground effect are shown in Table 1. We started off with a very streamlined symmetrical shape (model B) and tested it both in free air and in ground proximity. In free air the lift was approximately zero, but bringing the body down to the ground was equivalent to generating negative camber. A negative lift or down force developed ($C_L = -0.23$) and the drag coefficient rose from 0.16 to 0.18. Since negative lift adversely affected tire loads and positive lift was undesirable from the vehicle stability point of view, our objective was to achieve *zero* lift in ground proximity. We tried various amounts of

**Korff, W. H. (1966), The Aerodynamic Design of the Goldenrod – to Increase Stability, Traction, and Speed, SAE 660390.*

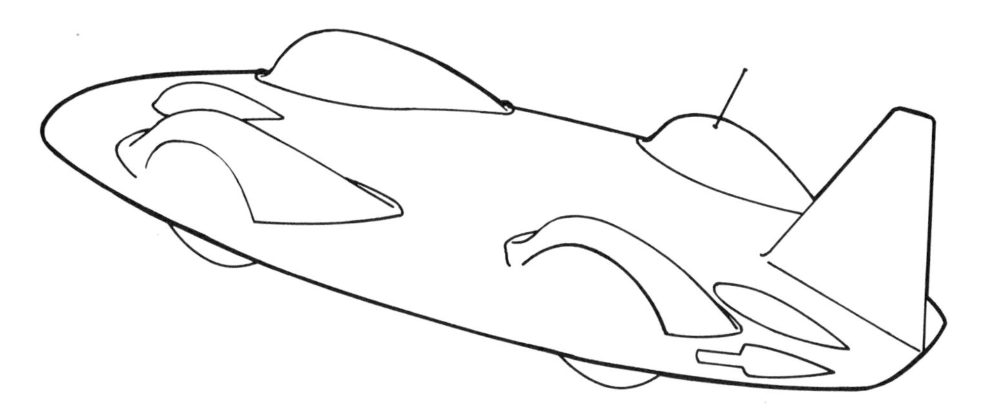

Fig. 2. The Donald Campbell land-speed-record car.

Table 1

Car Shape	Ground	Lift Coefficient	Drag Coefficient
B, Symmetric	No Yes	-0.02 -0.23	0.16 0.18
C, Large Positive camber (flat bottom)	Yes	+0.39	0.20
D, Medium Positive Camber	Yes	+0.14	0.17

positive camber, as given by models C and D. By adjusting the camber of the car body we were able to almost cancel the ground effect on lift, i.e. bring C_L back to approximately zero in the presence of ground, and nearly recover the original free-air drag coefficient of 0.16 that was measured for model B. Of course this car is not typical of passenger automobiles, but I feel confident of the general point that by adjusting the camber line to give $C_L = 0$ near the ground, then the same C_D can be obtained as in free-air conditions with C_L again zero. Whether such a cambered shape is acceptable to the automobile industry and the driving public is another question.

Moderator

This goes to the question of minimizing the vorticity drag.

G. W. Carr *(Motor Industry Research Association, England)*

I have some experimental results which cover the whole transition from a streamlined body to an actual car shape. The question of how the drag coefficient of a European saloon car is built up to its typical value of 0.45 from the value of about 0.05 for a streamlined body of revolution in free air has been investigated in model wind tunnel tests at MIRA. Some results*† are shown in Fig. 3, and part of them are in line with what John Stollery has just said.

(a) Aerodynamic Drag of Basic Shapes

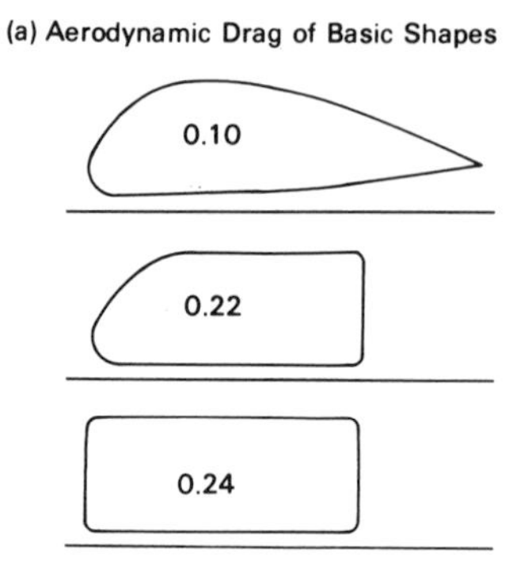

(b) Aerodynamic Drag of Saloon Car Components

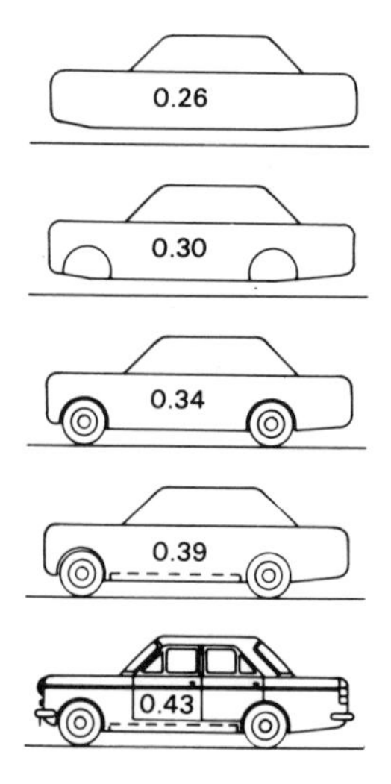

FIGURES ON MODELS DENOTE C_D VALUES

Fig. 3. Variation in drag as saloon car model is developed in stages from streamlined body.

*Carr, G. W. (1969), The Study of Road Vehicle Aerodynamics Using Wind Tunnel Models, Proc. of Symposium on Road Vehicle Aerodynamics, City University, London, Paper 14.

†Carr, G. W. (1973), Aerodynamic Lift Characteristics of Cars, Proc. I. Mech. E., Vol. 187 30/73, p. 333.

A drag coefficient of 0.08 was obtained in the absence of ground effect for a model of symmetrical streamlined side-elevation, rectangular planform and square maximum cross-section with rounded edges. This value increased to 0.11 with ground proximity typical of a road vehicle application, but was slightly reduced by camber or negative incidence, as shown in Fig. 3(a).

Replacing the streamlined rear section of the model with a parallel square-ended section increased C_D to 0.22, and a similar change to the front end caused a further increase to 0.24. The final modification to the basic shape of the model, incorporating notched front and rear ends as shown in Fig. 3(b), raised C_D to a value of 0.26.

The addition of the various protuberances and cavities essential to a practical road vehicle resulted in roughly equal increments in C_D of approximately 0.045 as wheel-arches, wheels, underbody details and upper-body details were added in turn to the basic car shape. The final value of 0.43 for C_D was obtained with a model fully representative of a typical car except for the engine cooling system. This would normally increase the drag coefficient further, to give a value of approximately 0.45 for the complete car in operating condition.

MIRA's work on drag reduction on cars bears out the statement that minimum drag requires zero lift, and results recently published† have shown that a drag coefficient approaching 0.30 can be achieved simply by eliminating the lift forces on a typical medium-sized European car.

W.-H. Hucho

Your comments, Professor Stollery, are part of the answer to the question I put. But please tell me, how can I design a passenger car without lift?

J. L. Stollery

With difficulty!

G. W. Carr

I would like to contradict that. We have designed cars in our wind tunnel which had zero and even negative lift, and they *are* practical.

W.-H. Hucho

Where is the show room where I can see them?

†Carr, G. W. (1976), Reducing Fuel Consumption by Means of Aerodynamic "Add-on" Devices, SAE 760187.

G. W. Carr

They are not being manufactured yet.

Moderator

You are saying they would not be saleable Dr. Hucho?

W.-H. Hucho

I don't know. I would have to see them before I could judge.

G. W. Carr

He is unduly pessimistic, I'm quite sure.

M. T. Landahl *(Massachusetts Institute of Technology)*

Maybe there is something about the rear wheels that should be looked into. They may shed longitudinal vorticity.

Discussion contributed after the symposium

J. E. Hackett *(Lockheed-Georgia Company)*

It may be helpful to explore the subject of vortex drag a little more deeply. If a net transverse force (either lift or side force) is present on a finite body moving through free air, then some vortex drag is inevitable. In an aircraft, its magnitude is reduced by the use of a large aspect ratio wing. For cars, this approach is obviously impractical and is also unnecessary. A car does not need aerodynamic lift, and even if lift occurs there is no necessity for a drag producing trailing vortex system.

The trailing vortex system of a car can be controlled in two ways. Low or zero lift and consequently low vortex drag are sometimes achieved for speed record purposes by adding appropriate camber to a conventional streamlined shape to compensate for ground effect. In this case, the flow is *attached* and closes smoothly behind the body; the vortex system forms *closed* loops lying within the boundary layers with their planes parallel to the local surfaces. These loops do not leave the body surface and trail downstream.

The second method is by means of ground clearance. In both passenger and racing cars, the reduction of ground clearance is currently receiving considerable attention. Though the immediate reason for this appears related to underbody drag *per se*, proximity to the ground provides an important option in reducing or eliminating vortex drag. By way of illustration, consider a hemisphere with completely attached flow (as might be produced, for example, by appropriate boundary layer control) placed with its equator on the ground. Because of the ground image, a lift is generated on the hemisphere (for example, the classical potential flow solution for a sphere gives a lift coefficient equal to 11/16 based on planform area). Despite this lift, there is no vortex drag, on either the whole sphere or the hemisphere. Transverse bound vortices are present but their ends loop down to the ground, forming arches transverse to the flow, rather than trail downstream. At the foot of each arch, the vorticity fans out radially and mixes with the ground boundary layer.

Returning now to practical shapes with *finite* ground clearance, it is evident that a gap is introduced between the foot of each vortex arch and the ground. Each vortex must therefore trail aft at an altitude slightly above the ground, rather than blend into the boundary layer, and a vortex drag component will develop which is roughly proportional to ground clearance. This vortex drag can be reduced by using small ground clearance to decrease the distance between the trailing vortices and their images in the ground plane. The vorticity can be eliminated by appropriately tailoring the pressure distribution on the body so that the bound-vortex arches run right down to the ground and join up with those from the image system to form *closed* loops. This does not necessarily require zero lift.

In summary, we see from the above that vortex drag can, in principle, be reduced to a low value for car-like shapes with *attached* flow: indeed, useful experiments could be performed to establish the effect of ground clearance on an idealized family of attached-flow body shapes. Such a study would complement the work described by Tom Morel on separated afterbody flow. This could pave the way for an integrated design approach.

SYMPOSIUM WRAP-UP

G. SOVRAN

General Motors Research Laboratories, Warren, Michigan

Through yesterday and today I have been thinking about what few on-the-spot comments I could make at this time that would be useful. First of all, and particularly for those not previously involved with road-vehicle aerodynamics, I hope we have opened up many new possibilities for future research. Two things should be evident. The problems are certainly significant and challenging from a technical point of view; furthermore, they are also very interesting. We all drive cars. As we drive around, the complex and fascinating aerodynamics that surround our vehicles naturally pique the curiosity of a fluid dynamicist and make him want to understand. As we have seen, there is much that is not well understood.

G. Sovran

The Symposium had certain objectives. Our hopes were probably greater than we could realistically expect to realize so we did not resolve all the problems posed. It may even have seemed at times, particularly in the final general discussion, that instead of opening up new vistas we just opened various cans of worms. But I think we did make progress, and we did effect some accomplishments.

We have explored the heretofore distinct possibility that answers to some of the key questions about the three-dimensional flow fields of road vehicles lie, for the taking, in the vast bluff body literature on simple, aeronautical and architectural configurations. Is it possible that the knowhow already exists and all that is necessary is to pose the proper questions to the right people? I think it is now clear that this is not the case. We might have hoped for better, but at least we can check off that possibility and get on with the specifically directed research that is required.

If the answers are not lying there, is there at least sufficient information and understanding to be of substantial assistance in generating them? To get the best evaluation we went to the researchers who have been leading contributors to bluff body technology. Most of that research has been either in two dimensions or on a restricted class of axisymmetric geometries. As the contributors from Cambridge have so clearly and forthrightly indicated, it is very limited in its application to the characteristically three-dimensional flows with vortical near-wakes that can exist around automobiles. Research on bluff bodies having significant amounts of trailing streamwise vorticity is required, and hopefully this Symposium has persuaded some of the active workers to direct at least a portion of their efforts to such flows.

The aerodynamic effects associated with the slanted upper-rear surfaces of bluff bodies has been a fascinating highlight of the Symposium. At the GM Research Laboratories, the instigation for the work reported by Tom Morel was the observations of Dr. Hucho on the effect of backlight angle on the lift and drag of hatchback cars. These results were very unusual, particularly if one thought about them from the perspective of two-dimensional flow. Depending on the slant angle, separation occurs either at the end of the roof or the end of the slanted surface. Rather surprisingly, the flow pattern with the smallest separation region produces the highest drag, the difference being of sufficient magnitude to be very significant to automobile fuel economy. It is clear that *two*-dimensional reasoning does not apply since one of the patterns is so highly *three*-dimensional. Tom Morel isolated the basic flow modules on a simple body whose geometry could be varied in a systematic manner. He examined in detail the dramatic and abrupt changes in drag and lift occurring at the critical geometry. We also find that Geoffrey Carr conducted related studies on automobile configurations at MIRA, and he reported those to us. As Mark Morkovin first brought to our attention, we can find essentially the same three-dimensional module, in an upside-down sense, on aircraft with upswept tails. The flow behind the hull of some ships appears to be another related piece of the puzzle. If we can bring all these pieces together it is possible that an even clearer understanding of the overall phenomenon will result, and that a design tool for its control will be developed.

In the case of tractor-trailer trucks and buses we had a remarkably well-matched trilogy of contributions that told a very comprehensive story. The Mason & Beebe paper presented a large amount of information on the overall and detailed aerodynamic characteristics of actual vehicle shapes, and it was condensed into a relatively small number of key bits of knowledge about their flow fields. The front end is clearly the place to expend most of the initial efforts on drag reduction. Furthermore, the aerodynamic size, AC_D, of a tractor relative to that of its trailer proves to be a considerably more significant parameter than its aerodynamic quality, C_D, *per se*. In effect, the most important message that emerged was that one designs tractor-trailer *systems,* not individual tractors or trailers.

Then came the contribution from Japan presented by Professor Nakaguchi. In their search for understanding, simplified geometries compared to those of actual trucks and buses were used. Although there were no wheels, some of the configurations looked like tractor-trailer trucks, and they were positioned at typical ground clearances. Some fundamental studies were also made out of ground effect on sharp-edged rectangular bodies representative of truck trailers and buses. The last piece provided by Roshko & Koenig took us to a very simple configuration in which the basic flow module of deflector-type drag reducers was captured and isolated. This group of contributions provided a systematic progression from actual configurations in which important flow modules were identified, to simple geometries suited for in-depth investigation. This represents basically the same type of approach and incisive research used so effectively for slanted bases.

It was of more than passing interest to me to learn how Dr. Saunders, the inventor of the Airshield™, and Professor Roshko and Mr. Koenig conceived their particular investigations. A number of years ago, upon seeing a configuration similar to that of Roshko & Koenig briefly treated in an aerodynamics text by von Mises, Dr. Saunders had the inspiration and engineering sense to realize that the principle involved could be adapted to tractor-trailer trucks for reducing drag. On the other hand, Roshko & Koenig's reasoning went in the other direction. Having as input the visual appearance of deflector-type devices and the knowledge that they are effective in reducing drag, they deduced the nature of the controlling flow pattern responsible for its aerodynamic performance and set out to investigate it in detail.

We explored the prospects for analyzing and computing the nature and magnitude of the aerodynamic drag nechanisms and forces associated with road vehicles. Seeking physical insight from the governing equations of fluid mechanics and from experimental evidence, Maskell and Landahl probed the three-dimensional near-wake flow patterns for their trendwise relationship to drag. Some information on the nature of drag emerged, and possibilities for managing the near-wake structure to reduce drag were suggested. Hirt & Ramshaw examined the ability to quantify drag by numerical computation of the Navier-Stokes equation. While substantial advances have been made in numerical fluid dynamics that will be helpful in predicting drag trends, there is little prospect that computations will replace wind tunnel testing for the complex shapes and flows of road vehicles in the foreseeable future.

Although we didn't answer all the questions, I think the Symposium has provided significant stimulus to the subject area – both the basic fluid dynamics and the road-vehicle applications. I hope it will be a source of inspiration for new research in the former that will be of utility to the latter.

PARTICIPANTS

Abernathy, F. H.
Harvard University
Cambridge, Massachusetts

Achenbach, E.
Institut für Reaktorbauelemente
Jülich, Germany

Agnew, W. G.
General Motors Research Laboratories
Warren, Michigan

Aldikacti, H.
Pontiac Motor Division, GMC
Pontiac, Michigan

Allison, R.
Penske Racing
Hueytown, Alabama

Amann, C. A.
General Motors Research Laboratories
Warren, Michigan

Barrows, T. M.
Department of Transportation
Cambridge, Massachusetts

Bearman, P. W.
Imperial College of Science and Technology
London, England

Beebe, P. S.
General Motors Research Laboratories
Warren, Michigan

Bennethum, J. E.
General Motors Research Laboratories
Warren, Michigan

Bensinger, K. W.
Fisher Body Division, GMC
Warren, Michigan

Bell, A. H.
Engineering Staff, GMC
Warren, Michigan

Bettes, W.
California Institute of Technology
Pasadena, California

Bevilaqua, P.
Rockwell International
Columbus, Ohio

Bidwell, J. B.
General Motors Research Laboratories
Warren, Michigan

Bohn, M. S.
General Motors Research Laboratories
Warren, Michigan

Buckley, Jr., F. T.
University of Maryland
College Park, Maryland

Buzan, L. R.
General Motors Research Laboratories
Warren, Michigan

Carey, V.
Harrison Radiator Division, GMC
Lockport, New York

Carr, G. W.
Motor Industry Research Association
Lindley, England

Charwat, A. F.
University of California
Los Angeles, California

Chenea, P. F.
General Motors Research Laboratories
Warren, Michigan

Chock, D. P.
General Motors Research Laboratories
Warren, Michigan

Cooper, K. R.
National Research Council
Ottawa, Ontario, Canada

Dalton, C.
University of Houston
Houston, Texas

Denkinger, R. A.
Adam Opel AG, GMC
Russelsheim, Germany

Denzer, R. E.
Engineering Staff, GMC
Warren, Michigan

Dolan, T. E.
Engineering Staff, GMC
Warren, Michigan

Doll, T. F.
Fisher Body Division, GMC
Warren, Michigan

Elliott, W. A.
Engineering Staff, GMC
Warren, Michigan

Ellis, M.
GMC Truck and Coach Division
Pontiac, Michigan

Emmelmann, H.-J.
Volkswagenwerk AG
Wolfsburg, Germany

Fink, M. R.
United Technologies Research Center
East Hartford, Connecticut

Frederiksen, G. A.
GMC Truck and Coach Division
Pontiac, Michigan

Frey, W. H.
General Motors Research Laboratories
Warren, Michigan

Gartshore, I. S.
University of British Columbia
Vancouver, British Columbia, Canada

Goetz, H.
Daimler-Benz AG
Sindelfingen, Germany

Gondert, T. R.
Engineering Staff, GMC
Warren, Michigan

Greenley, K.
Vauxhall Motors Limited
Luton, England

Gross, D. S.
University of Maryland
College Park, Maryland

Hackett, J. E.
Lockheed Georgia Company
Marietta, Georgia

Hakkinen, R. J.
McDonnell Douglas Corporation
St. Louis, Missouri

Hamsten, B.
AB Volvo
Gothenburg, Sweden

Hanson, E. K.
Buick Motor Division, GMC
Flint, Michigan

Harvey, J. K.
Imperial College of Science
and Technology
London, England

Heltsley, F. L.
ARO Incorporated
Tullahoma, Tennessee

Heskestad, G.
Factory Mutual Research Corporation
Norwood, Massachusetts

Hickling, R.
General Motors Research Laboratories
Warren, Michigan

Hirt, C. W.
Los Alamos Scientific Laboratory
Los Alamos, New Mexico

Hollasch, K. D.
Harrison Radiator Division, GMC
Lockport, New York

Hollyer, R. N.
General Motors Research Laboratories
Warren, Michigan

Hucho, W.-H.
Volkswagenwerk AG
Wolfsburg, Germany

Hunter, R.
Department of Transportation
Cambridge, Massachusetts

Ichimura, H.
Nissan Motor Company, Ltd.
Yokosuka, Japan

Jobe, D. W.
Engineering Staff, GMC
Warren, Michigan

Jones, R. T.
NASA-Ames Research Center
Moffett Field, California

Kaiser, F. G.
Transportation Systems Division, GMC
Warren, Michigan

Kelly, K. B.
Engineering Staff, GMC
Warren, Michigan

Kline, S. J.
Stanford University
Stanford, California

Klomp, E. D.
General Motors Research Laboratories
Warren, Michigan

Koenig, K.
California Institute of Technology
Pasadena, California

Korst, H. H.
University of Illinois
Urbana, Illinois

Krystoff, S. F.
American Motors Corporation
Detroit, Michigan

Landahl, M. T.
Massachusetts and Royal Institutes of Technology
Stockholm, Sweden

Larrabee, E. E.
Massachusetts Institute of Technology
Cambridge, Massachusetts

Lazurenko, L. B.
Engineering Staff, GMC
Warren, Michigan

Lea, G. K.
National Science Foundation
Washington, D.C.

Leonard, A.
NASA-Ames Research Center
Moffett Field, California

Liddle, S. G.
General Motors Research Laboratories
Warren, Michigan

Lissaman, P. B. S.
AeroVironment Incorporated
Pasadena, California

Mair, W. A.
Cambridge University
Cambridge, England

Marks, C.
Engineering Staff, GMC
Warren, Michigan

Marte, J. E.
Jet Propulsion Laboratory, Cal Tech
Pasadena, California

Martens, S. W.
Environmental Activities Staff, GMC
Warren, Michigan

Maskell, E. C.
Royal Aircraft Establishment
Farnborough, England

Mason, Jr., W. T.
General Motors Research Laboratories
Warren, Michigan

Matthews, C. C.
General Motors Research Laboratories
Warren, Michigan

Maull, D. J.
Cambridge University
Cambridge, England

McDonald, A. T.
Purdue University
West Lafayette, Indiana

McGill, R. N.
General Motors Research Laboratories
Warren, Michigan

Meyer, H. R.
Cadillac Motor Car Division, GMC
Detroit, Michigan

Morel, T.
General Motors Research Laboratories
Warren, Michigan

Morkovin, M. V.
Illinois Institute of Technology
Chicago, Illinois

Muench, N. L.
General Motors Research Laboratories
Warren, Michigan

Muto, S.
Japan Automobile Research Institute
Yatabe, Japan

Nagib, H. M.
Illinois Institute of Technology
Chicago, Illinois

Nakaguchi, H.
University of Tokyo
Tokyo, Japan

Nedley, A. L.
Chevrolet Motor Division, GMC
Warren, Michigan

Nishimura, Y.
National Research Council
Ottawa, Ontario, Canada

Oswald, L. J.
General Motors Research Laboratories
Warren, Michigan

Palmer, G. M.
Purdue University
West Lafayette, Indiana

Perry, A. E.
University of Melbourne
Parkville, Australia

Pershing, B.
The Aerospace Corporation
El Segundo, California

Peterka, J.
Colorado State University
Fort Collins, Colorado

Phillips, P.
Regie- Renault
Livonia, Michigan

Pilibosian, J. D.
General Motors Overseas Operations
Detroit, Michigan

Porter, F. C.
Chevrolet Motor Division, GMC
Warren, Michigan

Provencher, L. G.
Engineering Staff, GMC
Warren, Michigan

Przirembel, C. E. G.
Rutgers University
New Brunswick, New Jersey

Rainbird, W. J.
Carleton University
Ottawa, Ontario, Canada

Ramshaw, J. D.
Los Alamos Scientific Laboratory
Los Alamos, New Mexico

Rask, R. B.
General Motors Research Laboratories
Warren, Michigan

Romberg, G. F.
Chrysler Corporation
Detroit, Michigan

Rosenkrands, J.
Transportation Systems Division, GMC
Warren, Michigan

Roshko, A.
California Institute of Technology
Pasadena, California

Roussillon, G. L.
Peugeot
La Garenne-Colombes, France

Sagi, C. J.
General Motors Research Laboratories
Warren, Michigan

Sano, S.
Honda Research and Development Company, Ltd.
Wako, Japan

Saunders, S.
Airshield Division, Rudkin-Wiley Corporation
Stratford, Connecticut

Scharpf, G.
Engineering Staff, GMC
Warren, Michigan

Schenkel, F. K.
Engineering Staff, GMC
Warren, Michigan

Schilke, N. A.
General Motors Research Laboratories
Warren, Michigan

Schulze, P. T.
Design Staff, GMC
Warren, Michigan

Sedney, R.
U.S. Army Ballistic Research Laboratories
Aberdeen Proving Ground, Maryland

Shakespear, H.
Engineering Staff, GMC
Warren, Michigan

Shibakawa, H.
Toyota Motor Company, Ltd.
Susono, Japan

Singleton, R. E.
U.S. Army Research Office
Research Triangle Park, North Carolina

Skellenger, G. D.
General Motors Research Laboratories
Warren, Michigan

Smith, A. M. O.
University of California
Los Angeles, California

Smith, M. C.
Michigan State University
East Lansing, Michigan

Sovran, G.
General Motors Research Laboratories
Warren, Michigan

Stollery, J. L.
Cranfield Institute of Technology
Cranfield, England

Surry, D.
University of Western Ontario
London, Ontario, Canada

Szurpicki, J. J.
Engineering Staff, GMC
Warren, Michigan

Thompson, G.
Environmental Protection Agency
Ann Arbor, Michigan

Tipei, N.
General Motors Research Laboratories
Warren, Michigan

Tishkoff, J. M.
General Motors Research Laboratories
Warren, Michigan

Torner, C. H.
Design Staff, GMC
Warren, Michigan

Turunen, W. A.
General Motors Research Laboratories
Warren, Michigan

Wardlaw, R. L.
National Research Council
Ottawa, Ontario, Canada

White, R. A.
University of Illinois
Urbana, Illinois

Widnall, S. E.
Massachusetts Institute of Technology
Cambridge, Massachusetts

Williams, C. V.
Lockheed Georgia Company
Smyrna, Georgia

Williams, J. E.
Ford Motor Company
Dearborn, Michigan

Willmarth, W. W.
University of Michigan
Ann Arbor, Michigan

Wolffelt, K. W.
SAAB-SCANIA
Linköping, Sweden

Zielinski, N.
Cadillac Motor Car Division, GMC
Detroit, Michigan

Zlotnick, M.
Energy Research and Development Administration
Washington, D.C.

SUBJECT INDEX